Physical Electronics

Physical Electronics

An Introduction to the Physics of Electron Devices

J. Seymour PhD, CEng, MIEE, MInstP
Principal Lecturer
School of Electrical and Electronic
Engineering, Thames Polytechnic

A Halsted Press Book

John Wiley & Sons
New York

First Published in Great Britain in 1972
by Sir Isaac Pitman & Sons Ltd

Published in the U.S.A.
by Halsted Press
a division of John Wiley & Sons, Inc
New York

Library of Congress Catalog Card No 72–9029

ISBN 0 470–77851–2

Printed in Great Britain

Series Foreword

Professor W. A. Gambling

Electronics (or electronic engineering since the two names are synonymous) is still a rapidly-evolving subject not only in terms of its boundaries, which are still expanding, but also in the techniques being used. For example active devices have progressed from the electron tube through the bipolar transistor to field-effect devices and integrated circuits, while completely new types such as lasers, transferred-electron diodes and electro-acoustic devices have also emerged. In such a dynamic field it is difficult to keep up with the continually-changing scene and to obtain a modern, authoritative coverage of a topic at an introductory level. The days of the large comprehensive textbook, of which there have been some excellent examples in the past, have gone of course, since the gestation period of a work of this kind is such that it is bound to be out of date before it reaches the bookshelf.

The aim of the present series, therefore, is to produce a number of reasonably concise treatments each covering a specific aspect of electronics. Not only can these be written in a shorter time than can a large book but they are more easily revised and brought up to date. The student (using the term in its broadest sense) can thus select the appropriate volumes to suit the particular aspects of the subject he wishes to cover. Each book is written by a specialist and the academic level in the series as a whole ranges from the early undergraduate stage to postgraduate and professional standard. While potential readers will include both students and young professional engineers the authors have not written with any particular type of examination or teaching syllabus in mind. Indeed some of the titles cover material which students should be aware of but which is not normally classed as "examinable".

Finally a word should perhaps be said about the overall coverage. As stated above, the terms "electronics" and "electronic engineering" mean the same thing and were coined at a time when active devices consisted of electron tubes in which the flow of free electrons is controlled in such a way as to produce amplification of an electrical signal. Nowadays electronics engineers are equally at home using photons, phonons, valence and conduction electrons in solids, electron spin and other quantum states, etc., so that the terms have a less direct meaning than formerly. A better definition comes from a consideration of the function of electronics which is concerned with the transmission, storage, control and processing of information in all its different aspects. A more appropriate title might therefore be "information engineering" although the term "electronics" is now so widely accepted that it is unlikely to be displaced in the near future. In the broad meaning of the term, therefore, electronics involves the processing of information whether it appears in electrical, acoustic, optical or any other form. This is certainly the sense in which it is used in the present series.

Preface

This book provides an introduction to the physical principles underlying the operation of present-day electronic devices, and is based on lectures given in recent years to students at Thames Polytechnic studying for degrees awarded by the University of London and by the Council for National Academic Awards. It is intended to be of interest to electrical engineers and applied physicists and is written at about the level of the second or third year of an undergraduate course. It is also suitable for students preparing for the examinations of the Council of Engineering Institutions, the Institute of Physics and Higher National Certificates.

The emphasis is on solid-state devices, since they have now largely replaced thermionic valves. Chapters 1 and 2 therefore cover the basic physics of electrons in atoms and crystals. Wave mechanics is treated only descriptively in these chapters, but this approach is reinforced by a more analytical treatment in the Appendices. In Chapter 3 contacts between materials are discussed, leading to alloyed and diffused *p-n* junctions as the basis of transistors with uniform and graded bases, described in Chapter 4. Later semiconductor devices such as field-effect transistors are analysed in Chapter 5, which also includes integrated circuits and opto-electronic devices such as the electroluminescent diode. In Chapter 6 the four methods of electron emission are discussed and the thermionic diode is developed only far enough for an appreciation of the electron gun and the triode. Electrical conduction in gases and breakdown mechanisms are described in Chapter 7. Chapter 8 covers both vacuum and solid-state microwave devices, and Gunn and Read diodes are included, together with an introduction to electrical noise. Chapter 9 is concerned with the applications of stimulated emission in the maser and solid-state and gas lasers.

The general approach is to formulate initially a simple physical theory for the operation of a device. Then, where appropriate, an equivalent circuit for the device is evolved whose components depend on such external factors as applied voltage, current and temperature. The equivalent circuit can then represent the device in particular applications and the effect of the external factors can also be deduced. Thus the advantages and limitations of devices, both as discrete components and in the form of integrated circuits, are obtained, the applications of transistors being illustrated by simple amplifier and switching circuits. References are given both to more detailed theories and to wider applications, so that the book may serve as an introduction to more advanced studies.

SI units have been used throughout the text, with the exception of the electronvolt and the degree Celsius. However, measurements in CGS units are referred to in the problems so that practice is given in converting to the international system. The answers to the problems are entirely my responsibility and I am grateful for permission to reproduce questions from the examinations of the following bodies:

Institution of Electrical Engineers, Part III of the graduateship examination (*IEE*)

Institute of Physics, graduateship examination (*G. Inst. P.*)

University of London, B.Sc. (Engineering) examination, Parts II and III (*LU. BSc. (Eng.)*)

Finally, I would like to express my gratitude to all those who have helped to make this book possible. They include my colleagues in the School of Electrical and Electronic Engineering and the School of Materials Science and Physics at Thames Polytechnic for stimulating discussions and advice, and my students for critical comment. They also include the ladies who have typed the manuscript, especially Mrs. A. M. White and my wife, who has also shown great patience with my preoccupation while it was being written.

London 1971   J.S.

Contents

Foreword		v
Preface		vii
Chapter 1	Electrons in Atoms	1
Chapter 2	Electrons in Crystals	21
Chapter 3	Contacts between Materials and *p-n* Junctions	75
Chapter 4	Junction Transistors with Uniform and Graded Bases	121
Chapter 5	Field-effect Transistors and Other Semiconductor Devices	194
Chapter 6	Electron Emission and Vacuum Devices	234
Chapter 7	Electrons in Gases	297
Chapter 8	Microwave Devices and Electrical Noise	325
Chapter 9	Masers and Lasers	380
Appendix 1	Travelling Waves	399
Appendix 2	Wave Mechanics, an Introduction	402
Appendix 3	Density of Energy Levels in a Semiconductor	413
Appendix 4	Relativistic Mass Increase	420
Appendix 5	Electric Field between a Cylinder and a Coaxial Wire	423
Symbols and Abbreviations		426
Physical Constants		429
Index		430

1

Electrons in Atoms

INTRODUCTION

Physical electronics is concerned with the principles governing the behaviour of electric charge carriers. These principles underly the design of devices in which the flow of carriers is controlled by electric and magnetic fields.

Studies in the field of physical electronics may be said to originate with J. J. Thomson's experiment of 1897, which is analysed on page 267. He produced *cathode rays* by applying a voltage across a gas-filled tube (Fig. 1.1) and showed they could be deflected by an electric field and a

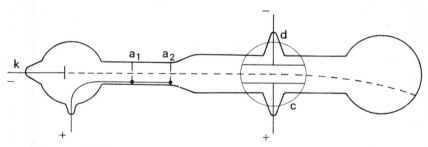

FIG. 1.1 Section through J. J. Thomson's apparatus

The tube contains gas at a low pressure and when a high voltage is applied between the cathode k and anodes a_1 and a_2 a discharge occurs and electrons are produced. They pass through the holes in the anodes and travel down the tube, striking the bulb at the end and causing it to fluoresce. The resulting bright spot is deflected by applying a voltage between the plates d or by means of a magnetic field acting across the tube within the area of the circle c.

magnetic field. They were found to be subject to the Newtonian laws of dynamics and he postulated that they consisted of minute negatively

1

charged particles, later called *electrons*. He determined their charge/mass ratio, e/m, and showed that this was the same when different gases were used.

Further studies showed that particles of the same charge/mass ratio were emitted from metal wires heated to a high temperature, a phenomenon called *thermionic emission*, and also when certain metallic surfaces were irradiated with ultraviolet light, called *photoemission*. The results suggested that the electron is universally present in matter. Thomson's value of e/m was close to the presently accepted value of $1·759 \times 10^{11}$ C/kg. This very high ratio suggests that the effect of gravitational force on the electron is negligible compared to the influence of an electric field.

In 1909 Millikan determined the charge on the electron, the accepted value of which is now $1·602 \times 10^{-19}$ C, so that the mass can be calculated to be $9·108 \times 10^{-31}$ kg. This may be compared with the mass of the hydrogen atom, which is the lightest atom known and yet is 1 837 times as heavy as an electron.

RUTHERFORD'S ATOMIC MODEL

If electrons are present in each atom of matter, which is itself electrically neutral, their combined negative charge must be balanced by an equal amount of positive charge. In 1911 Rutherford proposed a model of the atom in which all the positive charge and most of the mass was concentrated in a nucleus. The electrons rotated round it, so that the electrostatic attraction of the nucleus was balanced by the centrifugal force of their motion. He had determined by experiment that the diameter of a gold atom was about 10^{-10} m and the diameter of the nucleus about 10^{-14} m. Hence the atom consisted mainly of empty space, since its diameter was 10 000 times that of the nucleus.

As an electron moved in its orbit round the nucleus, its tangential direction of motion would change continuously so that it would experience acceleration. Theory shows that an accelerating electron radiates electromagnetic energy, a practical example being a radio transmitting aerial within which free electrons are accelerated to and fro, causing radiation from the aerial. Hence the electron in its orbit would radiate continuously, resulting in a continuous loss of energy which would allow the electrostatic attraction of the nucleus to overcome the centrifugal force of its motion. Eventually the electron would strike the nucleus, so that such an atom could not be stable. However, in practice many types of atom are stable, so that the model was not complete.

PLANCK'S QUANTUM THEORY OF RADIATION

In 1913 Bohr resolved the inconsistency of Rutherford's atom by assuming that Planck's *quantum theory* of radiation could be applied to atomic structure. His postulates were confirmed by experimental evidence and justified by the later wave-mechanical theory. In 1901 Planck had shown that when a source was radiating electromagnetic waves its energy was quantized. This means that its magnitude was always a multiple of a unit known as a *quantum* and so could not change continuously. For electromagnetic radiation the quantum is called a *photon*, which may be regarded as a particle representing the interaction of radiation with matter. However, when the radiation is travelling through free space with the velocity of light it retains its wave properties. If the frequency of the radiation is f, each photon has energy hf joules, where h is the Planck constant. Thus h relates the dual wave and particle properties of the radiation and has the value $6{\cdot}625 \times 10^{-34}$ Js.

The frequencies of electromagnetic radiation cover a very wide range, as shown in Table 1.1, extending from long radio waves of frequency 30 kHz to γ-rays with an upper frequency of about 3×10^{21} Hz. The corresponding wavelengths λ, obtained from the relationship $\lambda = c/f$ (Appendix 1), range from 10^4 to 10^{-13} m. One quantum thus represents an extremely small amount of energy, indeed only about 2×10^{-12} J for γ-rays of the highest energy. A practical source of radiation may radiate many joules per second; i.e. its power will be measurable in watts. Hence it is not surprising that energy appears to be continuously variable on a macroscopic scale.

However, on an atomic scale the energy of an individual quantum is no longer negligible. An electron moving through a potential difference of 1 V acquires an energy of $1{\cdot}602 \times 10^{-19}$ J, which is called 1 *electronvolt* (1 eV). This is comparable with the energy of a photon, and the energies of electrons and photons are normally expressed in electronvolts, as in Table 1.1.

BOHR'S MODEL OF THE HYDROGEN ATOM

Let us consider the application of Bohr's theory to the simplest atom, that of hydrogen, which has only one electron. The nucleus consists of a

3

Table 1.1 Electromagnetic Radiation

Velocity in free space, $c = 3 \times 10^8$ m/s

Type of radiation	Approximate wavelength	Frequency	Quantum energy
	m	Hz	eV
Radio	10^4–10^{-3}	3×10^4–3×10^{11}	$1\cdot2 \times 10^{-10}$–$1\cdot2 \times 10^{-3}$
Thermal			
infra-red	10^{-3}–7×10^{-7}	3×10^{11}–4×10^{14}	$1\cdot2 \times 10^{-3}$–$1\cdot7$
visible light	7×10^{-7}–4×10^{-7}	4×10^{14}–7×10^{14}	$1\cdot7$–3
ultra-violet	4×10^{-7}–10^{-9}	7×10^{14}–3×10^{17}	3–$1\cdot2 \times 10^3$
X-rays	10^{-9}–10^{-11}	3×10^{17}–3×10^{19}	$1\cdot2 \times 10^3$–$1\cdot2 \times 10^5$
γ-rays	10^{-11}–10^{-13}	3×10^{19}–3×10^{21}	$1\cdot2 \times 10^5$–$1\cdot2 \times 10^7$

particle called a *proton* which carries a positive charge of the same magnitude as that of an electron, but whose mass is 1 836 times that of the electron. Thus to a good approximation the proton may be considered fixed and unaffected by the rotation of the electron round it. Then the Coulomb electrostatic force of attraction between the charges is given by

$$F = \frac{e^2}{4\pi\epsilon_0 r^2} \tag{1.1}$$

where ϵ_0 is the permittivity of free space, which has a value of $8\cdot854 \times 10^{-12}$ F/m, and r is the distance between electron and proton (Fig. 1.2).

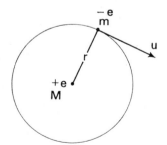

FIG. 1.2 Diagram of hydrogen atom

Suppose that the electron is moving in a circular orbit of radius r with velocity u so that a centrifugal force of mu^2/r is established. Then, at equilibrium,

$$\frac{mu^2}{r} = \frac{e^2}{4\pi\epsilon_0 r^2} \tag{1.2}$$

The energy associated with the atom will be the sum of the potential energy, V, of the two separated charges and the kinetic energy, K, of the electron rotating in its orbit. Let us assume for convenience that the potential energy is zero when the electron is an infinite distance from the nucleus. In effect this means a distance large compared with the atomic diameter of about 10^{-10} m. Then the work done in bringing the electron from infinity to distance r from the proton is

$$\int_{\infty}^{r} F\,dr = \int_{\infty}^{r} \frac{e^2}{4\pi\epsilon_0 r^2}\,dr = V$$

and

$$V = -\frac{e^2}{4\pi\epsilon_0 r} \tag{1.3}$$

The kinetic energy is $\frac{1}{2}mu^2$, which is obtained from eqn. (1.2), so that

$$K = \frac{e^2}{8\pi\epsilon_0 r} \tag{1.4}$$

The total energy of the system is then $W = V + K$, i.e.

$$W = -\frac{e^2}{4\pi\epsilon_0 r} + \frac{e^2}{8\pi\epsilon_0 r}$$

$$= -\frac{e^2}{8\pi\epsilon_0 r} \tag{1.5}$$

W is negative because zero energy is considered to be at infinity.

Up to this point Bohr's theory is the same as Rutherford's. But if the effect of radiation from the electron is included, W must decrease owing to conservation of energy, so that r decreases correspondingly and the electron spirals towards the nucleus. Hence Bohr first postulated that the electron could only move in orbits of certain radii corresponding to fixed amounts of energy, and while in one of these permitted orbits no radiation would occur. This idea is related to the quantum theory by considering the product of momentum and orbital circumference, which has the same dimensions as Planck's constant and gives

$$mu \times 2\pi r = nh$$

or

$$mur = n\frac{h}{2\pi} \tag{1.6}$$

mur is called the *moment of momentum*, or *angular momentum*. *n* is 0, 1, 2, 3, etc., and is a *quantum number*.

Squaring and substituting in eqn. (1.2) leads to

$$r = n^2 \frac{\epsilon_0 h^2}{\pi e^2 m} \tag{1.7}$$

and if this expression for *r* is substituted in eqn. (1.5) we have

$$W_n = -\frac{1}{n^2} \frac{e^4 m}{8 \epsilon^2_0 h^2} \tag{1.8}$$

for the energy of the *n*th orbit, which thus can be changed only in discrete steps, as illustrated in Fig. 1.3.

Inserting the values of the constants gives

$$W_n = -\frac{13 \cdot 58}{n^2} \text{ electronvolts} \tag{1.9}$$

and

$$r_n = n^2 \times 0 \cdot 053 \text{ nanometres} \tag{1.10}$$

The corresponding values of *n*, W_n and r_n for hydrogen are given in Table 1.2. The minimum energy is $-13 \cdot 58$ eV, which is the total energy of the stable or *ground* state of the atom. The higher permitted energy levels are unstable or *excited* states. The electron does not radiate while it is in any of the permitted levels so its energy remains constant. Fig. 1.3 has symmetry about the energy axis and is said to represent a *potential well*. At the bottom of the well the atom is in the ground state with the electron tightly bound to the nucleus.

The electron makes a transition from the ground state to a higher level when it receives energy corresponding to the difference between the levels.

Table 1.2 Energy Levels and Orbit Radii for Hydrogen Atom

n	$-W_n$	r_n
	eV	nm
1	13·58	0·053
2	3·39	0·212
3	1·51	0·477
4	0·85	0·848
5	0·54	0·133
10	0·14	0·532

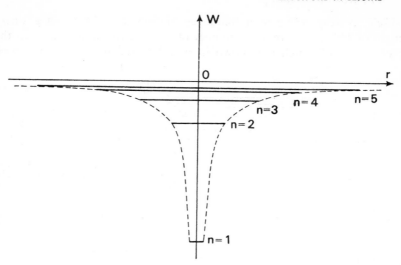

FIG. 1.3 Energy levels and orbit radii in hydrogen atom

Thus a transition from the ground state to level 2 requires 10·19 eV; from the ground state to level 3, 12·07 eV; and so on. If the electron were to receive 11 eV of energy, which does not correspond to a permitted level, it would not make any transition. An electron spends only a very short time in the excited state and then returns to the ground state, either directly or by way of intermediate levels.

Bohr's second postulate is that, when an electron jumps from an excited

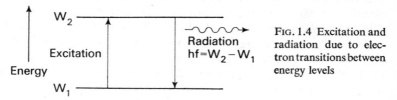

FIG. 1.4 Excitation and radiation due to electron transitions between energy levels

level to one of lower energy, a quantum of radiation is emitted (Fig. 1.4). The frequency of the radiation is given by

$$hf = W_2 - W_1 \qquad (1.11)$$

where $W_2 - W_1$ is the difference in energy between the two levels. For instance, in the transition from level 4 to level 2 the energy difference is

7

2·54 eV, corresponding to a frequency of $6·43 \times 10^{14}$ Hz and a wavelength of 467 nm. There are a number of possible transitions between the levels, some of which are shown in Fig. 1.5. When a sufficiently high

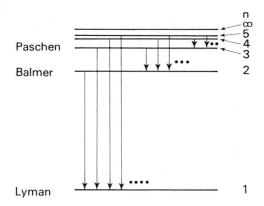

FIG. 1.5 Some of the possible transitions between the energy levels of a hydrogen atom

Paschen, Balmer and Lyman were spectroscopists who observed spectral lines at the wavelengths corresponding to transitions ending at the appropriate level. The values of wavelength were later given by Bohr's theory.

electric field is set up in the gas the energy of the atoms will be raised, as discussed in detail in Chapter 7, and the transitions will occur, each causing emission of radiation at the corresponding frequency. The wavelength of the radiation may be measured by means of a *spectroscope*, which employs a prism to separate the individual wavelengths into spectral lines, and it is found that these experimental values agree very closely with the theoretical values for hydrogen.

If the atom receives energy of 13·58 eV or more, the electron can become detached from the nucleus. This process is known as *ionization*, and 13·58 eV is the *ionization energy*, W_i, for hydrogen. Thus the larger the orbit radius, r, the smaller is the energy binding the electron to the nucleus, since the attractive force is reduced. As r is increased the energy difference between the levels becomes less and less until when the electron is freed from the nucleus its energy can change continuously.

Bohr's theory was extended by Sommerfeld in 1915, who showed that in general an orbit could be elliptical with the circular orbit as a special case having equal axes. The behaviour of the hydrogen atom, with only one electron, was then explained in considerable detail and attempts were made to extend the general principles to atoms with more than one electron. However, in these atoms repulsive forces occur between electrons, which means that a particular electron is affected by the repulsive forces of all the other electrons. This makes the calculation of the energy levels very difficult, and in addition Bohr's postulates are assumptions which do not follow from first principles. These difficulties were not resolved until it was realized that the electron possesses the properties of a wave as well as a particle.

WAVE PROPERTIES OF THE MOVING ELECTRON

The idea that a wave phenomenon, such as electromagnetic radiation, could also have particle properties was familiar from Planck's quantum theory, which had been confirmed by experiment. In 1924, de Broglie proposed that a moving particle such as an electron, and indeed all matter in motion, could possess wave properties and that the wavelength λ associated with a particle of momentum p ($=mu$) was given by

$$\lambda = \frac{h}{p} \tag{1.12}$$

as shown in Appendix 2. Thus the wave and particle properties of matter are linked by Planck's constant h, just as for the wave and particle properties of radiation, and in 1927 the wave nature of electrons was confirmed by experiments carried out by Davisson and Germer and also by G. P. Thomson, the son of J. J. Thomson.

It can easily be shown that Bohr's assumption of preferred orbits for an electron follows naturally from the concept of matter waves. Such a wave has an amplitude related to the probability of locating the electron at a point and is known as a *guiding wave*. For a non-preferred orbit there will not be a whole number of wavelengths round the orbit (Fig. 1.6(*a*)), which means that it will not be stable since the peaks will also travel round as the wave makes successive revolutions. However, for a preferred orbit Bohr's expression for momentum may be used and equated to eqn. (1.12), i.e.

$$p = \frac{nh}{2\pi r} = \frac{h}{\lambda}$$

so that

$$n\lambda = 2\pi r \tag{1.13}$$

Thus there are a whole number of wavelengths and the orbit is stable (Fig. 1.6(*b*)), which gives a physical interpretation of Bohr's first postulate

9

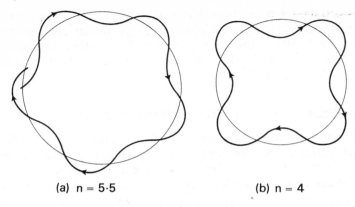

(a) n = 5·5   (b) n = 4

FIG. 1.6 De Broglie wave round a circular orbit

(page 5). The wave concept of electrons was expanded by Schrödinger in 1925 into *wave mechanics*, which leads automatically to preferred energy levels and has been used very successfully to obtain the structure of atoms with many electrons. Some of the basic ideas of wave mechanics are discussed in Appendix 2 and developed further in Ref. 1.1.

One result of wave mechanics is that the orbits are no longer precisely defined, and in a hydrogen atom, for example, the electron can traverse the whole of the space about the nucleus. However, it spends most of the time at a distance from the nucleus corresponding to a permitted Bohr radius; i.e. the probability of locating it is greatest at this distance (Fig. 1.7). The uncertainty in determining the properties of an electron was expressed by Heisenberg in 1927 in his *uncertainty principle*, which may be formulated as follows:

If the error in determining the momentum of a particle is Δp, then the error in determining its position is Δx, where

$$\Delta x \, \Delta p \geqslant \frac{h}{4\pi} \tag{1.14}$$

Putting $p = mu$,

$$\Delta x \, \Delta u \geqslant \frac{h}{4\pi m} \tag{1.15}$$

which shows that for a particle of very low mass, such as an electron, the

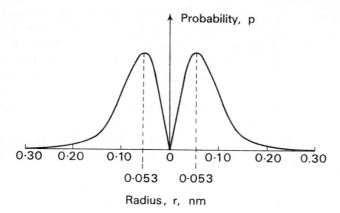

FIG. 1.7 Probability distribution for hydrogen atom in the ground state
The probability is zero of the electron being either at the nucleus or more than about 0·3 nm away from it. The probability of finding the electron is a maximum at the first Bohr orbit of radius 0·053 nm.

errors involved can be large. For example, if the position of an electron is determined within 2×10^{-10} m the error in its speed is

$$\Delta u \geqslant \frac{6 \cdot 63 \times 10^{-34}}{4\pi \times 9 \cdot 11 \times 10^{-31} \times 2 \times 10^{-10}}$$

$$\geqslant 2 \cdot 7 \times 10^{5} \, \text{m/s}$$

Thus, when the position of an electron is specified within a tolerance corresponding to the dimensions of an atom, the tolerance on its speed is very large and this uncertainty requires a statistical approach to events on an atomic scale. In fact, we can only find the probability that an electron will be in a certain place with a given speed, and this leads to the use of probability functions as described in the next chapter.

ATOMS WITH MANY ELECTRONS

The total number of electrons in an atom is always equal to the number of protons, so that the atom is electrically neutral; this is called the *atomic number*, Z. In addition, the nucleus may contain neutrons, which have no electrical charge, but have almost the same mass as the proton, namely $1 \cdot 67 \times 10^{-27}$ kg. Thus for most purposes the total mass of the atom may be considered to be in its nucleus.

11

The 105 elements can be arranged in ascending order of atomic number, starting with hydrogen and adding one electron at a time. Although it might appear that all the electrons should occupy the orbit of lowest energy, the number of electrons in any one orbit is limited. When this limit is reached a new orbit of greater radius is started, the heaviest atoms having seven orbits. It is still convenient to retain the idea of an orbit, even though it has been replaced by a charge distribution having a maximum value at the preferred distance from the nucleus.

The term *orbital* is used in wave mechanics to denote a particular energy state from which the probability of finding an electron at a given point can be calculated.

QUANTUM NUMBERS

Each possible electron orbital is uniquely defined by a set of four *quantum numbers*, which arise naturally from wave mechanics, and each number is related to a physical property of the electron. The first is the *principal quantum number*, n, which corresponds to the Bohr quantum number n (eqn. (1.6)). It is related to the total energy of the atom and hence specifies the number of orbitals, or orbits, so that its allowed values are

$$n = 1, 2, 3, \ldots 7$$

The second is the *orbital angular momentum quantum number*, l, which specifies the angular momentum in units of $h/2\pi$, and wave mechanical theory shows that the angular momentum is given by

$$p = \sqrt{[l(l + 1)]}\, \frac{h}{2\pi} \tag{1.16}$$

The allowed values of l are limited by the value of n being considered so that

$$l = 0, 1, 2, \ldots (n - 1) \tag{1.17}$$

Angular momentum has direction as well as magnitude, i.e it is a vector. This is illustrated in Fig. 1.8 for $l = 2$, with the z-axis taken as reference. The allowed directions of $\sqrt{[l(l + 1)]}$ (whose magnitude is 2·45 in this case) are determined by taking integral steps along the z-axis. The component of angular momentum along this axis is then

$$p_z = m\, \frac{h}{2\pi} \tag{1.18}$$

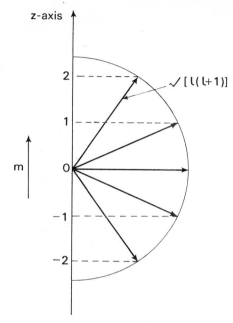

FIG. 1.8 Prefered directions
of the angular momentum
vector for $l = 2$

where m has the values $-2, -1, 0, 1, 2$ for $l = 2$. When $l = 0$, p_θ is also
zero, which corresponds to a random orientation of the vector.

In fact, m is the *magnetic quantum number*, which specifies the magnetic
moment produced along the z-axis by an electron moving round its
orbit. Consider a circular orbit of radius r with an electron moving round
it with velocity u (Fig. 1.9(a)). The time for one revolution is $2\pi r/u$ so that
the moving electron is equivalent to a current

$$i = \frac{eu}{2\pi r}$$

Now, a current i moving in a circle whose area is S is equivalent to a
magnetic dipole of moment iS, so that the magnetic moment due to the
electron is

$$iS = \frac{euS}{2\pi r} = \frac{eur}{2} \tag{1.19}$$

since $S = \pi r^2$. Using the Bohr condition for angular momentum (eqn.
(1.6)), $n = 1$ gives the smallest value, which is

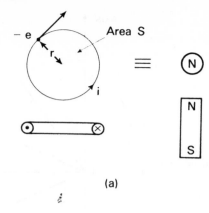

(a)

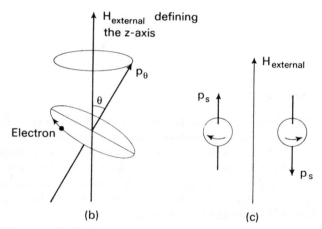

(b)　　　　　　　　(c)

FIG 1.9 Orbital and spin angular momentum

(a) Magnetic dipole due to the rotation of an electron in an orbit
(b) Precessional motion of the angular momentum vector p_θ about an external magnetic field
(c) Angular momentum due to an electron spinning round its axis. The left-hand vector is parallel and the right-hand vector is antiparallel to the magnetic field vector

$$mur = \frac{h}{2\pi} \tag{1.20}$$

Hence the magnetic moment becomes

$$\mu_B = \frac{eh}{4\pi m} \tag{1.21}$$

14

μ_B is the atomic unit of magnetic moment and is known as the *Bohr magneton*. When the values of the constants are substituted μ_B has a value of $9\cdot27 \times 10^{-24}\,Am^2$, and the actual value of magnetic moment is then $m\mu_B$. m can only have integral values, but these can be positive or negative corresponding to a clockwise or anticlockwise rotation of the electron. As shown above the allowed values of m are related to the value of l in that it can have any integral value between $-l$ and $+l$, so that

$$m = -l, -(l-1) \ldots -1, 0, +1 \ldots + (l-1), +l \qquad (1.22)$$

The z-axis is defined by the direction of an external magnetic field (Fig. 1.9(*b*)), and owing to the interaction of this field with the electronic magnetic moment the angular momentum vector rotates round the z-axis, or "precesses". The angle θ is quantized as shown in Fig. 1.8.

An electron can be visualized as a spherical charge spinning on its own axis (Fig. 1.9(*c*)), which is in addition to its orbital motion. This gives rise to a *spin angular momentum* given by

$$s = \sqrt{[s(s+1)]}\,\frac{h}{2\pi} \qquad (1.23)$$

where s is the *spin quantum number*, and the corresponding magnetic moment allows only two senses for p_s, parallel and antiparallel to the magnetic field. Thus the allowed values of s are $+\frac{1}{2}$ or $-\frac{1}{2}$, which differ by a whole quantum number. The total magnetic moment is then a combination of the values due to orbital motion and spin, the whole being expressed in Bohr magnetons.

Each electron is then specified by a set of the four quantum numbers and the number of electrons in an orbit is governed by Pauli's *exclusion principle*, introduced in 1925. This states that in any one atom no two electrons can have the same set of quantum numbers. Thus, when all the possible combinations of a particular set of numbers have been achieved with one value of n, a new value of n corresponding to a new orbit is required for any further electrons.

ELECTRONIC STRUCTURE OF ATOMS

The general ideas of atomic structure can now be illustrated by considering the elements of lowest atomic number and extending the results to those with higher values of Z. This assumes that the elements can be built up by starting with the simplest, hydrogen, and adding one electron at a time, which is only possible in theory.

Hydrogen, Z = 1
Hydrogen has only one electron and one orbit in the ground state. Hence $n = 1$, so that $l = 0$ and $m = 0$, and the electron may have either value of s (eqns. (1.17) and (1.22)).

Helium, Z = 2
The next in order is helium, which is an inert gas, and the two electrons can be in the same orbit but with opposite spins. Their quantum numbers are $n = 1$, $l = 0$, $s = +\frac{1}{2}$; and $n = 0$, $l = 0$, $m = 0$, $s = -\frac{1}{2}$. There are no other possible combinations of the numbers, so two electrons is the maximum number that can be contained in the orbit.

Lithium, Z = 3
Lithium is a metal, and the third electron has to go into a second orbit since the first one is full. The third electron is thus further from the nucleus than the two others and is less tightly bound to it, so that it can become detached giving rise to metallic properties (see page 24). The two inner electrons have the same quantum numbers as the helium electrons and the third has $n = 2$, $l = 0$, $m = 0$, $s = \frac{1}{2}$.

With elements of increasing atomic number further electrons can go into this orbit, as shown in Table 1.3 for the first 18 elements. This shows that only two electrons, with opposite spin, can have $n = 2$, $l = 0$, but the second orbit is not filled until a further six electrons have been added. These correspond to $n = 2$, $l = 1$, so that m can have the values -1, 0 and $+1$, each of which corresponds to two electrons with opposite spins.* Thus the orbit can contain a maximum of eight electrons, each with a different set of quantum numbers, and a completely filled second orbit corresponds to the inert gas neon. If this process is continued it is found that each filled orbit contains a maximum of $2n^2$ electrons. Hence for $n = 1, 2, 3, 4 \ldots$ the maximum numbers of electrons in each orbit are $2, 8, 18, 32 \ldots$ respectively. In addition when all the $l = 1$ states have been filled in a particular orbit the atom is exceptionally stable and these are the atoms of the inert gases neon, argon, krypton and xenon.

After neon the next element, sodium, requires three orbits and the

---

* An individual orbit may be conveniently referred to by the system at the top of Table 1.3. Here the number is the value of n, and the symbol is related to the value of l, as follows:

$l = 0, 1, 2, 3$
$\quad s \ p \ d \ f$

Thus the $3p$ orbit has $n = 3$, $l = 1$ and the $2s$ orbit has $n = 2$, $l = 0$, and in this context the symbol s has no connection with the spin quantum number. A superscript to each letter gives the number of electrons in that state. Thus the electronic configuration of silicon is given by $1s^2 2s^2 2p^6 3s^2 3p^2$, for example.

Table 1.3 Electronic Configurations of the First 18 Elements

		1s	2s	2p	3s	3p
	n =	1	2		3	
Z	Element	l = 0	0	1	0	1
1	Hydrogen	1				
2	Helium	2				
3	Lithium	2	1			
4	Beryllium	2	2			
5	Boron	2	2	1		
6	Carbon	2	2	2		
7	Nitrogen	2	2	3		
8	Oxygen	2	2	4		
9	Fluorine	2	2	5		
10	Neon	2	2	6		
11	Sodium	2	2	6	1	
12	Magnesium	2	2	6	2	
13	Aluminium	2	2	6	2	1
14	Silicon	2	2	6	2	2
15	Phosphorus	2	2	6	2	3
16	Sulphur	2	2	6	2	4
17	Chlorine	2	2	6	2	5
18	Argon	2	2	6	2	6

elements of higher atomic number are obtained as further electrons are added. As the orbits of lower energy are filled, according to the exclusion principle, electrons are arranged in an outer orbit. The electrons in the outermost orbit are known as the *valence* electrons and determine the physical and chemical properties of the element. An arrangement of elements in terms of their valence electrons forms the basis of the *periodic table*, and this is a striking confirmation of Mendeleeff's similar classification of the elements by purely chemical properties. Here elements in Groups I to VII have from one to seven valence electrons respectively, with the inert gases having eight outer electrons and forming a group on their own. Thus lithium and sodium are in Group I, boron and aluminium in Group III and nitrogen and phosphorus in Group V, as may be seen from Table 1.3. Further orbital arrangements are shown in Fig. 1.10.

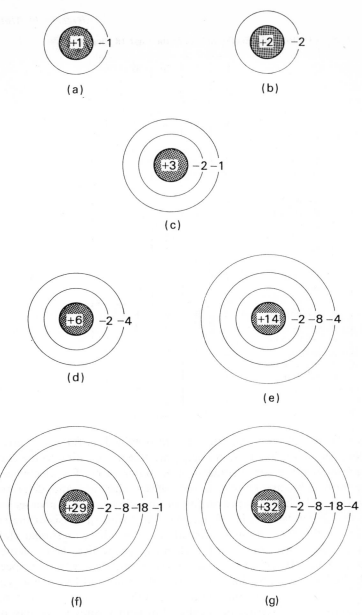

FIG 1.10 Electronic arrangement of some atoms

(a) Hydrogen, $Z = 1$ (d) Carbon, $Z = 6$
(b) Helium, $Z = 2$ (e) Silicon, $Z = 14$
(c) Lithium, $Z = 3$ (f) Copper, $Z = 29$
(g) Germanium, $Z = 32$

Both lithium and copper have one valence electron, which is typical of elements with strong metallic properties found in Group I. Carbon, silicon and germanium all have four valence electrons and are therefore in Group IV. Further details of the electronic configuration of all the elements are provided by References 1.2 and 1.3.

REFERENCES

1.1 LOTHIAN, G. F., *Electrons in Atoms* (Butterworth, 1963).
1.2 POHL, H. A., *Quantum Mechanics for Science and Engineering* (Prentice-Hall 1967).
1.3 HUME-ROTHERY, W., *Atomic Theory for Students of Metallurgy* (Institute of Metals, 1962).

FURTHER READING

BORN, M., *Atomic Physics* (Blackie, 1962).
CARO, D. E., *et al.*, *Modern Physics* (Arnold, 1964).
SPROULL, R. L., *Modern Physics* (Wiley, 1956).

PROBLEMS

1.1 Calculate the quantum energy in electronvolts of (*a*) radio waves of frequency 100 MHz, (*b*) infra-red radiation of wavelength 10 μm, (*c*) blue light of wavelength 488 nm, (*d*) X-rays of wavelength 1 pm.
(*Ans.* (*a*) $4\cdot14 \times 10^{-7}$ eV, (*b*) $0\cdot124$ eV, (*c*) $2\cdot55$ eV, (*d*) $1\cdot24 \times 10^{6}$ eV)

1.2 Obtain the uncertainty in the velocity of an electron confined within a volume of (*a*) 10^{-6} m^3, (*b*) 10^{-18} m^3, (*c*) 10^{-30} m^3.
(*Ans.* (*a*) $5\cdot8 \times 10^{-3}$ m/s, (*b*) 58 m/s, (*c*) $5\cdot8 \times 10^{5}$ m/s)

1.3 Show that the expression for Heisenberg's uncertainty principle

$$\Delta x \, \Delta p \geqslant \frac{h}{4\pi}$$

can take the form

$$\Delta W \Delta t \geqslant \frac{h}{4\pi}$$

for the tolerances in energy and time respectively.

1.4 The time taken for an electron to drop from an excited state to the ground state is about 10^{-8} s. If the energy difference is 1 eV find the uncertainty in (*a*) the energy of the emitted radiation, (*b*) the frequency of the radiation, (*c*) the wavelength of the radiation.
(*Ans.* (*a*) $3\cdot3 \times 10^{-8}$ eV, (*b*) 8 MHz, (*c*) $-4\cdot1 \times 10^{-14}$ m)

1.5 Determine the de Broglie wavelength of (*a*) 10 eV electrons, (*b*) 1 MeV protons, (*c*) neutrons at room temperature, having an energy of 0·025 eV, (*d*) An aircraft of mass 10^5 kg flying at a speed of 600 m/s.

(*Ans.* (*a*) $3·88 \times 10^{-10}$ m, (*b*) $2·87 \times 10^{-14}$ m, (*c*) $1·81 \times 10^{-10}$ m, (*d*) $1·1 \times 10^{-27}$ m)

1.6 Discuss the complementary wave and particle aspects in the atomic world. A parallel beam of electrons of $2·0 \times 10^3$ eV energy passes through a thin metal foil. If the first diffraction maximum ring is found at an angle of 5·0° with the incident beam, what is the lattice spacing of the crystal planes from which reflection occurred?

(*G.Inst.P., Part I*, 1967)

(*Ans.* $1·58 \times 10^{-10}$ m)

2

Electrons in Crystals

THE COMBINATION OF ATOMS

When large numbers of atoms are combined, as in a crystal, they are held in their relative positions by a balance between various types of inter-atomic force. In general the attractive forces are more important at large distances between atoms, and the repulsive forces between the nuclei predominate at close spacings. Thus there is an equilibrium distance between the nuclei corresponding to the actual spacing in the crystal, which is maintained to give a regular 3-dimensional array. This is illustrated

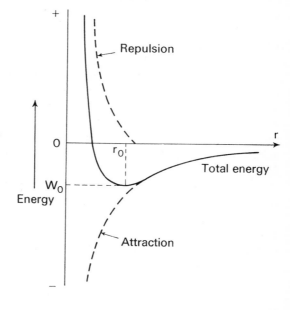

FIG. 2.1 Interatomic energies as a function of spacing

in Fig. 2.1, where the total energy is the sum of the attraction and repulsion energies. The corresponding forces are given by the slope of the energy curve, dW/dr, at a point. Thus at the equilibrium spacing r_0 the total force is zero since $dW/dr = 0$, and at this point the energy W_0 is known as the *binding energy*, since it corresponds to an attraction between the atoms. The energy has to be greatly increased to move atoms closer than the spacing r_0 against nuclear repulsion, and rather less increased to separate them against the binding energy.

There are two very important ways in which the binding energy is provided to form a stable structure. These are the *covalent bond* involved in crystals of carbon, germanium and silicon and the *metallic bond* involved in crystals of metals such as copper or silver.

The Covalent Bond

As an example of this bond let us consider the simplest, which is the formation of molecular hydrogen. The hydrogen atom can attain the stable structure of the inert gas helium if it can acquire one more electron. It achieves this by sharing its electron with another hydrogen atom, so that both electrons are associated with one of the nuclei for part of the time (Fig. 2.2(a)). Thus each atom has a share in a structure like helium and most of the time the electrons are between the nuclei, screening them from each other and reducing their electrostatic repulsion. This screening therefore constitutes an effective attractive force between the atoms, which is the covalent bond.

The covalent bond also occurs in crystals of carbon, germanium and silicon, where an atom of each element has four valence electrons. Each of these electrons follows an elliptical orbit and these orbits are symmetrically orientated so that they point to the corners of a regular tetrahedron (Fig. 2.2(b)). In a crystal there is another atom at each corner of the tetrahedron, so that each atom has a share in eight electrons and thus achieves the inert-gas type of structure for much of the time. This is illustrated in two dimensions in Fig. 2.2(c) for germanium and silicon, the nuclei plus completed inner electron orbits being drawn as atomic cores with a net charge of $+4e$, balanced by the valence electrons.

The Metallic Bond

Consider now the elements in the first group of the periodic table, such as copper, with one valence electron. The valence electrons of neighbouring copper atoms can be shared in a similar manner to the electrons of a

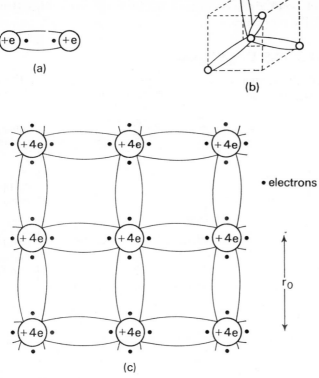

FIG. 2.2 Examples of covalent bonding

    (a) Molecular hydrogen
    (b) Tetrahedral arrangement of germanium or silicon atoms
    (c) Two-dimensional diagram of germanium or silicon crystal
    r_0 is 0·243 nm for germanium and 0·234 nm for silicon

covalent bond, but since it is not possible for each copper atom to have seven close neighbours an inert-gas structure cannot be achieved. However, electron sharing does occur with the electrons not restricted to particular atomic cores but wandering freely through the crystal. Nearly every atom contributes one valence electron to the "cloud" of charge which forms bonds between the atoms by screening nuclei from each other. The metallic bond is more flexible than the covalent bond, which leads to the ductile properties of a metal, and the presence of the freely moving cloud of electrons ensures good electrical and thermal conductivity. This type of bond also occurs with the elements in Groups II and III having metallic properties.

ENERGY LEVELS IN CRYSTALS

For isolated atoms such as occur in a gas at a low pressure the energy levels are discrete as shown in Fig. 2.3 for an isolated atom. If two such

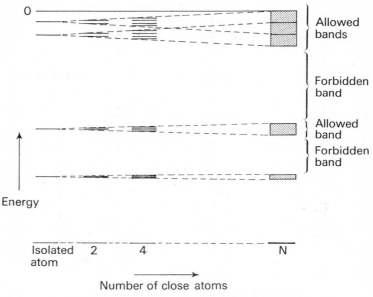

FIG. 2.3 Interaction of atoms leading to energy bands

atoms approach each other to form a molecule the electrons in the various levels cannot have the same energy, since this would mean that their quantum numbers were identical and the exclusion principle would be violated. However, this does not occur because each energy level splits into two, and for four atoms each level splits into four and so on. Hence the number of available levels equals the number of atoms involved. Each level can contain the same number of electrons as in the single atom, so that for the $2p$ level, for instance, there are six possible states (Table 1.3), and if there are N atoms there will be $6N$ states in the $2p$ group of energies.

In a crystal, where N is approximately 10^{29} atoms per cubic metre the energy difference between each level and the next becomes so small (approximately 10^{-8} eV) that the allowed energies may be considered continuous, forming an *allowed band*. In Fig. 2.3 the allowed bands are indicated by cross-hatching and are separated by regions with no allowed levels, which are known as *forbidden bands* (see also Appendix 2). It may be noted that the levels of highest energy, separated by small energy differences in a single atom, have overlapping bands in a crystal, while for the lower energy levels the width of the corresponding band is much less. This is because the inner electrons, which have the lowest energy, are tightly bound to their respective nuclei, so that they behave almost as though they were in a single atom. The potential energies of adjacent nuclei (Fig. 2.4(*a*)) combine to form a periodically varying potential through the crystal, which is shown in one dimension superimposed on the energy bands for a metal in Fig. 2.4(*b*). Here all the lower allowed bands or *core levels* are filled with the inner electrons, but the upper band is only partly filled with the higher energy levels unoccupied. The filled part is known as the *valence band* and extends above the periodic potential through the whole crystal. Thus electrons in this band are shared by all the atoms and form the metallic bond. The empty levels above the valence band are known as the *conduction band*, and if an electric field is applied across the crystal electrons from the upper levels in the valence band can easily gain enough energy to move into the vacant levels just above. Here they can move under the influence of the field and so give rise to the high electrical conductivity of a metal. For copper virtually all the valence electrons take part in conduction and since each atom has one valence electron the number density or concentration of free electrons is about 8×10^{28}/m^3. This value may be obtained from the number of atoms in the atomic weight, *Avogadro's number*,

$$N_A = 6 \cdot 023 \times 10^{26} \text{ atoms/kmol}$$

together with the atomic weight and density of copper given in Table 2.1:

Table 2.1

Element	Atomic weight	Density (kg/m³)	Atoms per cubic metre
		$\times 10^3$	$\times 10^{28}$
Copper	63·5	8.96	8.5
Germanium	72·6	5·3	4·4
Silicon	28·1	2·4	5·0

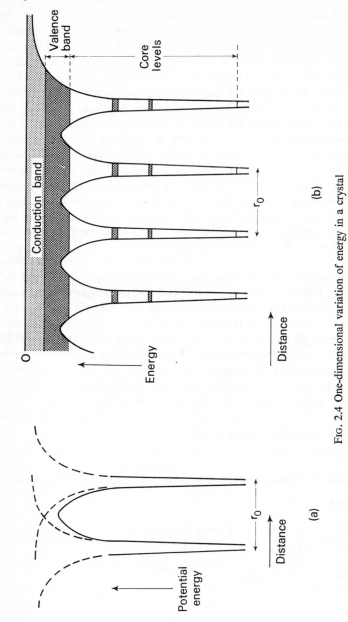

FIG. 2.4 One-dimensional variation of energy in a crystal

whence

$$\frac{(6 \cdot 023 \times 10^{26})(8 \cdot 96 \times 10^{3})}{63 \cdot 5} = 8 \cdot 5 \times 10^{28}/\text{m}^{3}$$

ELECTRICAL CONDUCTION IN SOLIDS

Solids may be classified according to their electrical resistivity, which is very high for insulators and very low for good conductors such as metals. A semiconductor has a resistivity between these two extremes as shown in Table 2.2, and this wide variation in resistivity may be explained in terms of the energy band structure of the materials.

Table 2.2

Material	Resistivity (Ω-m)
Metals	10^{-8}
(e.g. aluminium, copper, silver)	
Semiconductors	10^{-4}–10^{7}
Insulators	10^{12}–10^{20}
(e.g. glass, mica, polystyrene)	

The upper allowed bands of a metal, a semiconductor and an insulator are compared diagrammatically in Fig. 2.5, which shows that in both a semiconductor and an insulator the valence and conduction bands are separated by a forbidden band or *energy gap*, W_{g}. This is much larger for an insulator than for a semiconductor, typical values being $5 \cdot 2 \, \text{eV}$ for diamond, which is an insulator, and $0 \cdot 72 \, \text{eV}$ for germanium and $1 \cdot 09 \, \text{eV}$ for silicon, which are semiconductors. Even for a semiconductor the energy required to move an electron from the top of the valence band is too great to be acquired from a normal electric field: in fact, electrons are moved into the conduction band by excitation due to temperature or the absorption of radiation. It is shown on page 42 that at room temperature for germanium only about 1 in every 10^{9} of the valence electrons is free, giving a density of about $5 \times 10^{19}/\text{m}^{3}$. In silicon the energy gap is wider and the density of free electrons is only about $7 \times 10^{16}/\text{m}^{3}$. These densities are far less than for copper and account for the lower conductivity of semiconductors, since conductivity is proportional to the number of current carriers (eqn. (2.6(*b*))). For insulators, the carrier densities are even less than for semiconductors owing to their much

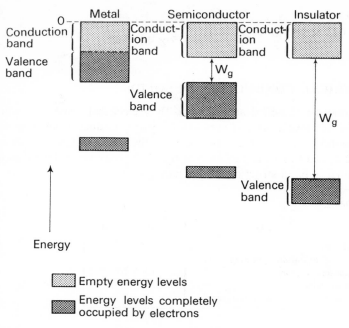

FIG. 2.5 Energy bands in a metal, a semiconductor and an insulator

larger energy gap, so that the conductivity of insulators is practically negligible.

Intrinsic Conduction

In any solid the atoms possess energy due to the temperature of the material, and this energy causes each atom to vibrate about its mean position in the crystal. Near and above room temperature this thermal energy is proportional to the temperature (see page 35). Some of the energy is shared with the valence electrons, and in a semiconductor a few are ejected from the covalent bonds and become free to take part in conduction. This corresponds to an electron acquiring energy W_g and moving into the conduction band (Fig. 2.6(a)), and the number becoming free electrons increases rapidly with temperature as discussed on page 43. When an electron is released the charge of $+4e$ on each atomic core is only compensated by three electrons, so that a *hole* is left in the bond with an effective charge of $+e$. If the free electron moves away from the hole

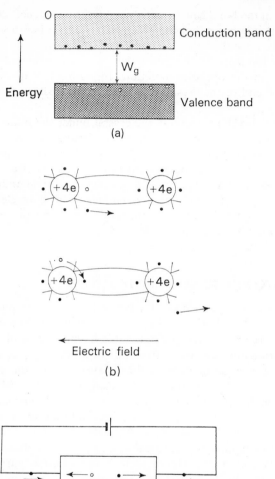

FIG. 2.6 Intrinsic semiconductor

    (*a*) Energy diagram
    (*b*) Electron-hole pair
    (*c*) Current flow

the positive charge in the bond can attract a neighbouring valence electron (Fig. 2.6(*b*)), so that it appears that the hole has moved. This process can continue through the crystal, so that the motion of valence electrons in one direction under the influence of an electric field can be considered as the movement of holes in the opposite direction (Fig. 2.6(*c*)). When a hole reaches the negative end of the crystal it is filled by an electron from the external circuit.

The free electrons also move through the crystal under the influence of the electric field, so that conduction takes place simultaneously in both the conduction band and the valence band, which is a phenomenon peculiar to semiconductors. The hole may be conveniently considered as a current carrier with charge $+e$ and a similar mass to the electron, and conduction is said to occur through the agency of *electron-hole pairs*. Two electrons flow in the external circuit for each electron-hole pair, and the process within the semiconductor is known as *intrinsic conduction* since it is a property of the pure material.

IMPURITY OR EXTRINSIC SEMICONDUCTORS

The conductivity of a pure semiconductor is too low for most purposes and it is increased by the addition of impurities, which are chosen from elements in Group III or V of the periodic table having atoms of nearly the same size as germanium or silicon atoms. Thus an impurity atom can easily take the place of a germanium or silicon atom during the growth of a crystal from the liquid state and without distorting the crystal structure. Typically 1 atom in 10^7 is replaced by an impurity atom. The material is then called an *extrinsic semiconductor* and the addition of impurities is known as *doping*.

Suitable elements from Group V are arsenic, antimony and phosphorus, each with five valence electrons. Hence there is one electron surplus to the band requirements, which is very loosely bound and easily becomes free even at low temperatures (Fig. 2.7(*a*)). Each impurity atom contributes one conduction electron to the crystal, so that Group V impurities are known as *donors* and the doped material becomes an *n-type semiconductor*. This extra electron can be regarded as similar to the single hydrogen electron, so that we can apply eqn. (1.9) to calculate its binding energy. However, since the electron is not in free space but in a crystal, eqn. (1.9) must be modified in two ways. Firstly, we must account for the effect of the medium on the force between the electron and its atomic core. This can be approximated by including the relative permittivity of the material, ϵ_r, which is 16 for germanium and 12 for silicon. Secondly, the electron experiences a force due to the periodic electric field of all the nuclei in the crystal. It is shown in Appendix 2 that, provided the field is perfectly periodic, the nuclei do not impede the motion of an electron having its energy in the conduction band and so it can move through the crystal as though it were in free space. However, the force experienced by an electron due to an external electric field is modified by the internal periodic field, and this effect may be included by saying that the electron has an *effective mass*, m,* which is different from its mass outside the material.

If m is the mass of the electron in free space, $m*$ is about $0 \cdot 25m$ for germanium and about $0 \cdot 8m$ for silicon. The binding energy of a donor electron in germanium may then be obtained from eqn. (1.8) with $n = 1$, $m = m*$ and ϵ_0 replaced by $\epsilon_0 \epsilon_r$, which gives

$$\frac{0 \cdot 25 \times 13 \cdot 6}{16^2} = 0 \cdot 013 \, \text{eV} \tag{2.1}$$

A similar calculation for silicon gives $0 \cdot 076$ eV. These figures may be compared with practical values of about $0 \cdot 01$ eV for germanium and $0 \cdot 05$ eV for silicon. Such low binding energies mean that the "extra" electron can be detached very easily. On the energy diagram (Fig. 2.7(b)) each impurity atom introduces an allowed energy level in the forbidden band and just below the conduction band. Thus the effective energy gap for donor electrons is very small, about $0 \cdot 01$ eV for germanium and $0 \cdot 05$ eV for silicon, so that at room temperature practically all the donors have lost their electrons. These are free to move in the conduction band, and since there is one for each donor atom the conductivity can be increased by an amount controlled by the donor concentration. The ionized donor atoms remain as fixed positive charges in the crystal lattice.

The corresponding elements from Group III are indium, gallium and boron, each with only three valence electrons. This causes a deficiency of one electron for each impurity atom, giving a hole in one bond (Fig. 2.7(c)). This hole can accept an electron from the valence band, so that impurity atoms from Group III are known as *acceptors* and the material becomes a *p-type semiconductor*. Each impurity atom introduces an allowed energy level in the forbidden band and just above the valence band (Fig. 2.7(d)), so that at room temperature practically all the acceptor levels are filled with electrons. This leaves a number of holes in the valence band equal to the number of acceptor atoms, and conduction occurs

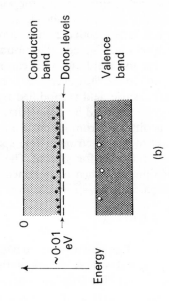

Conduction
band

Donor levels

Valence
band

0

~0·01
eV

Energy

(b)

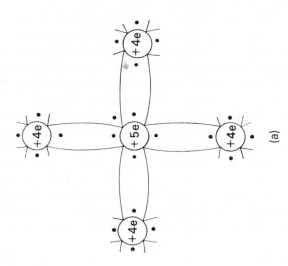

(a)

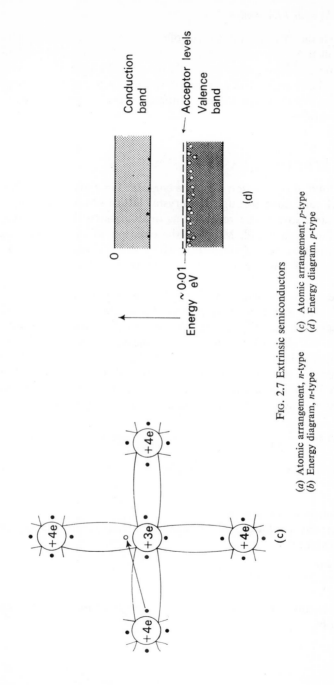

FIG. 2.7 Extrinsic semiconductors

(a) Atomic arrangement, n-type   (c) Atomic arrangement, p-type
(b) Energy diagram, n-type   (d) Energy diagram, p-type

through the effective movement of holes, the value of conductivity being controlled by the acceptor concentration. The ionized acceptor atoms remain as fixed negative charges in the crystal lattice, the ionization being due to the charge $-e$ of an electron thermally excited from a neighbouring covalent bond, which will be captured by the acceptor atom even at low temperatures.

THE ELECTRON GAS

The cloud of conduction electrons contained in a metal or a semiconductor crystal can move through the crystal lattice almost as easily as in free space and so may be regarded as analogous to the molecules of a gas contained in a vessel. Many of the concepts of the kinetic theory of an ideal gas have been applied successfully to the cloud of electrons, which is often called an *electron gas*. The more useful concepts are discussed below.

The most important measurable property of a gas is its pressure, which was formerly expressed as the height of a column of mercury exerting the same pressure. At standard temperature and pressure, namely 0°C and 101·3 kN/m², the volume occupied by the molecular weight in kilogrammes, or the kilomole, is 22·4 m³. Whatever the gas being considered, this volume must contain a definite number of molecules, which is *Avogadro's number*, N_A. Hence the concentration or number density of any gas at standard temperature and pressure is

$$n = \frac{6{\cdot}023 \times 10^{26}}{22{\cdot}4} = 2{\cdot}687 \times 10^{25} \text{ molecules/m}^3 \tag{2.2}$$

which is known as *Loschmidt's number*.

Boltzmann's Constant, k
Under any conditions of pressure and temperature the gas equation is applicable:

$$\frac{PV}{T} = \text{constant} \tag{2.3}$$

Thus at any pressure P and absolute temperature T the concentration is given by

$$n = 2 \cdot 687 \times 10^{25} \times \frac{P}{1 \cdot 013 \times 10^5} \times \frac{273}{T}$$

$$= 7 \cdot 29 \times 10^{22} \frac{P}{T} \text{ molecules/m}^3 \tag{2.4}*$$

This may be written in the form

$$P = nkT \tag{2.5}$$

where k is $1/7 \cdot 29 \times 10^{22}$ or $1 \cdot 38 \times 10^{-23}$ J/K, which is *Boltzmann's constant*.

It may be shown (Ref. 1.2) that the pressure of a gas is due to the transfer of momentum to the walls of the containing vessel and is given by

$$P = \tfrac{1}{3} n m \bar{u}^2 \tag{2.6}$$

where $\bar{u}$ is the average speed and m is the mass of each molecule. Hence, from eqn. (2.5),

$$\tfrac{1}{3} n m \bar{u}^2 = nkT \tag{2.7}$$

or

$$\tfrac{1}{2} m \bar{u}^2 = \tfrac{3}{2} kT \tag{2.8}$$

The left-hand side of this equation is the average kinetic energy of the gas molecules, which is directly related to the temperature of the gas through Boltzmann's constant. Thus the kinetic energy of a gas molecule, or any particle to which the kinetic theory can be applied, is given at room temperature (20°C or 293 K) by

$$\tfrac{3}{2} \times 1 \cdot 38 \times 10^{23} \times 293 = 6 \cdot 07 \times 10^{-21} \text{J}$$

$$= \frac{6 \cdot 07 \times 10^{-21}}{1 \cdot 6 \times 10^{-19}} = 0 \cdot 038 \text{ eV} \tag{2.9}$$

(The term kT on its own, which also occurs frequently, thus has a value of $0 \cdot 025$ eV). In fact, as a result of the temperature the molecules are in constant motion, in all directions, with a range of velocities above and below the mean value.

Mean Free Path, l

Due to this random motion collisions occur between the molecules as well as with the walls of the vessel. Hence there is a mean distance a

---

* The average atmospheric pressure is taken to be 760 mmHg which corresponds to $101 \cdot 3$ kN/m². Another unit of pressure is the bar, equal to 100 kN/m².

molecule can travel before colliding with another, the *mean free path, $\bar{l}$*. The molecules are considered as hard spheres of diameter d and it may be assumed that for a short time one molecule is moving with speed $\bar{u}$ while all the others are at rest. This molecule travels along the axis of an imaginary cylinder whose diameter is $2d$, and if any other molecule has its centre on or within this cylinder a collision occurs (Fig. 2.8(a)). In one

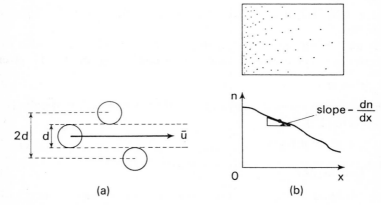

(a)   (b)

Fig. 2.8 The ideal gas

(a) Derivation of mean free path
(b) Diffusion and concentration gradient

second a volume $\pi d^2\bar{u}$ is swept out, and this volume contains $\pi d^2\bar{u}n$ molecules, so that this will also be the number of collisions per second. Thus the average time between collisions, t_c, known as the *relaxation time*, is given by

$$t_c = \frac{l}{\pi d^2\bar{u}n} \tag{2.10}$$

The distance covered by a molecule between two collisions is called its *free path* and its average value over a large number of collisions is its mean free path, $\bar{l}$. Since both t_c and $\bar{u}$ are average quantities, we can write

$$\bar{l} = t_c\bar{u} \tag{2.11}$$

so that

$$l = \frac{1}{\pi d^2 n} \qquad (2.12)$$

If allowance is made for the movement of the molecules within the cylinder it may be shown that

$$l = \frac{1}{\sqrt{2} \pi d^2 n} \qquad (2.13)$$

Thus the mean free path is inversely proportional to the number density of the molecules and hence also inversely proportional to the pressure.

Diffusion
Since the molecules are in continuous random motion, there may be at any instant a high concentration of molecules in one part of the vessel and a low concentration in another part. A transport process then occurs which restores equilibrium conditions, known as *diffusion*, in which movement of molecules takes place away from the region of high concentration. A familiar example is that of a balloon which is blown up to an internal pressure above atmospheric. If the neck of the balloon is opened air rushes out until the pressure is equal both inside and outside. A *concentration gradient* is set up under these conditions with a slope from high to low concentration (Fig. 2.8(*b*)). At any distance x the slope of the concentration/distance curve is $-dn/dx$, and the rate at which molecules diffuse down the gradient at this point is found to be proportional to the slope. The constant of proportionality is called the *diffusion coefficient*, D, so that the law of diffusion is

$$\frac{dn}{dt} = -D \frac{dn}{dx} \qquad (2.14)$$

We shall be concerned with numerical values of D when the idea of diffusion is applied to semiconductors (page 66).

The Boltzmann Factor
The velocities of the gas molecules cover a wide range which conforms to a distribution law evolved by Maxwell and Boltzmann (see Ref. 1.2 for a full discussion). We are concerned here with one result of the Maxwell–Boltzmann distribution which is of great practical importance in the operation of electronic devices.

If we choose any energy W the required result is that the number density of molecules with energies greater than W is given by

$$n_W = n \exp\left(\frac{-W}{kT}\right) \tag{2.15}$$

where n is the total number density of molecules. Since kT has the dimensions of energy, the exponential term is a pure number, known as the *Boltzmann factor*. Hence eqn. (2.15) becomes

$$\exp\left(-\frac{W}{kT}\right) = \frac{n_W}{n} = p_M(W) \tag{2.16}$$

Thus the Boltzmann factor gives the fraction of molecules with energy greater than W. This is also expressed as $p_M(W)$, the statistical probability that a molecule will have an energy greater than W. For example if $n = 10^{20}/m^3$, $n_W = 10^{12}/m^3$, then n_W/n is 10^{-8} and $p(W)$ is 1 in 10^8.

The Statistics of Electrons and Holes
The statistics of the molecules in an ideal gas, as exemplified by the Boltzmann factor, cannot be applied directly to the conduction electrons in a metal or a semiconductor. This is because the energy of gas molecules is not restricted to quantum values, but the energy of conduction electrons in a solid is governed by the number of available levels, which can only be filled according to the exclusion principle. The relevant statistics were first evolved by Fermi and Dirac and lead to a function in which the exponential form is retained, the *Fermi–Dirac function*, so that for solids

$$p_F(W) = \frac{1}{1 + \exp\left(\dfrac{W - W_F}{kT}\right)} \tag{2.17}$$

Here $p_F(W)$ is the probability that a level of energy W will be occupied by an electron, and W_F is the Fermi energy or *Fermi level*, which is a purely mathematical parameter. It may not correspond to an allowed level, but it provides a reference with which other energies can be compared. If we put $W = W_F$ in eqn. (2.17) the exponential term becomes unity and $p_F(W) = \frac{1}{2}$. Now, a probability of 1 implies that the event concerned is a certainty, while a probability of 0 implies that it can never happen. Hence a probability of $\frac{1}{2}$ can be interpreted in this case as meaning that an electron is equally likely to have an energy above the Fermi level as below it.

Let us now consider the effect of temperature on the Fermi function. At the absolute zero temperature, for all energies less than W_F, $p_F(W) = 1$, so that all these levels are empty. As T is increased above zero the function changes as shown in Fig. 2.9. Thus the probability of a level above W_F being occupied increases with temperature, and at the same

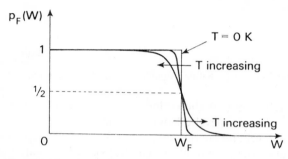

Fig. 2.9 The Fermi–Dirac function

time the probability of a level below W_F being occupied decreases, as electrons are transferred from lower to higher levels. For equal energies above and below W_F the probabilities of the level being filled and empty respectively are also equal (see Problem 2.1).

INTRINSIC SEMICONDUCTORS

In order to apply the Fermi function to an intrinsic semiconductor we have to decide where the Fermi level should be on the energy diagram. At $T = 0\,\text{K}$ the valence band is full of electrons and the conduction band is empty, so that we should expect the Fermi level to lie somewhere in the energy gap. As the temperature is increased electrons cross the gap to fill some of the levels in the conduction band, leaving an equal number of empty levels in the valence band. Hence the probabilities of a level being filled in the conduction band and empty in the valence band must be equal, and since the Fermi function changes symmetrically about W_F, the Fermi level should occur in the middle of the energy gap (Fig. 2.10). A detailed calculation (Appendix 3) shows that this is true provided that the effective masses of a hole and an electron are equal. It should be

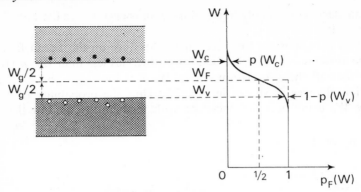

FIG. 2.10 Energy diagram and probability function for an intrinsic semiconductor

noted that in Fig. 2.10 energy is shown increasing upwards from an arbitrary zero. The actual energy level taken as zero is not required since W_F is now the reference level and energy differences from W_F are considered.

Let us now calculate the probability of an electron being excited into a state at the bottom of the conduction band at room temperature, 293 K. For germanium the energy gap, W_g, is 0·72 eV, so that $W_c - W_F = \frac{1}{2} W_g = 0·36$ eV, and $kT = 0·025$ eV, so that

$$p_F(W_c) = \frac{1}{1 + \exp 14·4}$$

$$= \frac{1}{1 + 1·78 \times 10^6} = 5·62 \times 10^{-7} \tag{2.18}$$

neglecting the 1 in the denominator. This is a very small probability but it does increase rapidly with temperature, becoming $2·15 \times 10^{-5}$ at 393 K. For silicon, with $W_g = 1·09$ eV, $p_F(W_c)$ at room temperature is $3·4 \times 10^{-10}$, much smaller than for germanium owing to the greater gap. These values of probability are the same for electrons in the conduction band as for holes in the valence band, since, because of the symmetry of the Fermi function about W_F, $p_F(W_c)$ in the conduction band has the same value as $1 - p_F(W_v)$ in the valence band (Fig. 2.10). $1 - p_F(W_v)$ is the probability of a level being empty, i.e. the probability of a hole being present in an otherwise filled band.

In the above examples the exponential term is much larger than unity so that we can write

$$p_F(W) \approx \frac{1}{\exp\left(\dfrac{W - W_F}{kT}\right)} = \exp\left(-\frac{W - W_F}{kT}\right) \qquad (2.19)$$

This is in a similar form to the Boltzmann factor (eqn. (2.16)), so that the Fermi–Dirac distribution has reduced to the Maxwell–Boltzmann distribution. This will be true provided that $W - W_F > 3kT$, which occurs in most intrinsic semiconductors and lightly doped extrinsic semiconductors, so that Maxwell–Boltzmann statistics are generally applicable to semiconductors. Here there are relatively few electrons available to fill a large number of empty states, so that the exclusion principle is comparatively unimportant. It may no longer hold at high temperatures or in extrinsic semiconductors with a large density of impurities (page 46).

The actual number density of electrons, n, will be limited by the density, N_c, of the energy levels in the conduction band, since each level can accommodate only one electron. Thus, as shown in Appendix 3,

$$n = N_c p_F(W_c) \qquad (2.20)$$

so that

$$n = N_c \exp\left(-\frac{W_c - W_F}{kT}\right) \qquad (2.21)$$

where W_c is the energy at the bottom of the conduction band, and then, for an intrinsic semiconductor,

$$n = N_c \exp -\frac{W_g}{2kT} \qquad (2.22)$$

Similarly the number density of holes, p, is governed by the density, N_v, of the energy levels in the valence band, since only one hole can appear in each level. The probability of a hole occurring is

$$1 - p_F(W) = 1 - \frac{1}{1 + \exp\left(\dfrac{W - W_F}{kT}\right)}$$

$$= \frac{1}{\exp\left(\dfrac{W_F - W}{kT}\right) + 1}$$

$$\approx \exp\left(-\frac{W_F - W}{kT}\right) \qquad (2.23)$$

Hence

$$p = N_v \exp\left(-\frac{W_F - W_v}{kT}\right) \tag{2.24}$$

where W_v is the energy at the top of the valence band. Thus, for an intrinsic semiconductor,

$$p = N_v \exp\left(-\frac{Wg}{2kT}\right) \tag{2.25}$$

For an intrinsic semiconductor $n = p = n_i$, the number density of electron-hole pairs, and an important result is obtained from the product of n and p, since

$$np = n_i{}^2 = N_c N_v \exp\left(-\frac{W_g}{kT}\right) \tag{2.26}$$

Thus

$$n_i = (N_c N_v)^{1/2} \exp\left(-\frac{W_g}{2kT}\right) \tag{2.27}$$

and n_i depends only on the type of semiconductor and the temperature. It is shown in Appendix 3 that

$$(N_c N_v)^{1/2} = GT^{3/2} \tag{2.28}$$

$$= 4 \cdot 83 \times 10^{21} T^{3/2} \tag{2.29}$$

This leads to values at room temperature of N_c and N_v near 10^{25} carriers/m³ The expression for n_i then becomes

$$n_i = GT^{3/2} \exp\left(-\frac{W_g}{2kT}\right) \text{ electron-hole pairs/m}^3. \tag{2.30}$$

G has been found experimentally to have values of $1 \cdot 76 \times 10^{22}$ for germanium and $3 \cdot 87 \times 10^{22}$ for silicon, both in units of m^{-3}-$K^{-3/2}$. At room temperature the corresponding values of n_i are 5×10^{19}/m³ for germanium and $6 \cdot 5 \times 10^{16}$/m³ for silicon, the very large difference being due to the small difference in energy gap between the two materials. Thus, for an insulator, where W_g is much larger than for silicon, n_i is negligible.

The temperature dependence of n_i can be obtained from eqn. (2.30) by taking logarithms of both sides:

$$\log_e n_i = (\log_e G + \tfrac{3}{2} \log_e T) - \frac{W_g}{2k} \frac{1}{T} \qquad (2.31)$$

Above about 200 K the bracketed term varies much more slowly with temperature than the $1/T$ term, so that $\log_e n_i$ increases almost linearly as $1/T$ decreases.

n- and *p*-TYPE SEMICONDUCTORS

The position of the Fermi level for an extrinsic semiconductor must depend on the density of the impurity atoms, since the greater this density the greater is the probability of electrons appearing in the conduction band, for *n*-type material. For an extrinsic semiconductor with an impurity density up to about $10^{23}/m^3$, i.e. about 1 atom in 10^6 replaced by an impurity atom, there are few enough carriers for Maxwell–Boltzmann statistics to be applied.

Thus for an *n*-type semiconductor we can use eqn. (2.21) to give the number density, n_n, of conduction electrons:

$$n_n = N_c \exp\left(-\frac{W_c - W_F}{kT}\right) \qquad (2.32)$$

Hence

$$W_c - W_F = kT \log_e \frac{N_c}{n_n} \qquad (2.33)$$

so that the position of the Fermi level is fixed relative to the bottom of the conduction band (Fig. 2.11(*a*)). Since the binding energy of the donor electrons is so low (page 31), near room temperature, practically all the donor atoms are ionized, so that $n_n \approx N_d$, the number density of the donor atoms. Thus as N_d is increased, $W_c - W_F$ is reduced and the Fermi level moves closer to the *bottom* of the conduction band.

Similarly for a *p*-type semiconductor the number density of holes, p_p, is given by

$$p_p = N_v \exp\left(-\frac{W_F - W_v}{kT}\right) \qquad (2.34)$$

and

$$W_F - W_v = kT \log_e \frac{N_v}{p_p} \qquad (2.35)$$

43

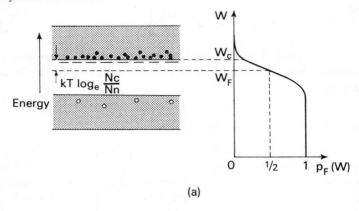

(a)

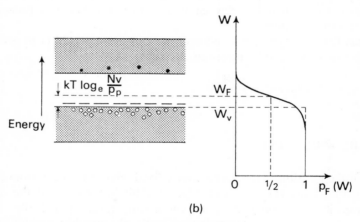

(b)

FIG. 2.11 Energy diagrams and probability functions

(a) *n*-type semiconductor   (b) *p*-type semiconductor

p_p is approximately equal to N_a, the number density of acceptor atoms, and as N_a is increased the Fermi level moves closer to the *top* of the valence band (Fig. 2.11(*b*)).

During the manufacture of semiconductor devices (page 183), *n*-type material may have to be changed to *p*-type by the addition of acceptor impurities, and the reverse may also be required. Hence both donor and acceptor impurities can be present at the same time, and if $N_d > N_a$ the material is *n*-type, while if $N_a > N_d$ it is *p*-type. In order to preserve electrical neutrality there must be equal densities of positive and negative charges in the material, so that

$$p + N_d = n + N_a \tag{2.36}$$

This assumes that all the impurity atoms are ionized, so that there is a density N_a of fixed negative charges in the lattice (page 34). Also eqn. (2.26) still holds, since it is independent of the impurity densities, so that

$$np = n_i^2 \tag{2.37}$$

Hence eqns. (2.36) and (2.37) must be solved simultaneously to find n and p. Putting $n - p = N_d - N_a = 2x$ and substituting from eqn. (2.37),

$$n - \frac{n_i^2}{n} = 2x$$

so that

$$n^2 - 2nx - n_i^2 = 0 \tag{2.38}$$

Hence

$$n = x + \sqrt{(x^2 + n_i^2)} \tag{2.39}$$

and similarly,

$$p = -x + \sqrt{(x^2 + n_i^2)} \tag{2.40}$$

where x is positive for n-type and negative for p-type material. When doping an intrinsic semiconductor, it is normal practice to make either N_a or N_d much larger than n_i at room temperature. Thus, applying the binomial theorem to eqns. (2.39) and (2.40),

$$n = x + \sqrt{x^2}\left[1 + \tfrac{1}{2}\left(\frac{n_i}{x}\right)^2\right] \tag{2.41}$$

and

$$p = -x + \sqrt{x^2}\left[1 + \tfrac{1}{2}\left(\frac{n_i}{x}\right)^2\right] \tag{2.42}$$

since $(n_i/x)^2 \ll 1$. Hence for n-type material $N_d \gg N_a$ and $x = N_d/2$, so that

$$n_n = N_d + \frac{n_i^2}{N_d} \approx N_d \tag{2.43}$$

and

$$p_n = \frac{n_i^2}{N_d} \tag{2.44}$$

Similarly, for p-type material, $N_a \gg N_d$ and $x = -N_a/2$, so that

$$n_p = \frac{n_i^2}{N_a}$$

and

$$p_p = N_a + \frac{n_i^2}{N_a} \approx N_a \tag{2.46}$$

taking $\sqrt{x^2}$ as always positive.

Majority and Minority Carriers

In n-type germanium with a donor density of $10^{22}/\text{m}^3$ and an electron-hole density of about $10^{19}/\text{m}^3$ at room temperature, the electron density will be $10^{22}/\text{m}^3$ and the hole density $10^{16}/\text{m}^3$, from eqns. (2.43) and (2.44). Thus there are 10^6 electrons for each hole, so that in an n-type semiconductor the electrons are called *majority* carriers and the holes, *minority* carriers. Similarly in a p-type semiconductor with a comparable impurity density the holes are the majority carriers and the electrons the minority carriers.

DEGENERATE SEMICONDUCTORS AND METALS

If the impurity density of an n-type semiconductor is increased, the Fermi level moves closer to the conduction band (page 43). When a density of about $5 \times 10^{23}/\text{m}^3$ is reached, N_c/n_n is about 20 and $\log_e(N_c/n_n) = 3 \cdot 0$, so that $W_c - W_F = 3kT$. Now, the approximate limit at which the Fermi function reduces to the Boltzmann factor is $W_c - W_F \geqslant 3kT$ (page 41), so that, for doping levels above about $5 \times 10^{23}/\text{m}^3$, $W_c - W_F < 3kT$ at room temperature and Maxwell–Boltzmann statistics are no longer valid. The semiconductor is then said to be *degenerate*, while a semiconductor with a lower doping level obeys the classical Maxwell–Boltzmann statistics and is said to be *non-degenerate*. Again similar considerations apply to p-type semiconductors where $W_F - W_v$ is considered.

With a further increase in impurity density above $5 \times 10^{23}/\text{m}^3$ the Fermi level continues to move until at a density of $10^{26}/\text{m}^3$ it is in fact *within* the conduction band in an n-type semiconductor and *within* the valence band in a p-type semiconductor (Figs. 2.12(a) and (b)). The semiconductor now has so many majority carriers that it has nearly degenerated into a metal, and some of the levels available for conduction are completely filled, both at room temperature and below. The energy diagram for a metal, given in Fig. 2.12(c), shows the Fermi level at the top

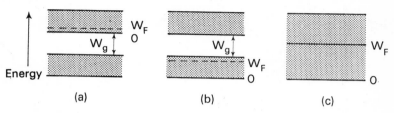

FIG. 2.12 Energy diagrams

(a) Degenerate n-type semiconductor
(b) Degenerate p-type semiconductor
(c) Metal

of the valence band. Virtually all the free electrons with energies near the top of the valence band can acquire sufficient energy from an applied field to appear in the adjacent levels in the conduction band. The distribution of energies obeys Fermi–Dirac statistics since only those electrons with energies close to the Fermi level will take part in conduction. For metals the value of W_F lies between 3 and 14 eV above the zero level at the bottom of the valence band, which corresponds to an average velocity near 10^6 m/s. The value of W_F for copper is 7 eV, which is much larger than the thermal energy at room temperature of 0·038 eV. Thus temperature has a negligible effect on both the number and energy of the current carriers in a metal. This contrasts with non-degenerate semiconductors, whose current carriers have energies of at least $3kT$ from the Fermi energy and hence obey Maxwell–Boltzmann statistics.

EFFECT OF TEMPERATURE ON n- AND p-TYPE SEMICONDUCTORS

In an n-type semiconductor at room temperature $n_i \ll N_d$, but n_i increases rapidly with temperature (eqn. (2.30)) until $n_i = N_d$ at the *transition temperature*. Then, from eqns. (2.43) and (2.44), $n_n = 2N_d$ and $p_n = N_d$. Similarly for a p-type semiconductor at the transition temperature, $p_p = 2N_a$ and $n_p = N_a$. Above this temperature the semiconductor reverts to the intrinsic type, so that $\log_e n$ or $\log_e p$ becomes proportional to $1/T$ (Fig. 2.13), while below it the carrier density is practically constant

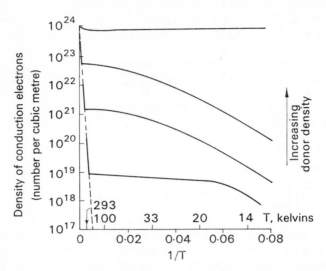

F IG. 2.13 Free electron concentration and temperature for n-type germanium
After P. P. Debye and E. M. Conwell, Ref. 2.3.

down to a temperature which depends on the density of impurities. As the temperature is further reduced fewer and fewer of the impurity atoms are ionized. The transition temperature increases with N_d or N_a and can extend up to about 400 K, while for a degenerate semiconductor it is even higher, so that over a useful operating range the carrier density of extrinsic semiconductors is constant. During the operation of semiconductor devices, the internal temperature must be kept well below the transition temperature, since if it is exceeded the sudden increase of current carriers can lead to further heating and rapid failure of the device.

LATTICE SCATTERING AND MOBILITY

For a single crystal with a perfectly periodic arrangement of atoms the effect of the periodic field of the crystal lattice has been included by giving the electron an effective mass (Appendix 2). A free electron can move through such a crystal as though it were in free space, so that the atoms will not impede its motion and the crystal will have zero electrical resistance. However, in practice no crystal is perfect and there are two main ways in which the perfectly periodic field may be disturbed. Firstly, there may be sites scattered at random through the lattice from which an atom is missing. Such a site may also be occupied by an impurity atom which will have a charge associated with it. In either case there is a local disturbance of the periodicity of the total electric field within the crystal. Secondly, at temperatures above absolute zero, the atoms are not stationary but are vibrating about their mean positions with an energy proportional to the temperature. These thermal vibrations result in the propagation of elastic waves through the crystal lattice at the speed of sound appropriate to the material. The energy of these waves can only change in discrete units of *hf* joules, known as *phonons*, where *f* is the frequency of the waves. The phonons can be considered as particles which are analogous to the photons of electromagnetic waves. The number of phonons flowing through the crystal will rise as the amplitude of the atomic vibrations is increased by a rise in temperature. An electron may be considered to collide with the phonons leading to a process known as *lattice scattering*. The resulting thermal motion of the electrons is completely random in direction, and an electron is supposed to move in a straight line at constant acceleration between collisions (Fig. 2.14(*a*)). In the absence of an applied field the

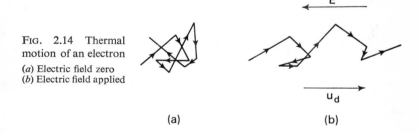

Fig. 2.14 Thermal motion of an electron
(*a*) Electric field zero
(*b*) Electric field applied

(a)   (b)

velocity of the electrons moving across any given plane at any instant is zero, owing to the random nature of the motion.

In a semiconductor the density of free electrons corresponds to the density of molecules in a gas at low pressure. Hence the electrons will have an average energy of $(3/2)kT$ due to their thermal motion, and their average speed may be obtained from eqn. (2.8), which gives

$$\bar{u} = \sqrt{\frac{3kT}{m}} \tag{2.47}$$

At room temperature, 293 K, and assuming that $m^* = m$, this leads to an average speed of about 10^5 m/s. The path of an electron may pass many atoms before scattering occurs and, as in a gas, the average distance travelled between collisions is known as the *mean free path*, l (page 36), which is about 10^{-7} m. Hence the *relaxation time*, t_c, between collisions is $l/\bar{u}$, about 10^{-12} s. For metals, although Maxwell–Boltzmann statistics are not applicable, a relaxation time can be found which is about 10^{-14} s; this is only 1 % of the value for semiconductors owing to the far higher density of free electrons (compare eqn. (2.10)).

If a constant electric field E is applied to the crystal the electron will experience a force eE moving it towards the positive terminal. This results in a *drift velocity*, u_d being superimposed on the thermal motion. (Fig. 2.14(b)). For small values of E the energy acquired by the electron from the field is dissipated by collision with the atoms. The amplitude of the atomic vibrations is thereby increased and the temperature of the crystal is raised, causing "joule heating" of the material.

The acceleration given to the electron by the field is eE/m, which is constant. Since the drift velocity is much less than the thermal velocity, as shown below, the average time between collisions is still t_c, so that the maximum velocity is eEt_c/m. Then in the simple case the average velocity, u_d, is half the maximum:

$$u_d = \frac{eEt_c}{2m} \tag{2.48}$$

This can be written in the form

$$u_d = \mu E \tag{2.49}$$

where

$$\mu = \frac{et_c}{2m} \tag{2.50}$$

μ is known as the *mobility* and represents the drift velocity per unit electric field, so that it is measured in metres per second divided by volts per metre. For small electric fields the value of μ is independent of the value of E.

Inserting the appropriate values of t_c in eqn. (2.50) yields mobilities of $0{\cdot}9 \times 10^{-1}$ m²/Vs for semiconductors and $0{\cdot}9 \times 10^{-3}$ m²/Vs for metals.

These figures are about right as may be seen from the experimental values in Table 2.3. μ_p for holes is less than μ_n for electrons since the

Table 2.3 Mobilities at 20°C

			m²/Vs
Aluminium			$1{\cdot}2 \times 10^{-3}$
Copper			$3{\cdot}2 \times 10^{-3}$
Germanium:	electrons	μ_n	$3{\cdot}9 \times 10^{-1}$
	holes	μ_p	$1{\cdot}9 \times 10^{-1}$
Silicon:	electrons	μ_n	$1{\cdot}35 \times 10^{-1}$
	holes	μ_p	$4{\cdot}8 \times 10^{-2}$

process of charge transfer for valence-band electrons is slower than for conduction-band electrons. A typical value of E in the operation of semi-conductors is 1 000 V/m, so in germanium at room temperature the drift velocity of electrons is 390 m/s and of holes 190 m/s, only about 0·1 % of the thermal velocity. Hence application of the field has a negligible effect on the thermal velocity.

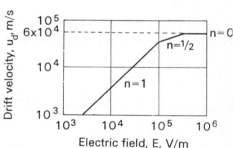

FIG. 2.15 Electron drift velocity and electric field, $u_d = \text{constant} \times E^n$

When E exceeds about 10^5 V/m (Fig. 2.15) the motion of an electron becomes much less random and more in the direction of the field. Since it moves through an effective potential $E\bar{l}$ between collisions it gains kinetic energy $e\bar{l}$ from the field. This energy imparts a drift velocity approaching the thermal velocity and is lost at each collision. Then the maximum velocity is u, given by

$$\tfrac{1}{2}mu^2 = Ee\bar{l} \tag{2.51}$$

and

$$u_d = \tfrac{1}{2}u = \tfrac{1}{2}\left(\frac{2Ee\bar{l}}{m}\right)^{1/2} = \left(\frac{e\bar{l}}{2m}\right)^{1/2}E^{1/2} \tag{2.52}$$

51

Thus in this region the drift velocity increases with the square root of the field and the mobility is dependent on the field, since

$$\mu = \frac{u_d}{E} = \left(\frac{e\bar{l}}{2mE}\right)^{1/2} \tag{2.53}$$

As E increases μ decreases until at fields above about 3×10^5 V/m the drift velocity reaches a limiting value of about 6×10^4 m/s at room temperature for germanium and silicon. At these very high fields the electron gives up so much energy to the crystal that new modes of atomic vibration are set up, so that an increase in field is compensated by a fall in mean free path.

Mobility and Temperature

The mobility in the low-field region is a function of temperature which can be deduced by putting $t_c = \bar{l}/\bar{u}$ in eqn. (2.50), which gives

$$\mu = \frac{e\bar{l}}{2m\bar{u}} \tag{2.54}$$

It has been found that the mean free path in semiconductors and metals is reduced as temperature is increased, so that approximately

$$\bar{l} = \frac{\text{const.}}{T} \tag{2.55}$$

The average velocity, given by eqn. (2.47), can be written

$$\bar{u} = \text{const.} \times T^{1/2} \tag{2.56}$$

for semiconductors, so that the mobility obtained from eqn. (2.54) becomes

$$\mu = \text{const.} \times T^{-3/2} \tag{2.57}$$

Experimental results confirm eqn. (2.57) for electrons in germanium, but for holes in germanium and both holes and electrons in silicon it is found that approximately

$$\mu = \text{const.} \times T^{-5/2} \tag{2.58}$$

near room temperature.

For metals only electrons with energies close to the Fermi energy take part in conduction, so that their energy and average velocity are mainly determined by the value of W_F. Hence $\bar{u}$ in eqn. (2.54) is virtually independent of temperature so that for metals

$$\mu = \frac{\text{const.}}{T} \qquad (2.59)$$

which is also confirmed by experiment.

Mobility and Ohm's Law

Consider a block of metal, of length L and cross-sectional area S, which contains n free electrons per cubic metre (Fig. 2.16). If a voltage V is

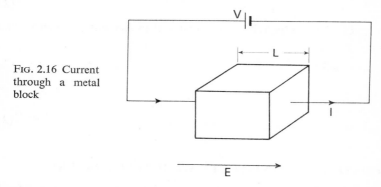

FIG. 2.16 Current through a metal block

applied across it a field $E = V/L$ will be set up, causing the electrons to move with a drift velocity u_d metres per second, and in one second they will sweep out a volume $u_d S$ cubic metres. Hence the total charge passing through a given plane in one second will be $neu_d S$ coulombs, which is also the current I, i.e.

$$I = neu_d S \qquad (2.60)$$

The current density, $J = I/S$ amperes per square metre, is therefore given by

$$J = neu_d = ne\mu E \qquad (2.61)$$

since $u_d = \mu E$. This may be shown to be a form of Ohm's law, corresponding to the familiar expression

$$I = \frac{V}{R} \qquad (2.62)$$

where R is the resistance of the block in ohms, given by

$$R = \rho \frac{L}{S} \tag{2.63}$$

ρ being the *resistivity* of the material in ohm-metres, or the resistance of a metre cube. Hence

$$J = \frac{V}{RS} = \frac{V}{\rho L} = \frac{E}{\rho} \tag{2.64}$$

This may be written in the form

$$J = \sigma E \tag{2.65}$$

where σ is the conductivity of the material in siemens per metre. Hence from eqn. (2.61),

$$\sigma = ne\mu \tag{2.66}$$

and

$$\rho = \frac{1}{\sigma} = \frac{1}{ne\mu} \tag{2.67}$$

VARIATION OF RESISTIVITY WITH TEMPERATURE

Metals
The resistivity of copper may be calculated using $n = 8\cdot5 \times 10^{28}/\text{m}^3$, $\mu = 3\cdot2 \times 10^{-3}\,\text{m}^2/\text{Vs}$ at room temperature, which gives $\rho = 2\cdot3 \times 10^{-8}\,\Omega\text{-m}$. The mobility of the free electrons in a metal varies inversely with temperature according to eqn. (2.59), so that the resistivity as a function of temperature becomes

$$\rho = \text{const.} \times T \tag{2.68}$$

In practice the resistivity of metals increases nearly linearly with temperature above about 100 K (Fig. 2.17).

Intrinsic Semiconductors
Both the electrons and the holes contribute to the total current (Fig. 2.6(c)), and although they flow in opposite directions their contributions add since their charges are of opposite sign. The electron current density is

54

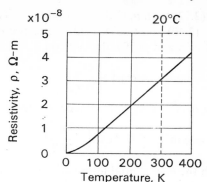

FIG. 2.17 Temperature dependence of the resistivity of a metal

$$J_n = n_i e \mu_n E \tag{2.69}$$

and the hole current density is

$$J_p = n_i e \mu_p E \tag{2.70}$$

since the densities of electrons and holes are both equal to the density of electron-hole pairs, n_i. Thus the total current density is

$$J_n + J = J = n_i e \mu_n E + n_i e \mu_p E$$

or

$$J = E n_i e (\mu_n + \mu_p) \tag{2.71}$$

and the intrinsic conductivity is

$$\sigma_i = \frac{J}{E} = n_i e (\mu_n + \mu_p) \tag{2.72}$$

For germanium at room temperature, using $n_i = 5 \times 10^{19}/\text{m}^3$ and the appropriate values of μ,

$$\sigma_i = 5 \times 10^{19} \times 1 \cdot 6 \times 10^{-19}(0 \cdot 39 + 0 \cdot 19)$$

$$= 4 \cdot 64 \, \text{S/m} \tag{2.73}$$

and

$$\rho_i = 1/\sigma_i = 0 \cdot 22 \, \Omega\text{-m} \tag{2.74}$$

3

In eqn. (2.72) both n_i and the mobilities vary with temperature, according to eqns. (2.30) and (2.57) or (2.58), respectively. Since σ_i depends on the product of n and μ, if μ depends on $T^{-3/2}$ the conductivity becomes

$$\sigma_i = \text{const.} \times \exp\left(-\frac{W_g}{2kT}\right) \tag{2.75}$$

and if μ depends on $T^{-5/2}$ the conductivity becomes

$$\sigma_i = \text{const.} \times T^{-1}\exp\left(-\frac{W_g}{2kT}\right) \tag{2.76}$$

in which the exponential term varies much more rapidly with temperature than the T^{-1} term. Hence in either case we may write

$$\rho_i = \text{const.} \times \exp\left(\frac{W_g}{2kT}\right) \tag{2.77}$$

so that the resistivity falls rapidly as the temperature rises. Taking logarithms in eqn. (2.77),

$$\log_e \rho_i = \log_e \text{const.} + \left(\frac{W_g}{2k} \times \frac{1}{T}\right) \tag{2.78}$$

and above about 200 K, $\log_e \rho_i$ varies almost linearly with $1/T$. The slope of the line is proportional to W_g and measurement of the resistance as a function of temperature is often used to determine the energy gap of an intrinsic semiconductor.

Eqn. (2.77) may be written in the form

$$R = R_0 \exp\left(\frac{b}{T}\right) \tag{2.79}$$

where R_0 and b are constants. This is the characteristic of a negative-temperature-coefficient device called a *thermistor* (Fig. 2.18), formed from material having the properties of an intrinsic semiconductor. The thermistor is widely used as a temperature-sensitive element in the measurement and control of temperature.

Extrinsic Semiconductors
In an extrinsic semiconductor the current is again carried by electrons and holes, so that

$$J = ne\mu_n E + pe\mu_p E \tag{2.80}$$

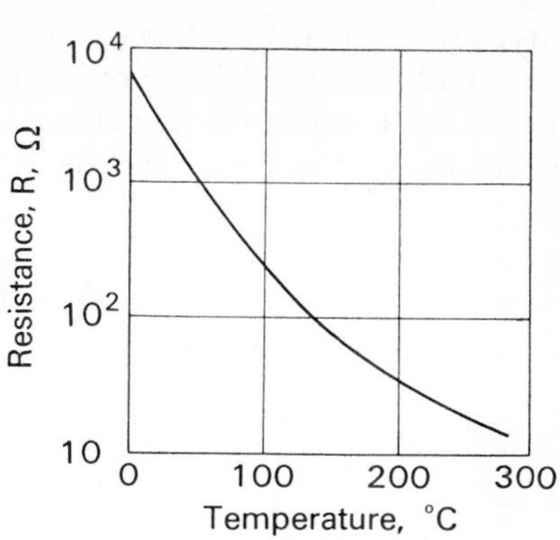

FIG. 2.18 Temperature dependence of the resistance of an intrinsic semiconductor (thermistor)

and

$$\sigma = e(n\mu_n + p\mu_p) \tag{2.81}$$

where n and p are no longer equal. The appropriate expressions for n and p are given on page 45, so that for an n-type semiconductor eqn. (2.81) becomes

$$\sigma_n = e(N_d\mu_n + \frac{n_i^2}{N_d}\mu_p) \tag{2.82}$$

$$\approx eN_d\mu_n$$

at room temperature; and for a p-type semiconductor

$$\sigma_p = e\left(\frac{n_i^2}{N_a}\mu_n + N_a\mu_p\right) \tag{2.83}$$

$$\approx eN_a\mu_p$$

at room temperature. These expressions are true down to the temperature above which nearly all the impurity atoms are ionized (Fig. 2.13). The corresponding expressions for resistivity are

$$\rho_n = \frac{1}{eN_d\mu_n} \tag{2.84}$$

for n-type semiconductors, and

$$\rho_p = \frac{1}{eN_a\mu_p} \tag{2.85}$$

57

for p-type semiconductors. Since μ decreases as temperature rises according to eqn. (2.57) or (2.58), the resistivity increases with temperature until the transition temperature is reached. At this point the semiconductor reverts to the intrinsic type and the resistivity falls again (Fig. 2.19), so

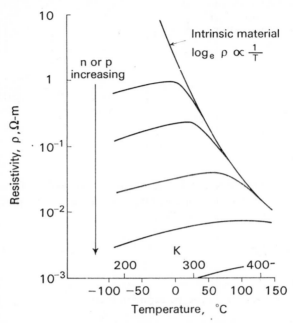

FIG. 2.19 Temperature dependence of the resistivity of an extrinsic semiconductor

After P. G. Herbart and J. Kurshan, Ref. 1.4

that the transition temperature is related to the temperature of maximum resistivity. Fig. 2.19 also shows the transition temperature increasing with the density of impurities and hence of current carriers.

THE HALL EFFECT AND MEASUREMENT OF SEMICONDUCTOR PROPERTIES

In 1879 Hall showed that a current flowing in a conductor would be deflected by a magnetic field, in a similar way to the deflection of a beam of electrons moving in the vacuum in a cathode-ray tube. The effect is very

58

small in metals, but may be easily observed in semiconductors and forms the basis of measuring the current carrier density and mobility. In addition the nature of the carriers, whether electrons or holes, may be established.

Consider a conducting strip whose length l is at least twice its breadth b to ensure uniform current flow across the plate (Fig. 2.20(a)). It has thickness t and carries a steady current I_x, and a uniform magnetic field

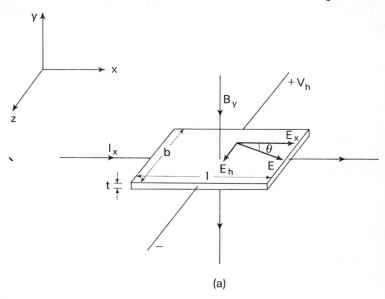

(a)

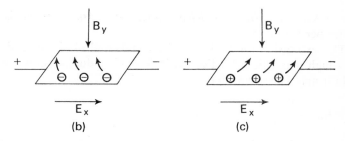

(b)   (c)

FIG. 2.20 The Hall effect

    (a) Arrangement of sample and positive axes
    (b) Deflection of electrons
    (c) Deflection of holes

B_y normal to its plane will deflect the current carriers according to the left-hand rule. This causes either type of carrier to crowd against the same side of the strip, (Figs. 2.20(*b*) and (*c*)), setting up an electric field E_H which will oppose further deflection of the carriers. E_H is called the *Hall field* and its establishment is the *Hall effect,* while as a result of the crowding the resistance of the conductor increases in the magnetic field.

Where the charge on each carrier is q, which will be $-e$ for an electron and $+e$ for a hole, we have at equilibrium

$$qE_H = B_y q u_x$$

or

$$E_H = B_y u_x \tag{2.86}$$

The current density is given by

$$\frac{J_x}{bt} = nqu_x$$

where there are n current carriers per cubic metre, so that

$$u_x = \frac{I_x}{btnq} \tag{2.87}$$

Hence

$$E_H = \frac{B_y I_x}{btnq}$$

$$= K_H \frac{B_y I_x}{bt} \tag{2.88}$$

where K_H is the *Hall coefficient*. Since $K_H = 1/nq$, its unit is the cubic metre per coulomb.

This expression for K_H is true for metals and extrinsic semiconductors in a high magnetic field. For extrinsic semiconductors in a low magnetic field it may be shown that

$$K_H = \frac{3\pi}{8} \frac{1}{nq}$$

The resistivity of the material is given by

$$\rho = \frac{1}{nq\mu} \tag{2.89}$$

and the *Hall mobility* is

$$\mu_H = \frac{1}{nqp} = \frac{K_H}{\rho} \tag{2.90}$$

K_H and hence μ_H may be determined by measuring the *Hall voltage*, given by

$$V_H = bE_H \tag{2.91}$$

when measured in the negative z-direction. Hence

$$V_H = K_H \frac{B_y I_x}{t} \tag{2.92}$$

and is directly proportional to the product of B_y and I_x. This property is also useful as the basis of a multiplying device, although the main application is the use of a Hall plate (or probe) to measure magnetic flux density.

The sign of V_H depends on the sign of K_H, which is negative for electrons and positive for holes. Hence the predominating current carrier may be identified and its density and mobility determined, provided that the density of the opposite carrier is negligible, which occurs in most n- or p-type specimens. The above expressions are true for the ratio $l/b \geqslant 2$.

The total electric field within the material is E, the vector sum of E_H and E_x (Fig. 2.20(a)). The angle θ between E and E_x is called the *Hall angle*:

$$\tan \theta = \frac{E_H}{E_x} = \frac{K_H B_y \dfrac{I_x}{bt}}{\dfrac{I_x}{bt} \rho}$$

$$= \frac{K_H B_y}{\rho} = \mu_H B_y \tag{2.93}$$

Thus the angle of deflection increases with the mobility of the charge carriers and the strength of the magnetic field. Since K_H is inversely proportional to n, it will be small for metals and large for semiconductors, whose current carriers thus experience a large deflection. They also produce a Hall voltage up to about 0.5 V as compared with about $1\,\mu$V for metals.

RECOMBINATION AND LIFETIME

We have seen that in a semiconductor at constant temperature there are fixed densities of electrons and holes, given by the expression $np = n_i^2$. This is in fact due to a dynamic equilibrium between the generation of carriers due to excitation and the *recombination* of holes and electrons, which may be expressed in terms of the equation.

$$\frac{dn_i}{dt} = g - r \tag{2.94}$$

Here g is the rate of generation and r the rate of recombination of electron-hole pairs, and at equilibrium $g = r$, so that $dn_i/dt = 0$, or n_i is constant.

Recombination through direct transition of a conduction electron to the valence band is unlikely in germanium or silicon, as explained on page 230. Not only must the electron lose energy W_g but also its momentum must be dissipated, which requires the simultaneous creation of many phonons. Instead, the electron returns to the valence band through one or more intermediate energy levels (Fig. 2.21). These are due to sites in the crystal lattice where there is a local discontinuity in the periodic potential caused by missing atoms, impurity atoms or other crystal defects. In an *n*-type semiconductor such a defect can lead to a localized level below the bottom of the conduction band which is normally empty and acts as an *electron trap*, with a high probability of capturing a free electron. A similar defect in a *p-type* semiconductor can lead to a localized level above the top of the valence band which is normally filled by an electron. When a free hole approaches the defect it will be filled by this electron, which thus returns to the valence band, so that the defect acts as a *hole trap* with a high probability of capturing a free hole.

Where the defect leads to a localized level nearer the middle of the energy gap it acts as a *recombination centre*, which has a high probability of capturing both an electron and a hole. If the centre is occupied by an electron it will attract a hole and recombination will occur, with the electron returning to the valence band. An electron-hole pair is removed and the centre is empty and ready to receive another electron. Before entering a recombination centre an electron or a hole may enter or leave several traps, energy being released as by radiation on entering and being absorbed from the surroundings on leaving the trap. The average time an electron or hole can exist in the free state is known as the *lifetime*, τ, and may be different for the two types of carrier.

EXCESS CURRENT CARRIERS IN SEMICONDUCTORS

The equilibrium density of current carriers may be increased, at constant temperature, in two main ways, resulting in either a uniform or a non-uniform distribution of carriers in the crystal. These are in addition to the carriers already present and so are known as *excess* carriers.

A uniform distribution may be obtained by illuminating the crystal with light of a suitable wavelength. In this case the excess carriers are due to excitation of electrons from the valence band to the conduction band, so that energy W_g is required. If the light has a frequency f, excitation will occur when $hf \geqslant W_g$ (Fig. 2.21), so that each photon of energy hf will generate one electron-hole pair and the rate of generation will be directly proportional to the number of photons being absorbed per second or the intensity of the light. A new dynamic equilibrium is set up with the total generation and recombination rates being equal, the number of electron-hole pairs generated optically being added directly to the number generated

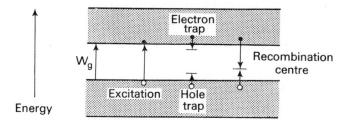

FIG. 2.21 Excitation, trapping and recombination in a semiconductor

thermally. The density of the excess conduction electrons Δn per cubic metre will be equal to the density of the excess holes, Δp, and will cause the conductivity to increase to a value $\sigma + \Delta\sigma$, where

$$\Delta\sigma = e\Delta n(\mu_n + \mu_p) \tag{2.95}$$

This increase in σ is known as *conductivity modulation*, and the excess conductivity $\Delta\sigma$ is proportional to the excess density of holes and electrons, which in this case depends on the light intensity. This is the principle of a *photoconductive cell*, the current through it being proportional to the intensity of light falling on it (page 224).

If the illumination is removed the dynamic equilibrium is disturbed and electron-hole pairs recombine until the equilibrium due to thermal generation is restored. The rate of recombination, $-d\Delta n/dt$, is proportional to the excess carrier density Δn and is given by

$$\frac{d\Delta n}{dt} = -\frac{1}{\tau}\,\Delta n \tag{2.96}$$

which defines the lifetime of the excess carriers. At time t after the illumination is removed the excess density, $\Delta n(t)$, is obtained by solving eqn. (2.96), which gives

$$\Delta n(t) = \Delta n(0)\exp\left(-\frac{t}{\tau}\right) \tag{2.97}$$

Here $\Delta n(0)$ is the excess density just before the illumination is removed, so that τ is the time taken for the excess density to fall to 37% of this initial value (Fig. 2.22). It may be shown that τ is also the average time

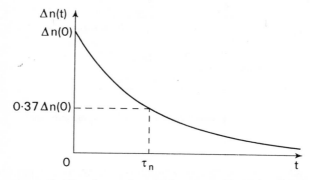

FIG. 2.22 Time decay of excess electron concentration due to recombination

any electron stays in the conduction band, whether excited there thermally, optically or owing to donor impurities. Similar considerations also apply to the lifetime of holes in the valence band.

In impurity semiconductors, illumination produces a large change in the density of the minority carriers, i.e. holes in *n*-type material and electrons in *p*-type, while hardly affecting the much larger numbers of majority carriers. Thus τ_n is the lifetime of electrons in p-type material and τ_p the lifetime of holes in n-type material. Practical values of τ are a few hundred microseconds and depend on the probability of recombination. This is increased by imperfections in the crystal structure such as grain boundaries. The surface of the material also has many recombination sites and has to be chemically cleaned and kept extremely dry to ensure that as little recombination occurs as possible.

Measurement of Lifetime

If a filament of extrinsic semiconductor is irradiated by a pulsed light source the lifetime of the excess carriers can be obtained by observing the growth and decay of current as the source is switched on and off. The principle of the method is illustrated in Fig. 2.23(a), where the light source

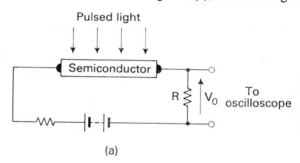

(a)

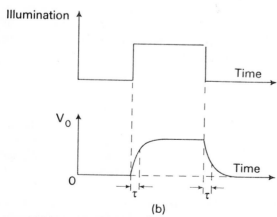

(b)

FIG. 2.23 Principle of lifetime measurement

65

may be pulsed by means of a fast-acting shutter. A current is passed through ohmic contacts along the filament, and this current is increased by the excess carriers to give a voltage waveform across the resistance R as shown at (b). Since the excess drift current ΔI at any instant is proportional to the excess carrier density, when the light source is switched off the current decays according to the expression

$$\Delta I = \Delta I_m \exp\left(-\frac{t}{\tau}\right) \tag{2.98}$$

where ΔI_m is the maximum excess current. When the light source is switched on the current rises exponentially, so that

$$\Delta I = \Delta I_m \left[1 - \exp\left(-\frac{t}{\tau}\right)\right] \tag{2.99}$$

τ is then the time for the current to rise to $0 \cdot 63 \Delta I_m$ or to fall to $0 \cdot 37 \Delta I_m$ and can be measured from the waveform by means of an oscilloscope.

The measured lifetime is due, not only to recombination in the bulk of the filament, but also to recombination at the surface. Since the surface contains many more imperfections than the bulk its recombination rate is higher and it is normally the controlling factor. If it is assumed that the rates of recombination are additive, then

$$\frac{1}{\tau_{measured}} = \frac{1}{\tau_{bulk}} \times \frac{1}{\tau_{surface}} \tag{2.100}$$

where $\tau_{bulk} \gg \tau_{surface}$. The lifetime at the surface is increased by etching. which has the effect of making it smoother and hence removing recombination sites. Ref. 2.5 has further details of the measurement of lifetime and other semiconductor parameters.

DIFFUSION OF MINORITY CARRIERS

If excess carriers are injected at one surface of a crystal, for instance at the contact between a semiconductor and a metal, or between two semiconductors, the carrier density near the contact will be higher than elsewhere in the crystal. This non-uniform distribution is similar to that occurring in a gas (page 37) and will cause diffusion of carriers away from the high-density region. Since they are charged they repel one another and a field is set up within the conductor also forcing the carriers away from the injecting contact. The number passing through unit area in one second is given by eqn. (2.14), so that for excess electrons this becomes

$$-D_n \frac{d}{dx} \Delta n \text{ electrons/m}^2\text{s}$$

where D_n is the *electron diffusion coefficient*, and for excess holes the corresponding number is

$$-D_p \frac{d}{dx} \Delta p \text{ holes/m}^2\text{s}$$

where D_p is the *hole diffusion coefficient*. The motion of the charged particles constitutes a *diffusion current*, which for electrons with charge $-e$ has a density of

$$J_n = eD_n \frac{d}{dx} \Delta n \text{ amperes per square metre} \tag{2.101}$$

and for holes with charge $+e$ has a density of

$$J_p = -eD_p \frac{d}{dx} \Delta p \tag{2.102}$$

It should be noted that no external field is required to cause a diffusion current to flow since there is an internal field set up due to the individual charges. This has the same effect on the current carriers as the pressure difference in a gas between regions of high and low density has on the gas molecules (Ref. 2.6). For a gas in the steady state, eqn. (2.5) gives

$$P = nkT \tag{2.103}$$

so that when a density gradient is set up the corresponding pressure gradient providing a force to expand the gas is

$$\frac{dP}{dx} = kT \frac{dn}{dx} \tag{2.104}$$

For a semiconductor with a non-uniform excess of electrons, at a certain distance from the injecting contact the excess density will be Δn per cubic metre. If at the same point the internal electric field is E the force on one electron is eE and the force on Δn electrons is

$$\Delta neE = \frac{\Delta neu}{\mu_n} = \frac{J_n}{\mu_n} \tag{2.105}$$

Here u is the corresponding velocity and Δneu is the current density J_n (eqn. 2.101)), so that we can write

$$\Delta neE = \frac{eD_n}{\mu_n} \frac{d}{dx} \Delta n \tag{2.106}$$

This force corresponds to the pressure gradient in a gas, so that, comparing eqns. (2.104) and (2.106), we have

$$kT = \frac{eD_n}{\mu_n}$$

or

$$\frac{D_n}{\mu_n} = \frac{kT}{e} \tag{2.107}$$

Similarly for a semiconductor with a non-uniform excess of holes,

$$\frac{D_p}{\mu_p} = \frac{kT}{e} \tag{2.108}$$

These are expressions of *Einstein's law* (1905) for the diffusion of charged particles. At room temperature, $kT/e = 0\cdot025\,\text{V}$, so that, using $\mu_n = 0\cdot4\,\text{m}^2/\text{Vs}$ and $\mu_p = 0\cdot2\,\text{m}^2/\text{Vs}$, we obtain $D_n = 0\cdot01\,\text{m}^2/\text{s}$ and $D_p = 0\cdot005\,\text{m}^2/\text{s}$.

Diffusion Length

Consider a long filament of p-type semiconductor with an excess concentration $\Delta n(0)$ of electrons maintained at one face. As the electrons diffuse away from the face they will recombine with the holes, but since they are continuously replaced the concentration gradient is maintained. Hence in the steady state there is a continuous flow of electrons through the filament, the rates of change due to diffusion and recombination being equal at any point. They may be obtained for an element of thickness δx and distant x from the injecting contact (Fig. 2.24(*a*)), with a cross-sectional area S and volume $S\,\delta x$. The rate at which electrons enter the left-hand side of the element is then

$$-D_n \frac{\partial}{\partial x} \Delta nS \tag{2.109}$$

and the rate at which they leave the right-hand side depends also on the rate of change of density gradient across the element and becomes

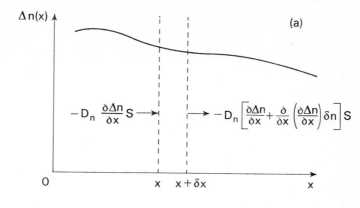

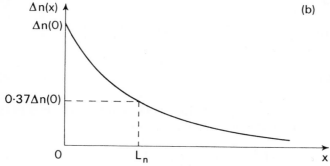

FIG. 2.24 Diffusion and recombination

(*a*) Excess electron flow through an element
(*b*) Fall of excess electron concentration

$$-D_n\left[\frac{\partial}{\partial x}\Delta n + \frac{\partial}{\partial x}\left(\frac{\partial}{\partial x}\Delta n\right)\delta x\right]S \qquad (2.110)$$

Hence the net rate of change of electrons is the difference between the rates of entering and leaving it, or the rate at which electrons are *gained* within the element is

$$D_n\frac{\partial^2}{\partial x^2}\Delta n\;\delta x\;S \qquad (2.111)$$

The rate of change of electron density due to recombination is given by eqn. (2.96), so that

$$\frac{\partial}{\partial t} \Delta n = -\frac{\Delta n}{\tau_n}$$

The rate at which electrons are *lost* within the element is then

$$\frac{\Delta n}{\tau_n} S \delta x \qquad (2.112)$$

Under equilibrium eqns. (2.111) and (2.112) are equal, so that

$$D_n \frac{\partial^2}{\partial x^2} \Delta n = \frac{\Delta n}{\tau_n} \qquad (2.113)$$

or

$$\frac{\partial^2}{\partial x^2} \Delta n - \frac{\Delta n}{D_n \tau_n} = 0 \qquad (2.114)$$

We can write $D_n \tau_n = L_n^2$, where L_n is the *diffusion length of electrons* in p-type material, which gives

$$\frac{\partial^2}{\partial x^2} \Delta n - \frac{\Delta n}{L_n^2} = 0 \qquad (2.115)$$

The general solution of eqn. (2.115) for the excess electron density at a point distant x from the injecting face is then

$$\Delta n(x) = A_n \exp\left(\frac{x}{L_n}\right) + B_n \exp\left(-\frac{x}{L_n}\right) \qquad (2.116)$$

A_n and B_n being determined by the boundary conditions. In the simple case considered at $x = 0$, $\Delta n(x) = \Delta n(0)$, and at $x = \infty$, $\Delta n(x) = 0$. Hence $A_n = 0$, $B_n = \Delta n(0)$, so that

$$\Delta n(x) = \Delta n(0) \exp\left(-\frac{x}{L_n}\right) \qquad (2.117)$$

Thus $\Delta n(x)$ decreases exponentially with distance (Fig. 2.24(*b*)), and L_n is the distance from the injecting face at which the density of excess electrons has fallen to 37% of its initial value. Similar considerations apply to the injection of excess holes into an n-type semiconductor, so that their density as a function of x is

$$\Delta p(x) = A_p \exp\left(\frac{x}{L_p}\right) + B_p \exp\left(-\frac{x}{L_p}\right) \qquad (2.118)$$

where L_p is the *diffusion length of holes*, and in the simple case as considered above,

$$\Delta p(x) = \Delta_p(0) \exp\left(-\frac{x}{L_p}\right) \qquad (2.119)$$

If we suppose that $\tau_n \approx \tau_p = 10^{-4}\,$s, then

$$L_n = \sqrt{(10^2 \times 10^{-4})} = 1\,\text{mm}$$

and

$$L_p = \sqrt{(5 \times 10^3 \times 10^{-4})} = 0.7\,\text{mm}$$

and using these values in eqns. (2.117) and (2.119) respectively, the corresponding carrier densities at any point in the filament can be obtained. Thus the diffusion length is a parameter which determines the length of any region in a device through which excess carriers are passing; it is discussed further for the *p-n* junction (page 99).

It may be noted that, when excess carriers are introduced uniformly into a material, the balance of charge is preserved since carriers of both signs are produced. It might appear that the injection of carriers of one sign would upset this balance. However, in a practical device this does not occur since an electric field is set up which attracts carriers of opposite sign from the surroundings, so that the space charge of the excess carriers is rapidly neutralized. Thus injection of excess holes is always accompanied by an equal flow of electrons, and vice versa. This is again discussed further in connection with the *p-n* junction (page 97).

REFERENCES

2.1 SHOCKLEY, W., *Electrons and Holes in Semiconductors* (Van Nostrand, 1950).

2.2 RAMEY, R. L., *Physical Electronics* (Prentice-Hall, 1961).

2.3 DEBYE, P. P., and CONWELL, E. H., "Electrical properties of *n*-type germanium", *Phys. Rev.*, **93**, p. 693 (Feb. 1954).

2.4 HERKART, P. G., and KIERSHAN, J., "Theoretical resistivity and Hall coefficient of impure germanium near room temperature", *RCA Rev.*, p. 427 (Sept. 1953).

2.5 EVANS, J., *Fundamental Principles of Transistors*, 2nd ed. (Heywood, 1962).

2.6 TEICHMANN, H., *Semiconductors*, (Butterworth, 1964).

FURTHER READING

HEMENWAY, C. L., *et al.*, *Physical Electronics*, 2nd ed. (Wiley, 1968).

Physical Electronics

PROBLEMS

2.1 If ΔW is a change in energy measured from the Fermi level show that

$$p_F(W_F + \Delta W) = 1 - p_F(W_F - \Delta W)$$

2.2 Explain briefly, in relation to a pure semiconductor, the terms energy gap and Fermi level.

A pure semiconductor has an energy gap of 1·0 eV. For temperatures of 0 K and 290 K respectively, calculate the probability of an electron occupying a state near the bottom of the conduction band. State, with reasons, whether the probabilities at each of these two temperatures will be increased if the semiconductor receives radiation of wavelength (a) 1·0 or (b) 2·0 μm. (L.U., B.Sc. (Eng.), 1967)

(Ans. 0, 2·06 × 10^{-9}; (a) Both probabilities increased since $\lambda < hc/Wg$; (b) Neither probability increased since $\lambda > hc/Wg$)

2.3 Describe the effect of incorporating atoms, like In, which carry three valence electrons, into a Ge lattice. Pay particular attention to the reason why only a small energy is needed to release carriers from these impurity atoms. A sample of Ge is doped with 3 × 10^{16} In atoms and 1·5 × 10^{16} As atoms per cm^3.

Assuming that all impurity atoms are ionized and that there is negligible intrinsic hole-electron pair contribution to carrier density at 27°C, calculate the position of the Fermi level, E_F, above the top of the valence band, E_F, given that the hole concentration p is controlled by

$$p = N_V \exp\left[-(E_F - E_V)/kT\right]$$

where N_V, the fictitious *number* of states all lying at the top of the valence band, has the value 2·5 × 10^{19} cm^{-3} at 27°C. (L.U., B.Sc.(Eng.), 1965)

(Ans. 0·19 eV)

2.4 Using an idealized energy-level diagram show that the conductivity of an intrinsic semiconductor is given by

$$\sigma = C \exp\left(-E_G/2kT\right)$$

where E_G is the energy gap between the valence band and the conduction band, C is a constant and T is the absolute temperature of the material.

For germanium $E_G = 0·72$ eV and the conductivity is 2·13 S/m at 300 K. Calculate the conductivity at $T = 400$ K.

Comment on the significance of E_G in determining the properties of a semi-conductor material—using the result of your calculation by way of illustration. (IEE, Nov. 1963)

(Ans. 70 S/m)

2.5 At room temperature the conductivity of a crystal of pure silicon is 5 × 10^{-4} S/m. If the electron mobility is 0·14 m^2/Vs and the hole mobility is 0·05 m^2/Vs, determine the density of electron-hole pairs in the crystal.

If doping with donor atoms to give an impurity density of 10^{22}/m^3 is carried out, calculate the new conductivity and the fraction of this conductivity due to holes at room temperature. Assume that all the donor atoms are ionized and that the electron and hole mobilities are unchanged.

(Ans. 1·64 × 10^{16}/m^3; 224 S/m; 9·65 × 10^{-13})

2.6 Explain the process of electrical conduction in (*a*) an intrinsic semiconductor at room temperature, (*b*) a heavily doped semiconductor, indicating briefly how the conductivity depends on temperature in each case.

An intrinsic silicon specimen at approximately $300\,K$ has a conductivity of $4\cdot3 \times 10^{-4}\,S/m$. What is the intrinsic carrier concentration? If a current is passed through the specimen what proportion of it is carried by the electrons? The same specimen is now doped to make it *n*-type. The donor concentration is $10^{21}/m^3$. Find the hole density of the doped specimen and also the proportion of current that would now be carried by electrons. Assume that the mobilities are substantially unchanged by the doping process.

Mobility of electrons in silicon at $300\,K = 0\cdot135\,m^2/Vs$. Mobility of holes in silicon at $300\,K = 0\cdot048\,m^2/Vs$. (*L.U.*, *B.Sc.* (*Eng.*), 1969)
(*Ans.* $1\cdot5 \times 10^{16}/m^3$; $0\cdot74$; $2\cdot25 \times 10^{11}/m^3$; $1 - (8 \times 10^{-11})$, so virtually all the current is carried by electrons)

2.7 Explain why the signs of the temperature dependence of the electrical resistivity of metals and intrinsic semiconductors are different.

Describe the physical conditions in which a slightly impure semiconductor could have negative temperature coefficients at high and low temperatures but a positive coefficient at intermediate temperatures. (*G.Inst. P.*, *Part II*, 1963)

2.8 Describe an experiment for determining the sign and density of the majority carriers in a semiconductor specimen. Derive any expressions necessary for the calculation of these parameters and comment on the validity of any assumptions made.

A sample of germanium has dimensions 1 cm long (*x*-direction), 2·0 mm wide (*y*-direction) and 0·2 mm thick (*z*-direction). A voltage of 1·4 V is applied across the ends of the sample and a current of 10 mA is observed in the positive *x*-direction. A Hall voltage of 10 mV is observed in the *y*-direction when there is a magnetic field of 0·1 T in the *z*-direction.

Calculate (*a*) the Hall constant, (*b*) the sign of the charge carriers, (*c*) the magnitude of the carrier density, and (*d*) the drift mobility of the carriers. (*L.U.*, *B.Sc.* (*Eng.*), 1967)
(*Ans.* (*a*) $-2 \times 10^{-3}\,m^3/C$; (*b*) Negative; (*c*) $3\cdot1 \times 10^{21}/m^3$; (*d*) $0\cdot35\,m^2/Vs$)

2.9 Explain in *physical* terms the mechanism of the Hall effect in a semiconductor.

A thin *p*-type germanium specimen of dimensions 2 cm × 1 cm is located in a plane normal to the earth's magnetic field. Calculate the potential difference which must be applied between the faces 2 cm apart if a Hall voltage of 10 mV is to arise. The earth's magnetic field may be taken as $44\mu\,T$. The mobility of positive holes in germanium is $0\cdot18\,m^2/Vs$.

Indicate on a diagram where this Hall voltage will appear, showing its direction, the direction of the magnetic field and of the applied voltage.

Explain fully what would be the effect of increasing this applied voltage by a factor of 200. (*IEE*, June, 1964)
(*Ans.* 2·52 kV)

2.10 A uniform excess of current carriers, Δn per m³, is introduced into a semiconducting crystal by illumination and the rate of recombination of electrons and holes is found to be proportional to Δn. Derive an expression for Δn at a time t after the light source has been suddenly extinguished.

If the equilibrium density of the minority carriers is 10^{20} per cubic metre and the initial rate of fall is $7 \cdot 1 \times 10^{23}$ per second, calculate (a) the minority carrier lifetime, and (b) Δn at a time 2 ms after the extinction of the source. What additional effects come into play if the excess carriers are injected into a localized region rather than uniformly into the whole sample? What effect will this have on the rate at which the carrier density approaches its equilibrium value? (*L.U.*, *B.Sc.* (*Eng.*), 1967)

(*Ans.* (a) 141 μs; (b) $6 \cdot 7 \times 10^{13}/m^3$)

2.11 Discuss the interdependence of electrical conductivity, carrier mobility, diffusivity, minority carrier lifetime and diffusion length in semiconductors.

The radiative recombination lifetime for germanium and indium antimonide may be taken as 1 s and $0 \cdot 5 \mu$s respectively. Experiments on the decay of photoconduction in monocrystalline InSb give decay times of order $0 \cdot 35 \mu$s. Experiments on the measurement of diffusion length of holes in a crystal of n-type germanium give values of 1 mm for this quantity. If the hole mobility may be taken as 1 750 cm^2/Vs and kT/e as $0 \cdot 025V$, what is the hole lifetime in the specimen of germanium?

Discuss the physical interpretation of the above observations and the probability or otherwise of obtaining lifetime values approaching 1 s for germanium. (*G.Inst.P.*, *Part II*, 1966)

(*Ans.* 228 μs)

3

Contacts between Materials and p-n Junctions

CONTACT BETWEEN TWO MATERIALS

When two crystals of different material are first placed in contact, there will be a flow of electrons from one to the other. This is because the electrons meeting the junction from one side will generally have more energy than those meeting it from the other side.

Consider two metals, A and B, with different Fermi levels W_{FA} and W_{FB} (Fig. 3.1(a)). Before they are brought into contact the energies at the surfaces of the metals will be equal and may be taken as zero. It is convenient to consider the energy difference between the Fermi level and the surface, which is known as the *work function*, ϕ. With metals this is also the depth of the conduction band. If $\phi_A < \phi_B$, filled states in the valence band of A will be at the same energy as empty states in the conduction band of B, so that, when the metals are brought into contact, electrons can flow from A to B. This results in the surface of A becoming positively charged, owing to the ionized atoms, and the surface of B becoming negatively charged (Fig. 3.1(b)). Flow proceeds until the two Fermi levels coincide (Fig. 3.1(c)), which implies that an electron has an equal probability of moving from A to B or from B to A. There is then a difference between the surface energy levels of $\phi_B - \phi_A$ and also a potential difference V_{AB} due to the surface charges (Fig. 3.1(d)).

If the ends of A and B not in contact are brought face to face without touching, work would have to be done in transferring an electron from A to B owing to the energy difference or *energy barrier* $\phi_B - \phi_A$, measured

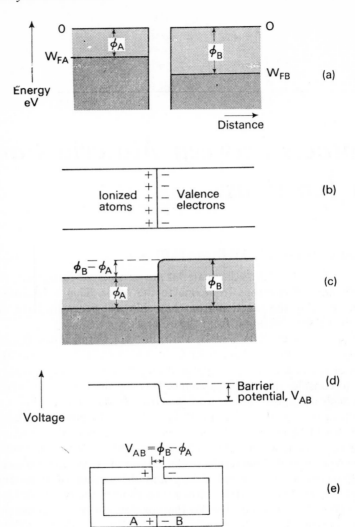

FIG. 3.1 Contact between two metals

(a) Energy diagram before contact
(b) Surface charges after contact
(c) Energy diagram after contact

(d) Barrier potential
(e) Contact potential

in electronvolts (Fig. 3.1(*e*)). This energy barrier must also equal V_{AB} electronvolts, or in terms of potential difference,

$$V_{AB} = \phi_B - \phi_A \text{ volts} \qquad (3.1)$$

V_{AB} is known as the *contact potential* between A and B and is given by the numerical difference between the two work functions.

If the circuit is completed by joining the free ends of A and B the contact potentials at the two junctions will be equal and opposite so that no current will flow after the initial connection. However, if a battery is inserted between A and B there will be a flow of electrons in the two conduction bands, which have become continuous. The Fermi levels will still coincide at the junction of A and B and the contact potential will be unaffected by the applied battery voltage. This is because the potential difference occurs in a very short distance on each side of the contact corresponding to the interatomic spacing, about 10^{-10} m, and the contact has a low resistance.

Other contacts between a metal and a semiconductor or between two semiconductors can be considered by applying the general principle of the alignment of Fermi levels.

METAL-TO-SEMICONDUCTOR CONTACTS

Consider a metal with work function ϕ_m and an extrinsic semiconductor with work function ϕ_s. There are four possibilities, depending on whether the semiconductor is *n*- or *p*-type and whether ϕ_m is less or greater than ϕ_s. In general, after the initial contact has been made, electrons flow from the material with the smaller work function.

n-type Semiconductor, $\phi_m < \phi_s$

Initially electrons flow from the metal to the semiconductor until the Fermi levels coincide, and as before, surface charges appear on each side of the junction (Fig. 3.2). Since the energy levels in the two conduction bands overlap, electrons can flow easily in either direction when $\phi_s - \phi_m$ is small, as often occurs. This type of contact between a metal and a semiconductor is called an *ohmic contact* and the contact potential $\phi_s - \phi_m$ is again unaffected by an applied voltage.

n-type Semiconductor, $\phi_m > \phi_s$

In this case the initial flow of electrons is from the semiconductor to the metal. When the Fermi levels coincide, the metal has acquired a negative

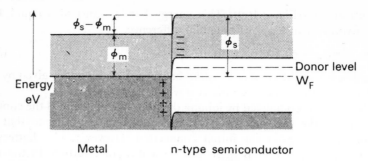

Metal   n-type semiconductor

FIG. 3.2 Energy diagram for contact between a metal and an
n-type semiconductor, $\phi_m < \phi_s$

surface charge, but the positive charge on the semiconductor is due to uncompensated donor atoms and so is distributed for a distance d from the junction (Fig. 3.3(b)). After contact, electrons can diffuse from semiconductor to metal if they have sufficient thermal energy to overcome the contact potential barrier $\phi_m - \phi_s$. This would result in an increase in the height of the barrier, but there is also diffusion of electrons from the metal over the somewhat larger barrier $\phi_m - \chi$, where χ is the depth of the conduction band of the semiconductor. At equilibrium these two currents are equal and opposite, so that the contact potential remains constant and $\phi_m - \phi_s$ is called the *diffusion potential*, ψ (Fig. 3.3(c)). The region of width d is called the *depletion layer*, since there are no free charges within it, the uncompensated atoms being fixed and the diffusing electrons passing through quickly. The resistance of the depletion layer is therefore much greater than that of the bulk of the metal or semiconductor.

The rate of diffusion of electrons across the depletion layer depends on the height of the potential barrier across it. An external bias voltage V applied between metal and semiconductor will appear almost entirely across the relatively high resistance of the depletion layer, so that this voltage adds algebraically to the diffusion potential ψ and current flow depends on both the magnitude and polarity of V. In general, the current has two components, I_F from semiconductor to metal, and I_0 in the reverse direction. The total current I is the difference between them, i.e. $I = I_F - I_0$, and the effect of the bias may be found by considering the voltage in the junction region.

For zero bias, $V = 0$, there is no net current since $I_F = I_0$ and $I = 0$. If the semiconductor is biased negatively with respect to the metal the

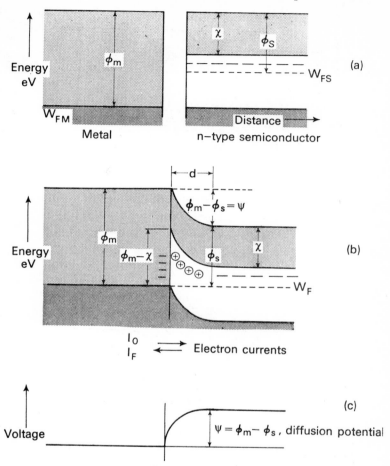

FIG. 3.3 Contact between a metal and an *n*-type semiconductor, $\phi_m > \phi_s$
  (*a*) Energy diagram before contact
  (*b*) Energy diagram after contact
  (*c*) Diffusion potential

energy of all the electrons in the semiconductor is raised and so the potential barrier is reduced to $\psi - V$ (Fig. 3.3(*d*)). Thus more electrons can diffuse from the semiconductor and I_F becomes greater than I_0, since the height of the barrier $\phi_m - \chi$ is unchanged. Fewer donors are

79

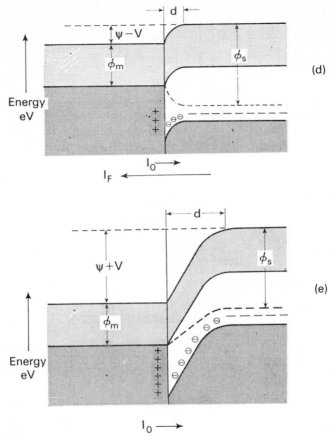

FIG. 3.3 (*contd.*)

(*d*) Energy diagram for forward bias
(*e*) Energy diagram for reverse bias

uncompensated and so *d* is also reduced. Any slope of the energy levels in the semiconductor is due to its resistance, which causes an *IR* voltage drop at high currents. The junction is said to be *forward biased* and when $I_F \gg I_0$, $I = I_F$.

If the bias is reversed, so that the semiconductor is positive with respect to the metal, the energy of the electrons in the semiconductor is lowered and the potential barrier is raised to $\psi + V$ (Fig. 3.3(*e*)). This greatly reduces the diffusion of electrons from the semiconductor and I_F tends to zero. Electrons are drawn away from the junction, so that more donor atoms are uncompensated and d increases. However, the barrier $\phi_m - \chi$ is unaffected so that I_0 is still unchanged. The junction is said to be *reverse biased* and when $I_F \to 0$, $I = I_0$. Since there is heavy current flow with forward bias and light flow with reverse bias, the junction acts as a rectifier, so that with $\phi_m > \phi_s$ a *rectifying contact* is formed. Practical applications of this effect occur in the so-called "hot carrier", or *Schottky-barrier diode*, in which platinum silicide makes contact with *n*-type silicon.

p-type Semiconductor, $\phi_m > \phi_s$
Again the initial flow of the electrons is from the semiconductor to the metal. This results in a surface charge of electrons on the metal and a surface charge of holes on the semiconductor (Fig. 3.4). This occurs

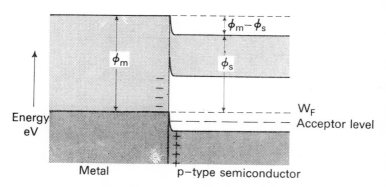

FIG. 3.4 Energy diagram for contact between a metal and a
p-type semiconductor, $\phi_m > \phi_s$

because holes are the majority carriers in *p*-type material and so are free to collect at a surface. There is no depletion layer and external bias does not affect the small contact potential, $\phi_m - \phi_s$. There is a free flow of holes (due to electrons moving within the valence band) in either direction so that the contact is ohmic.

p-type Semiconductor, $\phi_m < \phi_s$
Here the initial flow of electrons is from metal to semiconductor, so that a positive charge is formed on the metal. The electrons are captured by

81

acceptor atoms near the junction and a depletion layer of width d is formed (Fig. 3.5(b)). The action of the contact is similar to that of the

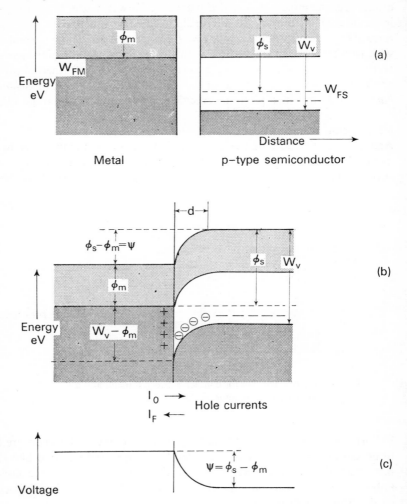

FIG. 3.5 Contact between a metal and p-type semiconductor: $\phi_m < \phi_s$

(a) Energy diagram before contact
(b) Energy diagram after contact
(c) Diffusion potential
(d) Energy diagram for forward bias
(e) Energy diagram for reverse bias

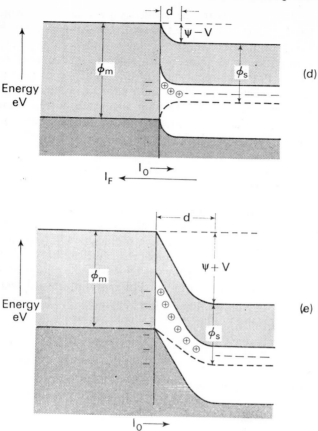

FIG. 3.5 (*contd.*)

metal-to-*n*-type contact with $\phi_m > \phi_s$, but with holes as current carriers. With zero external bias, equal and opposite currents I_F and I_0 are set up, where I_F is due to holes from the semiconductor which possess sufficient thermal energy to cross the barrier $\phi_s - \phi_m$. I_0 is due to holes from the metal crossing the larger barrier $W_v - \phi_m$, where W_v is the depth of the top of the valence band in the semiconductor.

Thus the contact is rectifying and forward bias occurs with the semiconductor positive, which reduces the potential barrier to $\psi - V$, where $\psi = \phi_s - \phi_m$ (Fig. 3.5(*d*)). This allows I_F to increase, while I_0 remains constant since the somewhat larger barrier $W_v - \phi_m$ is unaffected by the

83

bias. Reverse bias occurs with the semiconductor negative, which increases the barrier to $\psi + V$ so that $I_F \to 0$, but I_0 is unaffected (Fig. 3.5(e)). It should be noted that, for contacts with both n- and p-type materials, I_0 increases with temperature since it is due to thermally generated carriers. Practical application of this rectifying contact are the cuprous-oxide and selenium rectifiers. Both cuprous oxide and selenium are p-type semiconductors, and a rectifier is formed by the contact of copper with cuprous oxide, while an alloy of tin, cadmium and bismuth form a metallic contact with selenium.

CURRENT/VOLTAGE CHARACTERISTIC OF A RECTIFYING CONTACT

The relationship between the current and voltage of a rectifying contact may be obtained by considering the probability that a current carrier will have sufficient energy to cross a barrier of height W. Assuming that Maxwell–Boltzmann statistics are applicable, the number of carriers crossing the barrier per second from the semiconductor is $n(t)$, which is proportional to $\exp(-W/kT)$. For a rectifying contact with zero bias $W = e\psi$ and the current flowing in one direction is $I_F = en$, which is balanced by I_0 flowing in the opposite direction. Hence

$$I_F = \text{constant} \times \exp\left(-\frac{e\psi}{kT}\right) = I_0 \tag{3.2}$$

With forward bias, V is positive and $W = e(\psi - V)$, so that

$$I_F = \text{constant} \times \exp\left[-\frac{e(\psi - V)}{kT}\right] \tag{3.3}$$

$$= I_0 \exp \frac{eV}{kT} \tag{3.4}$$

The total current is $I = I_F - I_0$, so that

$$I = I_0\left[\exp\left(\frac{eV}{kT}\right) - 1\right] \tag{3.5}$$

for the characteristic of a contact with negligible IR drops on either side. This is illustrated in Fig. 3.6. With reverse bias V is negative and the exponential term becomes much less than unity, so that $I = -I_0$. For the metal-to-n-type-semiconductor contact I_0 is an electron current, while

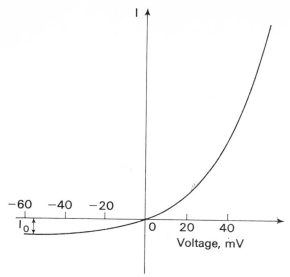

Fig. 3.6 Current/voltage characteristic of a rectifying contact
The current scale is determined by the value of I_0. The curve corresponds to $T = 293$ K, giving $e/kT = 40$/V.

for the metal-to-p-type-semiconductor contact it is due to holes. It will be shown in a later section that for a junction between p- and n-type semiconductors I_0 is due to both electrons and holes.

THERMOELECTRIC EFFECTS

When electrons flow from one material to another, energy is also transported in the form of heat, the *Peltier effect* (Fig. 3.7(a)). It is found that the quantity of heat transferred is proportional to the quantity of electricity flowing. The constant of proportionality is the differential *Peltier coefficient*, $\alpha_{P\,ab}$, given by

$$\alpha_{P\,ab} = \frac{W}{Q} = \frac{P}{I} \text{ volts} \tag{3.6}$$

where W is the energy in joules transferred to or from the junction between two materials, a and b, by a charge of Q coulombs. $\alpha_{P\,ab}$ is often more

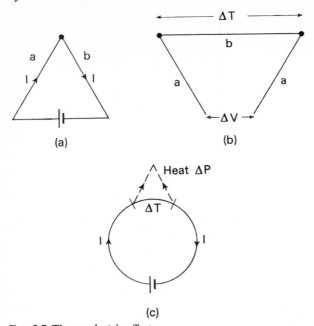

FIG. 3.7 Thermoelectric effects

(a) Peltier effect   (b) Seebeck effect   (c) Thomson effect

conveniently expressed in terms of the power P (watts) transferred by a current I (amperes).

If the two materials are joined at two points held at different temperatures, an open-circuit potential difference ΔV is produced as a result of a temperature difference ΔT between the junctions, the *Seebeck effect* (Fig. 3.7(*b*)). This leads to the differential *Seebeck coefficient*, $\alpha_{S\,ab}$, given by

$$\alpha_{S\,ab} = \lim_{\Delta T \to 0} \frac{\Delta V}{\Delta T} \text{ volts per degree Celsius} \tag{3.7}$$

The e.m.f. generated when $\Delta T = 1°C$ is sometimes called the *thermoelectric power*. The two coefficients are related by Kelvin's law:

$$\alpha_{S\,ab} = \frac{\alpha_{P\,ab}}{T} \tag{3.8}$$

where T is the absolute temperature of the cold junction (Ref. 3.1).

Finally, where there is a temperature difference ΔT over part of a *single* conductor the passage of current I leads to thermal power P being generated (Fig. 3.7(c)). This is the *Thomson effect*, related to the Peltier and Seebeck effects, but of small practical importance and not considered in this book.

The junction of two metals to form a thermocouple has been used for a long time as a method of measuring temperature, with copper-constantan or iron-constantan couples having values of α_S up to about $50\,\mu V/°C$. Correspondingly low values of α_P occur, so that little energy is transferred when a current is passed through the junction, with a consequently small cooling effect. This is because the conduction electrons all have energies close to the Fermi level, and very small energy changes occur when a current flows through the junction. However, for the ohmic contact between a metal and a non-degenerate semiconductor α is$_P$ much larger and a significant cooling effect may be obtained.

Consider an n-type semiconductor sandwiched between two metals (Fig. 3.8(a)). If a potential difference is applied as shown, only the higher-energy electrons in metal 1 will be able to move over the potential barrier $\phi_s - \phi_m$ into the semiconductor (Fig. 3.8(b)). Thus in metal 1 the average electron energy is reduced, while in metal 2 it is increased, so that heat is transferred from metal 1 to metal 2. If a p-type semiconductor is substituted and the same voltage applied (Fig. 3.8(c)), a hole current will flow due to movement of electrons in the valence bands under the potential barrier $\phi_m - \phi_s$. Thus low-energy electrons are removed from metal 1, increasing its average energy and reducing the average energy of metal 2, so that in this case heat is transferred from metal 2 to metal 1. The Peltier coefficients may be obtained from the energy diagrams, since the electrons crossing from a metal to an n-type semiconductor possess potential energy ($\phi_s - \phi_m$) and mean kinetic energy $\bar{w}$, which is proportional to temperature. Thus the energy transported per unit charge is

$$\alpha_{P\,mn} = -\frac{\bar{w} + (\phi_s - \phi_m)}{e} \tag{3.9}$$

the minus sign indicating removal of energy from the metal. Similarly, for a metal-to-p-type-semiconductor contact,

$$\alpha_{P\,mp} = +\frac{\bar{w} + (\phi_m - \phi_s)}{e} \tag{3.10}$$

the plus sign indicating energy transfer to the metal.

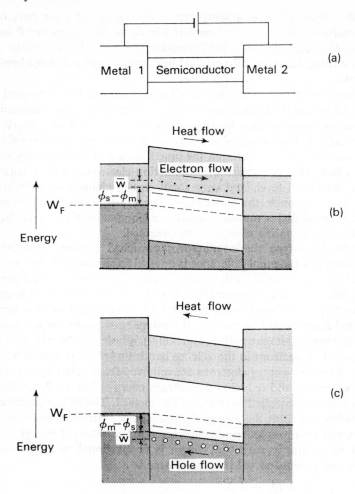

FIG. 3.8 Thermoelectric cooling

(a) Metal-semiconductor-metal ohmic contacts
(b) Energy diagram using *n*-type semiconductor
(c) Energy diagram using *p*-type semiconductor

A thermoelectric cooling device is obtained by arranging *n*- and *p*-type materials in couples (Fig. 3.9). The passage of current due to the indicated applied voltage will cause all the top metal surfaces to be cooled and the

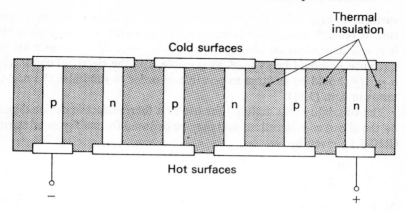

FIG. 3.9 Thermoelectric cooling device

lower ones to be heated, while reversal of the current will cause reversal of the direction of heat flow. Thus if one side of the device is fixed to a suitable heat sink maintained at room temperature, refrigeration can be carried out of an article attached to the other side. A *p-n* bismuth-telluride couple has a Seebeck coefficient of about $400 \, \mu\text{V}/^\circ\text{C}$, and for a well heat-insulated device with 16 couples, for example, a current of 10 A will cause a heat flow of about 3 W, maintaining a temperature difference of about 30°C between the two surfaces. From eqn. (3.6) the higher the current passed through the device the greater will be the rate of heat flow, but a limit is set by the heat dissipation due to the electrical resistance of the device and by the heat flowing in from the surroundings. It may be shown that the Joule heat produced in the resistance flows equally to the hot and cold surfaces, so that for a cooling unit of resistance R with the cold surface at temperature T_c, the equation governing the thermal condition of the load is

$$P_c \quad = \quad \alpha_S I T_c \quad - \quad \tfrac{1}{2} I^2 R \quad - \quad K \Delta T$$

| Net heat absorbed at cold junction | Peltier heat transferred from cold junction | Joule heat flowing to cold junction | Heat conducted from surroundings and hot junction |

(3.11)

K is the thermal conductance of the device which is reduced by efficient thermal insulation, and ΔT is the temperature difference between the

89

surfaces. A high value of α_S is desirable to give as large a drop in temperature as possible for a given current; α_S is used in the above equation since it is less dependent on temperature than is $-\alpha_P$. Also the thermal conductance K should be as small as possible, and the electrical conductivity $\sigma = 1/\rho$ should be as large as possible in order to minimize the negative terms in eqn. (3.11).

The suitability of a material for use as a thermoelectric device depends on the above considerations and may be deduced from a figure of merit, Z (Ref. 3.1), given by

$$Z = \frac{\alpha_s^2 \sigma}{K} \text{ per kelvin} \qquad (3.12)$$

At room temperature, for metal junctions Z is about $0 \cdot 1 \times 10^{-3}/\text{K}$, while for bismuth telluride it is about $2 \times 10^{-3}/\text{K}$, which indicates that semiconductors are better than metals for thermoelectric applications (see Problem 3.1).

The coefficient of performance or the efficiency of a thermoelectric refrigerator is defined as P_c/P_I, where P_I is the input power supplied to the device. P_I is partly dissipated in the resistance R to give an I^2R term, but the useful part of P_I is obtained from eqn. (3.7). It is expressed as the product of the current and the Seebeck voltage, ΔVI, so that

$$P_I = \alpha_S I \Delta T + I^2 R \qquad (3.13)$$

and

$$\frac{P_c}{P_I} = \frac{\alpha_S I T_c - \frac{1}{2} I^2 R - K \Delta T}{\alpha_S I \Delta T + I^2 R} \qquad (3.14)$$

THE *p-n* JUNCTION

The contact between p- and n-type semiconductors is particularly important since it forms the basis of electronic devices such as semiconductor diodes and transistors. It is a rectifying contact and is formed in a single crystal whose impurity atoms are changed from donors to acceptors at the junction.

An *abrupt* junction, as opposed to a graded one, discussed later in this chapter, is formed by *alloying* a Group III material and an n-type semiconductor at high temperature. The type of impurity atom then changes at a well-defined cross-section of the composite crystal, called the metallurgical junction.

In practice the densities of acceptors and donors are not equal, and in Fig. 3.10(*a*) an abrupt junction is illustrated in which the acceptor density, N_a, is considerably greater than the donor density, N_d. At the junction, electrons and holes have recombined, forming the depletion layer of width *d* in which there are no free charges. Since equal numbers of electrons and holes have recombined there must remain equal numbers of ionized impurity atoms on each side of the junction. Thus the areas enclosed under the charge-density diagrams are equal on each side of the junction (Fig. 3.10(*b*)), since they correspond to the numbers of charges, and the depletion layer extends further into the *n*-region than the *p*-region. It is shown on page 109 that the greater the charge density on either side of the junction, the narrower is the width of the depletion layer, *d*. As a result of the change from fixed negative to fixed positive charges a potential difference occurs across the junction, the diffusion potential ψ volts (Fig. 3.10(*c*)). On the energy diagram there is a corresponding fall in energy from the *p*- to the *n*-region of ψ electronvolts, arising as a result of the alignment of the Fermi levels. There are then equal probabilities of electrons and holes diffusing across the junction with zero applied bias.

In fact there are now two electron currents and two hole currents. The hole current consists of (*a*) minority holes from the *n*-region drifting into the *p*-region under the influence of the electric field *E* at the junction, and (*b*) majority holes from the *p*-region diffusing up the energy barrier. Similarly, the electron current consists of minority carriers drifting from the *p*-region and majority carriers diffusing from the *n*-region. The corresponding current densities are

$$J_p = ep\mu_p E - eD_p \frac{dp}{dx} \tag{3.15}$$

and

$$J_n = en\mu_n E + eD_n \frac{dn}{dx} \tag{3.16}$$

for holes and electrons respectively, from eqns. (2.80), (2.101) and (2.102).

Equilibrium Conditions, Zero Bias

Under equilibrium conditions both J_p and J_n are zero. Thus for holes,

$$ep\mu_p E = eD_p \frac{dp}{dx}$$

91

Physical Electronics

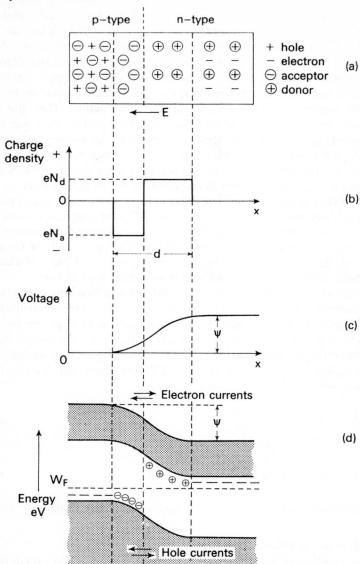

FIG. 3.10 The *p-n* junction

(*a*) Impurities and current carriers
(*b*) Charge density
(*c*) Junction potential
(*d*) Energy diagram

so that

$$\frac{p_p}{D_p} E \, dx = \frac{dp}{p} \qquad (3.17)$$

Similarly for electrons,

$$\frac{\mu_n}{D_n} E \, dx = -\frac{dn}{n} \qquad (3.18)$$

Substituting from eqns. (2.107) and (2.108),

$$\frac{\mu_p}{D_p} = \frac{\mu_n}{D_n} = \frac{e}{kT}$$

which gives

$$\frac{e}{kT} E \, dx = \frac{dp}{p} = -\frac{dn}{n} \qquad (3.19)$$

Considering two points, 1 in the *p*-region and 2 in the *n*-region outside the junction, and integrating eqn. (3.19),

$$-\frac{e}{kT}(V_2 - V_1) = \log_e \frac{p_2}{p_1} = \log_e \frac{n_1}{n_2} \qquad (3.20)$$

Now, for an abrupt junction, $p_1 = p_p \approx N_a$, and $p_2 = p_n \approx n_i^2/N_d$, where N_a and N_d are the densities of the acceptor and donor atoms respectively, which are all assumed to be ionized. Also $n_1 = n_p \approx n_i^2/N_a$ and $n_2 = n_n \approx N_d$. Finally $V_2 - V_1 = \psi$, the diffusion potential.

Hence, from eqn. (3.20),

$$\exp -\frac{e\psi}{kT} = \frac{p_n}{p_p} = \frac{n_p}{n_n} = \frac{n_i^2}{N_a N_d} \qquad (3.21)$$

and

$$\psi = \frac{kT}{e} \log_e \frac{N_a N_d}{n_i^2} \qquad (3.22)$$

ψ may thus be calculated from the doping densities and the energy gap which occurs in the expression for n_i (eqn. (2.30)). Using the value of n_i obtained on page 42 and taking as typical values $N_a = 3 \times 10^{23}$, $N_d = 3 \times 10^{22}$, $kT/e = 25 \, \text{mV}$ at room temperature gives $\psi = 0.38 \, \text{V}$ for germanium and $0.77 \, \text{V}$ for silicon.

The relationship between the electron densities on each side of the junction and also the corresponding relationship between the hole densities may be obtained from eqn. (3.21):

$$\frac{n_p}{n_n} = \frac{p_n}{p_p} = \exp\left(-\frac{e\psi}{kT}\right)$$

Thus

$$p_n = p_p \exp\left(-\frac{e\psi}{kT}\right) \tag{3.23}$$

and

$$n_p = n_n \exp-\left(\frac{e\psi}{kT}\right) \tag{3.24}$$

These equations specify the equilibrium conditions in which drift and diffusion currents balance in the depletion layer. If low-resistance ohmic contacts are made to the *p*- and *n*-regions (Fig. 3.11(*a*)) there will be changes in potential at these contacts exactly compensating the junction potential, so that the potential difference between the contacts is zero (Fig. 3.11(*b*)). Thus when they are joined no net current flows.

Forward Bias

Suppose the *p*-region is now made positive with respect to the *n*-region (Figs. 3.11(*c*) and (*d*)). The height of the junction potential will be reduced, so that more electrons can diffuse across the junction from the *n*-region and more holes from the *p*-region, increasing the total diffusion current I_F. The current I_0 will remain unchanged, since this is due to the drift of holes from the *n*-region and electrons from the *p*-region across the junction under the influence of the field *E*. This field is sufficient to extract all the minority carriers from each region, even when the junction potential is reduced.

If the current I_F is large there will be a voltage drop in the bulk of the semiconductor, away from the junction region. This will be mainly in the *n*-region when the *p*-region is the more heavily doped. This voltage drop may often be ignored in comparison with the drop across the depletion layer, and it may then be assumed that the whole of the applied voltage *V* reduces the junction potential to $\psi - V$. Consider first the effect of lowering the junction potential on the hole component of I_F. The effect on the electron component can then be deduced by comparison.

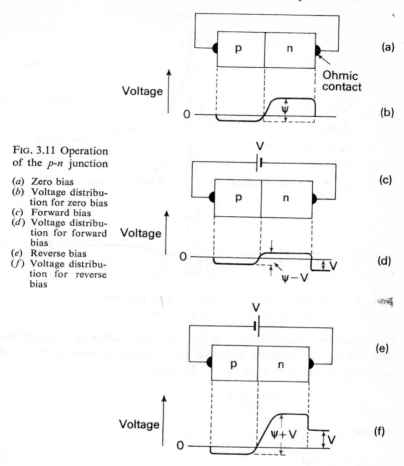

Fig. 3.11 Operation of the *p-n* junction

(*a*) Zero bias
(*b*) Voltage distribution for zero bias
(*c*) Forward bias
(*d*) Voltage distribution for forward bias
(*e*) Reverse bias
(*f*) Voltage distribution for reverse bias

The equilibrium hole density in the *n*-region is p_n, which is increased to a new value p_e due to the injection of excess holes from the *p*-region (Fig. 3.12(*a*)). It is assumed in the following analysis that the density of holes injected into the *n*-region is small compared to that of the majority carriers, which are electrons. This is known as *low-level injection* in which the charge neutrality in the *n*-region is undisturbed and the holes move only by diffusion to the ohmic contact. Recombination with the electrons also occurs, but since the holes are replaced continuously a constant excess density $p_e - p_n = \Delta p(0)$ is maintained at the boundary between

95

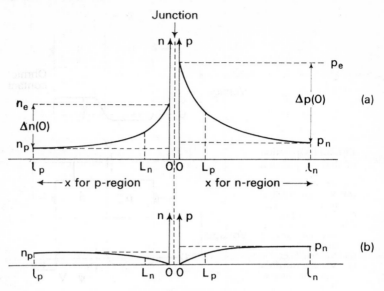

FIG. 3.12 Excess carrier distributions for the *p-n* junction

(a) Forward bias   (b) Reverse bias

the depletion layer and the *n*-region. It is reasonable to suppose that p_e depends on the junction potential in a similar manner to p_n, so that, replacing ψ with $\psi - V$ in eqn. (3.23), gives

$$p_e = p_p \exp\left[-\frac{e(\psi - V)}{kT}\right] \qquad (3.25)$$

Hence

$$p_e = p_n \exp\frac{eV}{kT} \qquad (3.26)$$

again using eqn. (3.23). Then

$$\Delta p(0) = p_e - p_n = p_n\left[\exp\left(\frac{eV}{kT}\right) - 1\right] \qquad (3.27)$$

which does not depend on p_p but on p_n, the minority carrier density in the *n*-region.

The current density carried by the injected holes at the edge of the depletion layer is then

$$J_p(0) = -eD_p\left(\frac{\partial \Delta p}{\partial x}\right)_{x=0} \tag{3.28}$$

Recombination occurs in the *n*-region, which has a length $l_n > L_p$, the diffusion length for holes. Thus an exponential decay of excess hole density will occur with increasing values of x as given by eqn. (2.119), and

$$\Delta p = \Delta p(0)\exp\left(-\frac{x}{L_p}\right)$$

so that

$$\frac{\partial \Delta p}{\partial x} = \frac{1}{L_p}\Delta p(0)\exp\left(-\frac{x}{L_p}\right) \tag{3.29}$$

and

$$\left(\frac{\partial \Delta p}{\partial x}\right)_{x=0} = -\frac{\Delta p(0)}{L_p} \tag{3.30}$$

Hence

$$J_p(0) = \frac{eD_p}{L_p}\Delta p(0) \tag{3.31}$$

and, substituting from eqn. (3.27),

$$J_p(0) = \frac{eD_p p_n}{L_p}\left[\exp\left(\frac{eV}{kT}\right) - 1\right] \tag{3.32}$$

which is of the same form as eqn. (3.5).

The electrons in the *n*-region which recombine with the injected holes are replaced by electrons from the external circuit, entering through the ohmic contact. At the junction the current is carried mainly by holes, but as x increases into the *n*-region the current is increasingly carried by electrons. At the ohmic contact all the current is carried by electrons and the hole density has fallen to the equilibrium value, p_n, which is one way of defining an ohmic contact. Thus the *total* current is constant throughout the *n*-region, at the value given by eqn. (3.32), but the proportion carried by holes or electrons changes as x is increased. Under low-current conditions the space charge carried by the injected holes is always neutralized by an equal number of excess electrons, which is small compared with the equilibrium density n_n.

The effect of lowering the junction potential on the flow of electrons from the n- to the p-region may be deduced in a similar way. An excess electron density, n_e, is set up at the boundary between the depletion layer and the p-region. n_e is obtained from eqn. (3.24) by replacing ψ with $\psi - V$, so that

$$n_e = n_n \exp\left[-\frac{e(\psi - V)}{kT}\right] \tag{3.33}$$

$$= n_p \exp\left(\frac{eV}{kT}\right) \tag{3.34}$$

and

$$\Delta n(0) = n_e - n_p = n_p\left[\exp\left(\frac{eV}{kT}\right) - 1\right] \tag{3.35}$$

Recombination occurs in the p-region, so that

$$\Delta n = \Delta n(0)\exp\left(-\frac{x}{L_n}\right) \tag{3.36}$$

with $x = 0$ taken at the depletion layer boundary and x increasing positively to the left of the boundary. These equations lead to the expression

$$J_n(0) = \frac{eD_n n_p}{L_n}\left[\exp\left(\frac{eV}{kT}\right) - 1\right] \tag{3.37}$$

for the electron current density in the p-region. This current remains constant, but an increasing proportion of it is carried by holes as x increases into the p-region, until at the ohmic contact it is entirely a hole current.

The processes of hole and electron injection occur simultaneously and independently so that the total current density, J, is the sum of the two components. Putting $I = JS$, where S is the junction area, the junction current is given by eqns. (3.32) and (3.37) and

$$I = eS\left(\frac{D_p p_n}{L_p} + \frac{D_n n_p}{L_n}\right)\left[\exp\left(\frac{eV}{kT}\right) - 1\right] \tag{3.38}$$

where for forward bias, V is positive. This is the basic equation for the I/V characteristic of a p-n junction diode, and by comparison with eqn. (3.5) it may be seen that

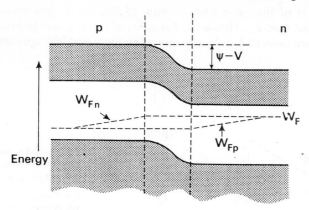

Fɪɢ. 3.13 Quasi-Fermi levels

$$eS\left(\frac{D_p p_n}{L_p} + \frac{D_n n_p}{L_n}\right) = I_0 \tag{3.39}$$

This is the current that will flow when reverse bias is applied, i.e. with V negative.

In practice two changes may be made in the structure of the device. Firstly, in order to keep the forward voltage drop low at high currents the diode is shortened so that the length of one or both regions is less than the diffusion length (Fig. 3.14(*a*)). This also reduces the transit time through the device. The excess carrier distributions then become approximately linear, since they correspond to the initial part of an exponential decay. The slopes of the distribution are determined by the lengths l_n and l_p respectively, so that

$$\frac{\partial \Delta p}{\partial x} = -\frac{\Delta p(0)}{l_n} \tag{3.40}$$

and

$$\frac{\partial \Delta n}{\partial x} = -\frac{\Delta n(0)}{l_p} \tag{3.41}$$

giving

$$I = eS\left(\frac{D_p p_n}{l_n} + \frac{D_n n_p}{l_p}\right)\left[\exp\left(\frac{eV}{kT}\right) - 1\right] \tag{3.42}$$

99

This equation is of the same form as eqn. (3.38), but I_0 has increased since $l_n < L_p$ and $l_p < L_n$. Thus a smaller value of V is required to produce the same forward current I at the expense of an increased leakage current.

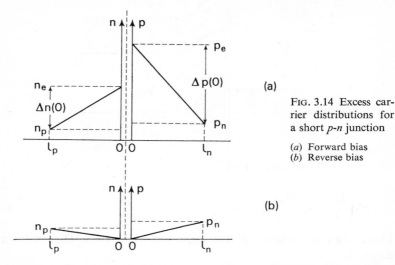

Fig. 3.14 Excess carrier distributions for a short *p-n* junction

(a) Forward bias
(b) Reverse bias

Secondly, a diode is usually constructed by starting with a material of relatively high resistivity and then doping with impurity to convert part of it to a lower-resistivity region of the opposite type. If the starting material is *n*-type this gives a p^+-*n* structure, which corresponds to the distributions of Figs. 3.12 and 3.14; and if the starting material is *p*-type an n^+-*p* structure results. For a p^+-*n* structure p_p will be much greater than n_n and consequently n_p will be much less than p_n, since

$$n_p = \frac{n_i^2}{p_p} \quad \text{and} \quad p_n = \frac{n_i^2}{n_n}$$

from eqns. (2.45) and (2.44). Hence the electron component in eqns. (3.38) and (3.42) will be much less than the hole component, so that the current will be carried mainly by holes.

Quasi-Fermi Levels*

When injection of minority carriers occurs the product of the electron and hole concentrations is no longer n_i^2, since the excess minority carriers

---

* The term "imref" for a quasi-Fermi level has been coined from the word Fermi spelt backwards.

are compensated by an equal number of excess majority carriers drawn into the region. For low-level injection the relative increase in the density of the majority carriers is very small, but the density of the minority carriers is greatly increased. On the energy diagram this is shown by using different Fermi levels for electrons and holes, which are illustrated in Fig. 3.13 for a forward-biased *p-n* junction. They are called *quasi-Fermi levels*, since they apply when the equilibrium conditions have been disturbed, and express the increased probability of finding minority carriers in otherwise empty levels.

In the *p*-region remote from the junction the Fermi level corresponds only to holes, since all the injected electrons have recombined. Towards the junction the hole Fermi level remains constant, but since the probability of an electron occupying a given level increases with W_F (eqn. (2.21)) and the electron density also increases exponentially with distance (Fig. 3.12(b)), the electron Fermi level rises linearly. In the depletion layer the quasi-Fermi levels remain constant since it is assumed that no recombination is occurring here and the excess electron and hole densities are determined by the applied voltage. In the *n*-region the probability of a hole occupying a given level falls as W_F increases (eqn. (2.24)) and the hole density also decreases exponentially away from the junction. Thus the hole Fermi level rises linearly, finally reaching the equilibrium level for electrons only when all the injected holes have recombined.

Reverse Bias

If the polarity of the bias is reversed, so that the *p*-region is negative with respect to the *n*-region, the junction potential is increased (Figs. 3.11(e) and (f)). This will prevent the diffusion of holes into the *n*-region and electrons into the *p*-region, but will increase the field E across the junction. Thus the extraction of minority carriers from each region, which constitutes the current I_0, will continue and diffusion of holes from the *n*-region and electrons from the *p*-region will occur to maintain this current across the junction. E is large enough to extract all the minority carriers near the junction even at low reverse bias voltages, so that the densities of minority carriers are zero at the edges of the depletion layer (Figs. 3.12(b) and 3.14(b)). Since diffusion is occurring away from the junction an exponential distribution is set up on each side, defined by the same diffusion lengths L_p and L_n as for forward bias, and corresponding linear distribution for a short diode. Hence on average all the minority carriers which are generated thermally within distances L_p and L_n from the edge of the depletion layer are extracted by the field across the junction.

The reverse bias current I_0 will be independent of the voltage across an ideal diode for voltages above a small value. At room temperature kT/e is about 25 mV so that if V is -75 mV, $\exp(eV/kT)$ is 1/20. Hence in eqn. (3.5) the exponential term is much less than unity and $I = -I_0$ for values of reverse bias greater than about 75 mV. However, a rise in temperature will cause both components of I_0 to increase, since both p_n and n_p are proportional to n_i^2 (eqns. (2.44) and (2.45)), and n_i^2 is approximately proportional to $\exp - W_g/kT$ from eqn. (2.30). This results in an almost exponential increase of I_0 with temperature.

The Effect of Temperature on Diode Characteristics

Differentiating eqn. (2.30) gives

$$\frac{dn_i^2}{dT} = \frac{W_g}{kT^2}\, n_i^2 + \frac{3}{T}\, n_i^2$$

or

$$\frac{dn_i^2}{n_i^2} = \left(\frac{W_g}{kT} + 3\right)\frac{dT}{T} = \frac{dI_0}{I_0} \tag{3.43}$$

Putting $W_g = 0.72$ eV for germanium and 1.01 eV for silicon and taking $kT = 0.025$ eV at room temperature,

$$\frac{dI_0}{I_0} = 32\,\frac{dT}{T} \text{ for germanium} \tag{3.44}$$

$$= 43\,\frac{dT}{T} \text{ for silicon} \tag{3.45}$$

As a guide to the rate of change from room temperature, 293 K, we can consider the case where $dI_0 = I_0$, or the current has doubled itself. The corresponding change in temperature, dT, then becomes 9.2°C for germanium and 6.6°C for silicon.

The change in I_0 will also result in a decrease in the value of V required to keep the forward current constant. From eqn. (3.5), for forward bias $eV/kT \gg 1$, so that

$$\log_e \frac{I}{I_0} = \frac{eV}{kT} \tag{3.46}$$

$$V = \frac{kT}{e} \log_e \frac{I}{I_0}$$

and

$$\frac{dV}{dT} = \frac{k}{e} \log_e \frac{I}{I_0} - \frac{kT}{e} \frac{1}{I_0} \frac{dI_0}{dT} \tag{3.47}$$

$$= \frac{V}{T} - 32 \frac{k}{e} \text{ for germanium} \tag{3.48}$$

$$= \frac{V}{T} - 43 \frac{k}{e} \text{ for silicon} \tag{3.49}$$

using eqns. (3.44) and (3.45). If a forward bias of 0·2 V is taken for germanium and 0·6 V for silicon, where the higher value is necessary for the same forward current due to I_0 being much less for silicon, then at room temperature the rate of change of V to keep I constant is about -2 mV/°C rise in temperature for germanium diodes and $-1·5$ m V/°C for silicon diodes. Thus the effect of temperature is to cause the reverse current to rise, and forward current will also rise in applications where the forward bias may be considered constant. Both of these effects are also important in the operation of transistors at high temperatures.

REVERSE BREAKDOWN MECHANISMS

As the reverse voltage across a junction is increased, I_0 remains constant until a voltage V_B is reached, the *reverse breakdown* voltage. Here I_0 increases very rapidly due to *avalanche multiplication* of the reverse current. This occurs because the field across the junction has become so large, owing to the applied bias, that an electron or a hole can acquire sufficient energy to ionize a lattice atom on collision and form a new electron-hole pair for each carrier. This process is repeated by the new carriers and is illustrated in Fig. 3.15. It is analogous to the mechanism of the Townsend gas discharge, described on page 311. The current I_0 is then multiplied by a factor M given by

$$M = \frac{1}{1 - (V_R/V_B)^n} \tag{3.50}$$

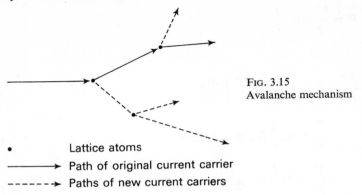

FIG. 3.15
Avalanche mechanism

•   Lattice atoms

————▸   Path of original current carrier

------▸   Paths of new current carriers

so that the reverse current is

$$I_R = \frac{I_0}{1 - (V_R/V_B)^n} \tag{3.51}$$

n is a constant whose value depends on whether the material is germanium or silicon and whether the current is carried mainly by electrons or holes. It lies between 2 and 6 and controls the rate of change of current with voltage as V_R approaches V_B (Fig. 3.16). Eqn. (3.51) suggests that $I_R = \infty$

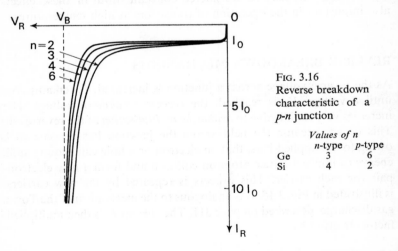

FIG. 3.16
Reverse breakdown characteristic of a *p-n* junction

| | *Values of n* | |
	n-type	*p*-type
Ge	3	6
Si	4	2

when $V_R = V_B$, but in fact I_R is limited by the resistance external to the junction. As shown on page 110, V_B is inversely proportional to the impuirty density, and thus increases with the resistivity of the side with fewer impurity atoms. In practice V_B may be as high as 500 V in germanium and 1 kV in silicon diodes. Thus when breakdown does occur there will be a power dissipation within the diode of $I_R V_B$, which may be sufficient to cause destruction of the diode by overheating if I_R is not restricted. However, if the dissipation is kept within safe limits the breakdown curve is reversible, and when V_R is less than V_B, $I_R = I_0$ once more.

Zener Breakdown

If the doping density is increased to above about $10^{24}/m^3$, or 1 in 10^5 atoms replaced, a different mechanism is responsible for breakdown. The depletion layer becomes very narrow and so a very high electric field exists across the junction (page 110) which can ionize the lattice atoms by removing electrons directly from their valence bonds, causing *field emission* (see also page 256). This is known as *Zener breakdown* and occurs for reverse voltages below about 5 V. The corresponding minimum field for Zener breakdown is about 1.2×10^8 V/m, which compares with about 5×10^7 V/m for avalanche breakdown in silicon. For breakdown voltages between about 5 and 8 V both avalanche and Zener effects occur simultaneously, while in diodes breaking down above 8 V the avalanche effect predominates. It should be noted that avalanche breakdown requires a sufficient mean free path for carriers to acquire the energy needed for ionization, while this is not necessary for Zener breakdown, which therefore occurs mainly within narrow depletion layers. Another difference between the two mechanisms is in the temperature coefficient of V_B, which is *negative* for Zener breakdown and *positive* for avalanche breakdown, being zero for a diode breaking down at a reverse voltage of about 5.2 V. The coefficient is of the order of 0.01 % per kelvin and in view of this stability and the constancy of V_B for large changes in reverse current, a diode working at its breakdown voltage may be used as a reference voltage source. Such a *breakdown diode* is designed to withstand a fairly large reverse current without damage and is commonly called a *Zener diode*, even though avalanche multiplication is occurring.

Further increases in doping density, up to the limit of the solubility of impurities in the semiconductor, will cause V_B to move closer to zero, and eventually breakdown occurs for small forward bias voltages. This results in another device, the *tunnel diode*, which is described on page 345.

PROPERTIES OF THE DEPLETION LAYER

The charge densities on each side of the depletion layer are shown in Fig. 3.17(a), where $x = 0$ at the junction between the p- and n-regions.

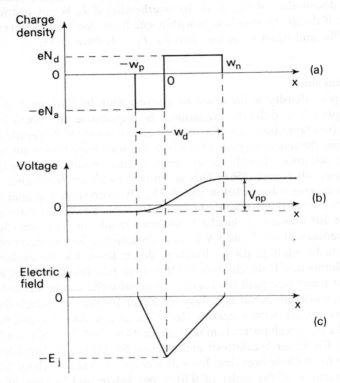

FIG. 3.17 Depletion layer of an abrupt junction

(a) Charge density distribution
(b) Voltage distribution
(c) Electric field

In order to simplify the analysis it is assumed that there is a constant density of ionized donor atoms N_d extending a distance w_n into the n-region and a constant density of ionized acceptor atoms N_a extending a distance $-w_p$ into the p-region. These charge densities may be related to the junction potential, the width of the depletion layer $w_n + w_p$, and the field E existing within it. In addition, the depletion layer will have a capacitance since it contains fixed charges of opposite sign separated by a high-resistance region, as in a parallel-plate capacitor.

Consider a surface within the depletion layer normal to the x-axis, where the electric flux density is D coulombs per square metre. The electric field due to this surface is

$$E = \frac{D}{\epsilon} \tag{3.52}$$

where $\epsilon = \epsilon_r\epsilon_0$ is the permittivity and ϵ_r the relative permittivity. Hence

$$\frac{dE}{dx} = \frac{d}{dx}\frac{D}{\epsilon} \tag{3.53}$$

and in the p-region of the depletion layer,

$$\frac{dE}{dx} = -\frac{eN_a}{\epsilon} \qquad (-w_p < x < 0) \tag{3.54}$$

while in the n-region of the depletion layer,

$$\frac{dE}{dx} = \frac{eN_d}{\epsilon} \qquad (0 < x < w_n) \tag{3.55}$$

which are forms of Poisson's equation for a region containing space charge. Integrating eqns. (3.54) and (3.55),

$$E_p = -\frac{eN_a x}{\epsilon} + C_1 \text{ in the } p\text{-region} \tag{3.56}$$

and

$$E_n = \frac{eN_d x}{\epsilon} + C_2 \text{ in the } n\text{-region} \tag{3.57}$$

where C_1 and C_2 are constants.

The electric field is continuous and exists only within the boundaries of the depletion layer. Hence $E_p = 0$ when $x = -w_p$, and $E_n = 0$ when $x = w_n$, so that

$$C_1 = -\frac{N_a w_p}{\epsilon} \tag{3.58}$$

and

$$C_2 = -\frac{eN_d w_n}{\epsilon} \tag{3.59}$$

Then

$$E_p = -\frac{eN_a}{\epsilon}(x + w_p) \tag{3.60}$$

$$E_n = \frac{eN_d}{\epsilon}(w_n - x) \tag{3.61}$$

Since x is negative in the p-region and positive in the n-region, the field in each region increases as x approaches zero (Fig. 3.17(c)) until, when $x = 0$,

$$E_p = E_n = E_j = -\frac{eN_a w_p}{\epsilon_r \epsilon_0} = -\frac{eN_d w_n}{\epsilon_r \epsilon_0} \tag{3.62}$$

E_j is thus the maximum field existing in the depletion layer, and is a controlling factor in reverse breakdown. Also from eqn. (3.62),

$$N_a w_p = N_d w_n \tag{3.63}$$

which expresses the fact that the total negative charge on one side of the junction equals the total positive charge on the other side, and also that the depletion layer penetrates a shorter distance into the more heavily doped region.

Junction Potential

The junction potential as a function of the width of the depletion layer may be found by integrating eqns. (3.60) and (3.61), since $V = -\int E \, dx$. Thus

$$V_p = \frac{eN_a}{2\epsilon}x^2 + \frac{eN_a}{\epsilon}w_p x + C_3 \tag{3.64}$$

$$V_n = -\frac{eN_d}{2\epsilon}x^2 + \frac{eN_d}{\epsilon}w_n x + C_4 \tag{3.65}$$

and the variations of V_p and V_n with x are shown in Fig. 3.17(b).

Since the voltage is continuous across the junction it must be zero at $x = 0$, so that $C_3 = C_4 = 0$. Also the voltage will be constant for values of $x \leqslant -w_p$ and for $x \geqslant w_n$. Putting $x = -w_p$ in eqn. (3.64) and $x = w_n$ in eqn. (3.65),

$$V_p(-w_p) = -\frac{eN_a}{2\epsilon} w_p^2 \tag{3.66}$$

$$V_n(w_n) = \frac{eN_d}{2\epsilon} w_n^2 \tag{3.67}$$

The junction potential is then $V_{np} = \psi - V$, where V is the external bias voltage, so subtracting eqn. (3.66) from eqn. (3.67),

$$V_{np} = \psi - V = \frac{\epsilon}{2\epsilon_r\epsilon_0} (N_a w_p^2 + N_d w_n^2) \tag{3.68}$$

Width of Depletion Layer
Since $N_a w_p = N_d w_n$ eqn. (3.68) yields

$$\psi - V = \frac{e}{2\epsilon} N_a^2 w_p^2 \left(\frac{1}{N_a} + \frac{1}{N_d}\right)$$

$$= \frac{e}{2\epsilon} \frac{N_a^2 w_p^2}{N_j} \tag{3.69}$$

and

$$\psi - V = \frac{e}{2\epsilon} \frac{N_d^2 w_n^2}{N_j} \tag{3.70}$$

where

$$\frac{1}{N_j} = \frac{1}{N_a} + \frac{1}{N_d} \tag{3.71}$$

Hence

$$w_p = \frac{1}{N_a}\left[\frac{2\epsilon N_j(\psi - V)}{e}\right]^{1/2} \tag{3.72}$$

$$w_n = \frac{1}{N_d}\left[\frac{2\epsilon N_j(\psi - V)}{e}\right]^{1/2} \tag{3.73}$$

and the width of the depletion layer is

$$w_d = w_n + w_p = \left[\frac{2\epsilon_r\epsilon_0(\psi - V)}{eN_j}\right]^{1/2} \tag{3.74}$$

As mentioned on page 91, w_d decreases as the doping density is increased. In a p^+-n diode, where $N_a \gg N_d$, $N_j \approx N_d$ and in a n^+-p diode with $N_d \gg N_a$, $N_j \approx N_a$, so that in each case w_d is controlled by the impurity density of the high-resistivity side. w_d also depends on the applied bias and increases with the reverse voltage (V negative). This effect is particularly important in transistors (page 134).

Electric Field at the Junction
The field at the junction is obtained by substituting the expression for w_p or w_n into eqn. (3.62), which gives

$$E_j = \left[\frac{2eN_j(\psi - V)}{\epsilon_r \epsilon_0} \right]^{1/2} \tag{3.75}$$

E_j increases with the reverse voltage and also with the doping density. Thus, the voltage applied to achieve a given field for breakdown of the junction will be reduced as N_j is increased.

Depletion Layer Capacitance
The depletion layer capacitance per unit area of junction, C_j, is given by the ratio of the change of the charge per unit area to the change of applied junction potential. Thus

$$C_j = \frac{dQ}{dV_{np}} \tag{3.76}$$

where $Q = eN_a w_p$ on the p-side. Hence

$$C_j = eN_a \frac{d_{wp}}{dV_{np}} \tag{3.77}$$

and differentiating eqn. (3.72) with respect to ($\psi - V$),

$$C_j = \left[\frac{e\epsilon_r \epsilon_0 N_j}{2} \right]^{1/2} (\psi - V)^{-1/2} \text{ farads/m}^2 \tag{3.78}$$

Thus C_j increases with the doping density and for a rectifier diode has a value in the order of $10\,\mathrm{pF/mm^2}$. Also $C_j \propto V_R^{-1/2}$ for the abrupt junction (Fig. 3.18), where V_R is the reverse voltage, and so the junction may be used as a voltage-dependent capacitor for remote tuning purposes or as a parametric amplifier (page 360). However, if the diode is to be used as a rectifier at very high frequencies, C_j degrades its performance by

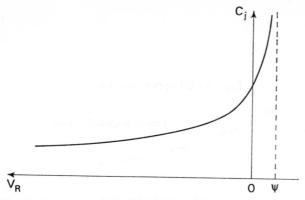

Fɪɢ. 3.18 Depletion layer capacitance and reverse voltage (abrupt
junction)

by-passing the junction, so that a diode with a small contact area between
a metal and an *n*-type semiconductor is preferred (page 77).

GRADED JUNCTIONS

The abrupt junction model considered above is most applicable to a *p-n*
junction produced by the alloying process. Another common method of
production involves the evaporation of impurity atoms to change one type
of semiconductor into the other to form a *graded* junction, and this
method is now widely used in the manufacture of high-frequency transistors and integrated circuits. In a graded junction there is a gradual transition from one type of impurity to the other across the junction, which may
be approximated by a linear distribution or *grade* in the region of the
junction.

The manufacturing process results in the diffusion of impurities into
the base material to give a distribution $N(x)$ as shown in Fig. 3.19. The
impurity density is N_0 at the surface and falls approximately exponentially
with penetration. At the depth x_j the impurity density of the diffused
material equals the impurity density of the base material, N_b. Thus, if
p-type impurities have been diffused into an *n*-type semiconductor at high
temperature, the material changes from *p*- to *n*-type at $x = x_j$, which
defines the junction. An approximate expression for the distribution is

$$N(x) = N_0 \exp\left(-\frac{ax}{N_b}\right) \tag{3.79}$$

111

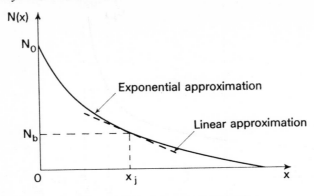

Fig. 3.19 Impurity distribution in a graded junction

where a is the *grade constant* in units of atoms/m^4. At $x = x_j$,

$$N(x_j) = N_b = N_0 \exp\left(-\frac{ax_j}{N_b}\right) \tag{3.80}$$

and the slope of the distribution is

$$\left.\frac{dN(x)}{dx}\right|_{x=x_j} = -a\,\frac{N_0}{N_b}\exp\left(-\frac{ax_j}{N_b}\right) \tag{3.81}$$

If we assume that the distribution at $x = x_j$ varies approximately linearly with distance, then

$$N(x) = -ax \tag{3.82}$$

and the slope is $-a$. Putting eqn. (3.81) equal to $-a$ and taking logarithms leads to

$$a = \frac{N_b}{x_j}\log_e\frac{N_0}{N_b} \tag{3.83}$$

for the grade constant.

The corresponding charge density in the region of the junction is illustrated in Fig. 3.20(a) and is also represented by a linear function

$$Q(x) = eax \tag{3.84}$$

The number of negative charges due to ionized acceptors in the p-region is equal to the number of positive charges due to ionized donors in the

112

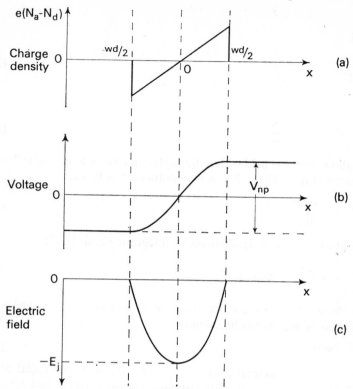

FIG. 3.20 Depletion layer of a graded junction
(*a*) Charge density distribution
(*b*) Voltage distribution
(*c*) Electric field

n-region. Then, taking x_j as the origin, the depletion layer extends an equal distance $w_d/2$ into each region owing to the linear distribution. When Gauss's theorem is applied,

$$\frac{dE}{dx} = \frac{eax}{\epsilon}$$

(3.85)

and

$$E = \frac{eax^2}{2\epsilon} + C_1$$

(3.86)

When $x = \pm w_d/2$, $E = 0$, since $Q(x) = 0$,

so that

$$C_1 = -\frac{eaw_d^2}{8\epsilon} \qquad (3.87)$$

and

$$E = \frac{eax^2}{2\epsilon} - \frac{eaw_d^2}{8\epsilon} \qquad (3.88)$$

which shows that the field within the depletion layer is a parabolic function of distance (Fig. 3.20(*c*)). It is a maximum at $x = 0$, where

$$E_j = -\frac{eaw_d^2}{8\epsilon_r\epsilon_0} \qquad (3.89)$$

The junction voltage is obtained by integrating eqn. (3.81):

$$V = -\frac{eax^3}{6\epsilon} + C_1x + C_2 \qquad (3.90)$$

which shows that it is a cubic function of distance (Fig. 3.20(*b*)). The voltage across the graded junction is

$$V_{np} = V_n - V_p \qquad (3.91)$$

where V_n is the potential at $x = w_d/2$, and V_p is the potential at $x = -w_d/2$. If we assume that all the applied voltage is developed across the depletion layer so that $V_{np} = V$, evaluation of eqn. (3.85) leads to

$$w_d = \left(\frac{12\epsilon_r\epsilon_0 V}{ea}\right)^{1/3} \qquad (3.92)$$

Thus, as the grade constant a is increased, the maximum junction field is increased and the depletion layer width is decreased.

Finally, the capacitance per unit area of the depletion layer is obtained from

$$C_j = \frac{\epsilon}{w_d} \qquad (3.93)$$

the expression for the equivalent parallel-plate capacitance. Substituting for w_d from eqn. (3.92) leads to

$$C_j = \left(\frac{\epsilon_r^2 \epsilon_0^2 ea}{12} \right)^{1/3} V^{-1/3} \tag{3.94}$$

which indicates that C_j is proportional to $V_R^{-1/3}$ for a reverse biased graded junction. This is confirmed by experiment for low reverse voltages, but as the voltage is increased C_j becomes proportional to $V_R^{-1/2}$ as for the abrupt junction (Ref. 3.3).

The diffusion potential ψ does not appear in eqns. (3.91), (3.92) and (3.94), so that V refers to the magnitude of the reverse bias voltage V_R. The reason for excluding ψ is that it is not a constant but depends on the bias voltage in a graded junction. An approximate expression for ψ may be obtained from the zero bias condition for a p-n junction, which led to eqn. (3.20). In the graded junction the impurity density varies according to eqn. (3.84), so that

$$N_a - N_d = -ax \tag{3.95}$$

and the impurity changes from an effective acceptor to an effective donor type as x is increased from a negative value. Then, assuming that all the impurity atoms are ionized, the hole density at the edge of the depletion layer on the p-side is

$$p_1 = \frac{aw_d}{2} \tag{3.96}$$

The hole density at the corresponding point on the n-side is related to the effective donor density by eqn. (2.37), so that

$$p_2 = \frac{2n_i^2}{aw_d} \tag{3.97}$$

Then rearranging eqn. (3.20) and putting $V_2 - V_1 = \psi$,

$$\psi = \frac{kT}{e} \log_e \frac{p_1}{p_2} = \frac{kT}{e} \log_e \frac{a^2 w_d^2}{n_i^2} \tag{3.98}$$

which suggests that ψ decreases with the magnitude of the bias since w_d depends on V (eqn. (3.92)). An exact analysis of the behaviour of the graded junction is difficult, but a more rigorous approach is given in Ref. 3.4, where inhomogeneous impurity distributions are considered in general.

CHARACTERISTICS OF A PRACTICAL *p-n* JUNCTION

With forward bias a practical *p-n* junction will have a greater voltage drop across it at a given current than is predicted by eqn. (3.5). One reason for this is the voltage developed across the *p*- and *n*-regions, which is due to the ohmic resistance of the semiconductor materials and of their contacts with the external circuit. Thus, if the total resistance outside the junction is R, the voltage across the junction is $V - IR$, so that a larger voltage must be applied to compensate for the IR drop. Another effect occurs when the density of injected carriers is comparable to the density of majority carriers. Thus in a p^+-n diode at high currents both the density of holes injected at the junction and the density of electrons drawn in at the ohmic contact with the *n*-region are no longer negligible compared with the density of majority electrons in the *n*-region. This results in a potential difference V_E being set up across the *n*-region in opposition to the applied voltage (Fig. 3.21) and leads to an I/V characteristic of the form

$$I = 2I_0 \left[\exp\left(\frac{eV}{2kT}\right) - 1 \right] \tag{3.95}$$

This equation is obtained in Ref. 3.3 and confirmed experimentally; it predicts a slower rise of current with voltage than eqn. (3.5). The slope of the $\log_e I$ against V graph is thus halved at high currents (Fig. 3.22), and

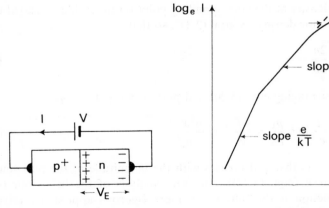

FIG. 3.21 High level injection in
a *p-n* junction

FIG. 3.22 Forward characteristic
of a *p-n* junction

in practice this may occur at currents lower than those for which the IR drop is significant.

With reverse bias the current I_R is not independent of voltage V_R but shows a slow rise with V_R. This is mainly due to current leakage across the outer surface of the diode, which may be represented by a resistance in parallel with it. It is reduced by chemical treatment of the surface and by hermetically sealing the device. Finally, when $V_R = V_B$, I_R rises sharply at a constant voltage and breakdown occurs (Fig. 3.23).

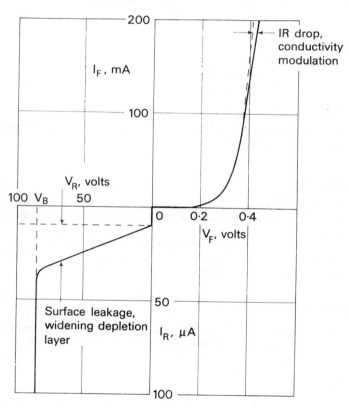

FIG. 3.23 I/V characteristic of a practical *p-n* junction

The forward characteristics of a germanium and a silicon diode are compared in Fig. 3.24. The diodes are assumed identical in construction, but $I_0 = 8\,\mu A$ for the germanium and $20\,nA$ for the silicon diode. This

117

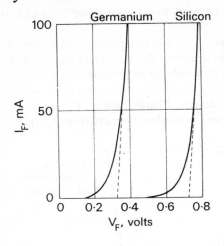

Fig. 3.24
Forward characteristics of
germanium and silicon *p-n*
junctions

causes the forward voltage for $I_F = 50$ mA to be 0.3 V for the germanium
and 0.7 V for the silicon diode.

REFERENCES

3.1 GOLDSMID, H. J., *Applications of Thermoelectricity* (Methuen, 1960).
3.2 ESAKI, L., "Avalanche breakdown in germanium", *Phys. Rev.* **99**, p. 1234 (Aug., 1955).
3.3 PHILLIPS, A. B., *Transistor Engineering* (McGraw-Hill, 1962).
3.4 LINDMAYER, J. and WRIGLEY, C. T., *Fundamentals of Semiconductor Devices* (Van Nostrand, 1965).

FURTHER READING

TAUC, J., *Photo and Thermoelectric Effects in Semiconductors* (Pergamon, 1962).
MORANT, M. J., *Introduction to Semiconductor Devices*, (Harrap, 1964).

PROBLEMS

At room temperature $\dfrac{e}{kT} = 40/\text{V}$

For germanium $\epsilon_r = 16$ and for silicon $\epsilon_r = 12$

3.1 Define the three main thermoelectric effects. By the analysis of a single thermo-junction, treated *either* as a refrigerator *or* as a generator, show how the appropriate expression for the coefficient of performance or efficiency may be derived. How does this expression lead to the idea of a thermoelectric figure of merit from which a choice of thermoelectric materials for application in the above devices may be made?

Two materials, A and B, have the following properties at room temperature:

Material	Electrical resistivity (Ω-cm)	Thermal conductivity (J/s-m-K)	Seeback coefficient (μV/K)
A	$1 \cdot 4 \times 10^{-2}$	4×10^{-2}	200
B	7×10^{-1}	3×10^{-3}	85

Which material is likely to be of greater use in the construction of a thermoelectric refrigerator? (*G.Inst. P., Part II*, 1966)

(*Ans.* $Z_A = 71 \times 10^{-6}$/K, $Z_B = 3 \cdot 3 \times 10^{-6}$/K, so that A is the more useful material)

3.2 A *p-n* junction is made of intrinsic germanium with 10^{13} free electrons per cm³ doped with 10^{17} and 5×10^{16} ionized impurity atoms per cm³ on the *p*- and *n*-sides respectively. The diffusion constants for the minority electrons and holes are 100 and 50 cm²/s respectively, the diffusion lengths being 0·08 cm in each case.

Estimate the value of the energy barrier assuming that the majority concentration in the material is proportional to exp ($-E/kT$), where E is the energy difference between the Fermi level and the conduction band (for electrons) or the valence band (for holes).

Calculate the saturation current density assuming that the deviation from the equilibrium concentration of minority carriers varies exponentially with distance from the junction. (*L.U., B.Sc. (Eng.)*, 1966)

(*Ans.* 0·44 eV; 4×10^{-7} A/cm²)

3.3 An abrupt silicon *p-n* junction is formed from *p*-type material with a resistivity of $1 \cdot 3 \times 10^{-3}$ Ω-m and *n*-type material with resistivity of $4 \cdot 6 \times 10^{-3}$ Ω-m at room temperature. The lifetimes of the *p*- and *n*-materials are 100 μs and 150 μs repectively, and the junction area is 1·0 mm².

If $\mu_p = 4 \cdot 8 \times 10^{-2}$ m²/Vs, $\mu_n = 0 \cdot 135$ m³/Vs and $n_i = 6 \cdot 5 \times 10^{16}$/m³, calculate the reverse bias leakage current, assuming the *p*- and *n*-regions are much longer than the diffusion length.

If a similar junction to the above is formed, except that the lengths of the *p*- and *n*-regions are each 50 μm, what is the new leakage current?

(*Ans.* $2 \cdot 7 \times 10^{-13}$ A; $2 \cdot 1 \times 10^{-12}$ A)

3.4 Calculate the forward voltage of an ideal *p-n* junction for a diode current of 10 mA when I_0 is (*a*) 1·0 μA, (*b*) 1·0 pA at room temperature. Suggest with reasons which of these values of I_0 would correspond to a germanium diode and which to a silicon diode, assuming the same physical dimensions in each case.

(*Ans.* (*a*) 0·23 V; (*b*) 0·58 V)

3.5 Describe briefly the physical processes of (i) avalanche and (ii) Zener breakdown in *p-n* junctions.

Physical Electronics

A certain abrupt *p-n* junction is made from germanium doped with 5×10^{23} and 1×10^{23} impurity atoms per m^3 on the *p-* and *n*-sides respectively. Using Gauss's theorem, or otherwise, obtain the width of the depletion layer when the maximum electric field within the junction has a magnitude of $10^7 \, V/m$ under reverse-bias conditions. (*L.U.*, *B.Sc.* (*Eng.*), 1968)
(*Ans.* 106 nm)

3.6 Using an idealized model, show that the depletion-layer capacitance of an abrupt plane *p-n* junction can be expressed as

$$C = \frac{K}{\sqrt{(V + V_b)}}$$

where K is a constant, V is the bias voltage and V_b is the intrinsic barrier potential difference.

When an alternating voltage of 0·5 V peak amplitude is applied across such a junction, the peak value of the junction capacitance is found to be 2 pF. If the capacitance of the junction at zero bias is 1 pF find (*a*) the intrinsic potential difference across the barrier, (*b*) the minimum value of the capacitance. State any assumptions made in the calculation. (*IEE June*, 1965)
(*Ans.* (*a*) 0·67 V; (*b*) 0·75 pF)

3.7 An abrupt silicon *p-n* junction is formed from a *p*-type semiconductor with an acceptor density of $3 \times 10^{23}/m^3$ and an *n*-type semiconductor with a donor density of $3 \times 10^{22}/m^3$. If the junction area is 1·0 mm² and the reverse bias is 10 V calculate (*a*) the width of the depletion layer, (*b*) the maximum field within the depletion layer, (*c*) the depletion layer capacitance.

Assuming that a field of $5 \times 10^7 \, V/m$ is required for avalanche breakdown, what is (*d*) the breakdown voltage of the junction?
(*Ans.* (*a*) 0·7 μm; (*b*) 2·86 × 10⁷ V/m; (*c*) 270 pF; (*d*) 99·6 V)

3.8 A graded silicon *p-n* junction is formed by diffusing boron into *n*-type silicon having a resistivity of 0·015 Ω-m. The surface density of boron atoms is $5 \times 10^{25}/m^3$ and the transition from *p-* to *n*-type material occurs 1·5 μm below the surface. If the junction area is 1·0 mm² and the reverse bias is 10 V, calculate for the depletion layer (*a*) its width, (*b*) its maximum field, (*c*) its capacitance, (*d*) its breakdown voltage, assuming the same breakdown field as in Problem 3.7.
(*Ans.* (*a*) 1·6 μm; (*b*) 9·65 × 10⁶ V/m; (*c*) 311 pF; (*d*) 120 V)

4

Junction Transistors with Uniform and Graded Bases

THE JUNCTION TRANSISTOR

A junction transistor consists of two *p-n* junctions formed back-to-back in a single crystal of germanium or silicon. Two arrangements are possible, *p-n-p* and *n-p-n*. A *p-n-p* transistor consists of a thin *n*-region sandwiched between two *p*-regions, while in the *n-p-n* transistor the types of semiconductor are reversed. In both types the central region is called the *base* and the outer regions are called the *emitter* and the *collector* respectively. In normal operation the emitter–base diode is forward biased and the collector–base diode is reverse biased, and under these conditions current from the emitter flows across the base to the collector. This is illustrated in Fig. 4.7(*b*), which shows a *uniform-base* transistor manufactured by alloying the emitter and collector onto the base. A more recent manufacturing technique uses diffusion processes to form a *graded base* transistor illustrated in Fig. 4.37, and described at the end of this chapter. In both types the area of the collector is greater than that of the emitter, and they are both available in *p-n-p* and *n-p-n* versions.

In the *p-n-p* type the emitter current is carried by holes and this type will be considered in detail. Similar considerations apply to the *n-p-n* type, but the emitter current is carried by electrons and the polarities of the applied voltages are reversed. When no potentials are applied the potential barriers and energy levels for the uniform-base *p-n-p* transistor are as shown in Fig. 4.1, while for the normal applied potentials the potential barriers are shown in Fig. 4.2, the collector region being more heavily doped than the base region. The emitter–base voltage V_{EB} is normally

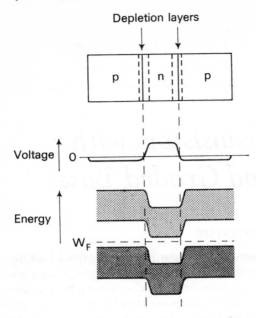

Depletion layers

Voltage

Energy

FIG. 4.1 Potential and energy through an unbiased uniform-base transistor

between 0·2 and 0·5 V, while the collector–base voltage V_{CB} is about −5 V. Since the collector–base junction is reverse biased the collector current contains a leakage component I_{CBO}, which is mainly due to electrons from the collector. I_{CBO} is the current flowing between collector and base when the emitter is open-circuited.

The emitter p-region is much more heavily doped than the n-region, so that holes are injected into the base and an excess density p_e is set up at the edge of the emitter–base depletion layer (Fig. 4.3). These holes diffuse across the base towards the collector, and if the effective width of the base region w_b is much less than the diffusion length for holes, very little recombination occurs. At the collector the field across the collector–base junction sweeps *all* the holes out of the base, whether they are the original minority carriers in the n-region or excess carriers injected from the emitter. Thus the hole density is zero at the base side of the collector–base depletion layer.

An important feature of the transistor is the fraction of the emitter current that reaches the collector. This can be expressed as the ratio of

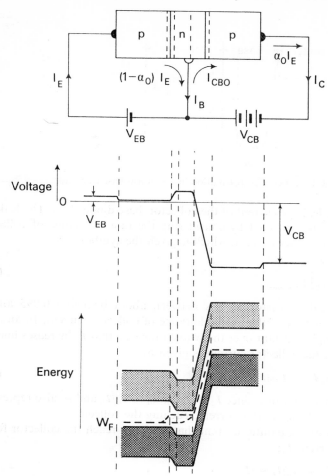

Fig. 4.2 Potential and energy through a uniform-base transistor biased
with the base common to input and output

the direct collector and emitter currents, I_C/I_E*. However, when the
transistor is connected as shown in Fig. 4.2, the measured collector current

---

* This ratio is given the symbol h_{FB}, because it is a large-signal *h*-parameter (see page 174),
so that

$$h_{FB} = \frac{I_C}{I_E} \bigg|_{V_{CB}\text{constant}}$$

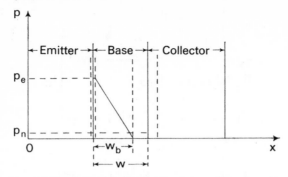

FIG. 4.3 Excess carrier distribution in a uniform-base transistor

includes the leakage current of the collector–base diode, I_{CBO}. The leakage current may be excluded by considering the rate of change of collector current with emitter current, which is given the symbol α_0:

$$\alpha_0 = \frac{\partial I_C}{\partial I_E}\bigg|_{V_{CB}\text{constant}} \tag{4.1}$$

In a practical transistor α_0 lies between about 0·95 and 0·995 and is constant over part of the available range of collector current. In an ideal transistor α_0 is constant over the whole range, so that I_C increases linearly with I_E and the collector current is given by

$$I_C = \alpha_0 I_E + I_{CBO} \tag{4.2}$$

Under these conditions, since $I_C \gg I_{CBO}$, $\alpha_0 \approx I_C/I_E$ and so also represents the fraction of the emitter current reaching the collector.

The part of the emitter current that does not reach the collector forms the base current I_B:

$$I_B = (1 - \alpha_0)I_E - I_{CBO} \tag{4.3}$$

as shown in Fig. 4.2. Owing to the negative bias on the collector in a *p-n-p* transistor, $\alpha_0 I_E$ is entirely a hole current.

Hole Emitter Efficiency

The first reason for α_0 being less than unity is that the emitter current will have an electron component, since the forward current of a *p-n* diode consists of both electrons and holes. This is described by the *hole emitter efficiency*, γ:

$$\gamma = \frac{I_p}{I_p + I_n} \tag{4.4}$$

I_p is the hole component and I_n the electron component of the current crossing the emitter–base junction. I_n is made very small by doping the p-region much more heavily than the n-region, so forming a p^+-n-p transistor. From eqn. (4.4),

$$\gamma = \frac{1}{1 + \dfrac{I_n}{I_p}} = \left(1 + \frac{I_n}{I_p}\right)^{-1}$$

$$\approx 1 - \frac{I_n}{I_p} \tag{4.5}$$

if $I_p \gg I_n$ and using the binomial theorem. I_n may be obtained from eqn. (3.34):

$$I_n = \frac{eD_n n_p S}{L_{ne}} \left[\exp\left(\frac{eV}{kT}\right) - 1\right] \tag{4.6}$$

S being the effective cross-sectional area of the base and L_{ne} the diffusion length of electrons in the emitter. The width of the base is normally much less than the diffusion length of holes, so that I_p is obtained from eqn. (3.41) with $l_n = w_b$, which gives

$$I_p = \frac{eD_p p_n S}{w_b} \left[\exp\left(\frac{eV}{kT}\right) - 1\right] \tag{4.7}$$

Dividing eqn. (4.6) by eqn. (4.7) and substituting into eqn. (4.5),

$$\gamma = 1 - \frac{D_n n_p w_b}{D_p p_n L_{ne}} \tag{4.8}$$

Since $n_p p_p = n_n p_n$ from eqn. (2.37) which gives $n_p/p_n = n_n/p_p$, and also $D_n/D_p = \mu_n/\mu_p$ from eqns. (2.106) and (2.107), we can write

$$\frac{D_n n_p}{D_p p_n} = \frac{\sigma_n}{\sigma_p} \tag{4.9}$$

using eqn. (2.66). Then

$$\gamma = 1 - \frac{\sigma_n w_b}{\sigma_p L_n e} \tag{4.10}$$

and γ approaches unity when the second term is small. This is achieved by making the base region narrow and also by making $\sigma_p \gg \sigma_n$ which occurs when $p_p \gg n_n$. It is common to make the conductivity of the p-region about 100 times that of the n-region, so that γ is normally greater than 0·99 since w_b in a uniform-base transistor is typically $10\,\mu$m, which is much less than L_{ne}, the diffusion length of minority carriers in the emitter.

Base Transport Factor, δ

The second reason for α_0 being less than unity is that some of the holes diffusing across the base will recombine with the electrons present as majority carriers. This results in a component of base current I_{BF} due to electrons replacing those that have recombined and leads to the *base transport factor*, δ, which is the ratio of the hole current entering the collector to the hole current leaving the emitter.

The distribution of excess holes in the base is shown in Fig. 4.4, and

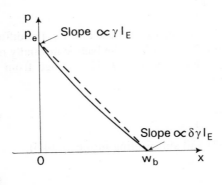

FIG. 4.4 Distribution of excess charge in the uniform base

since holes move across the base by diffusion, the current at any point is proportional to the slope of the hole distribution curve at that point. Thus the hole current leaving the emitter is

$$\gamma I_E = -eD_pS\frac{dp}{dx}\bigg|_{x=0} \tag{4.11}$$

and the hole current entering the collector is

$$\delta\gamma I_E = -eD_pS\frac{dp}{dx}\bigg|_{x=w} \tag{4.12}$$

The difference between these two currents constitutes the base current I_{BF} by which electrons which have recombined with holes are replaced from the external circuit, so that

$$I_{BF} = (1 - \delta)\gamma I_E \tag{4.13}$$

The component of the base charge due to leakage current is negligible since in a small transistor this current would be about 10^{-7} A as compared with an emitter current of about 10^{-3} A and a base current of about 10^{-5} A. Also δ is only just less than unity, so that to a first approximation

$$\left.\frac{dp}{dx}\right|_{x=0} \approx \left.\frac{dp}{dx}\right|_{x=w_b} \tag{4.14}$$

and the distribution of holes in the base may be considered linear, as indicated by the dotted line. The effective volume of the base is Sw_b, and the average density of the excess base charge is $\frac{1}{2}ep_e$, so that the actual excess base charge, q_B, is given by

$$q_B = \frac{ep_e w_b S}{2} \tag{4.15}$$

and

$$\gamma I_E = \frac{eD_p p_e S}{w_b} = q_B \frac{2D_p}{w_b^2} \tag{4.16}$$

This equation may be written in the form

$$\frac{q_B}{I_E} = \tau_C \tag{4.17}$$

since $\gamma \approx 1$ and putting $w_b^2/2D_p = \tau_C$.

The excess hole distribution of Fig. 4.4 is maintained in the base by holes entering from the emitter and leaving for the collector, so that τ_C represents the average transit time of holes across the base.

I_{BF} may also be related to q_B by the expression

$$I_{BF} = \frac{q_B}{\tau_B} \tag{4.18}$$

Since I_{BF} is due to the process of recombination in the base, we can see by comparing eqns. (4.18) and (2.96) that τ_B represents the lifetime of the minority carriers in the base. Combining eqns. (4.13), (4.16) and (4.18),

$$1 - \delta = \frac{w_b{}^2}{2D_p\tau_B} = \frac{w_b{}^2}{2L_p{}^2} \qquad (4.19)$$

where L_p is the diffusion length of holes in the base. Hence

$$\delta \approx 1 - \frac{w_b{}^2}{2L_p{}^2} \qquad (4.20)$$

in view of the original approximations and $\delta \to 1$ when $w_b \ll L_p$. This is achieved by making the base width small during manufacture and by making the diffusion length large by means of a long minority carrier lifetime in the base region.

In practice L_p is about 10^{-4} m, so that, if w_b is about 10^{-5} m, δ is about 0·995.

Collector Efficiency, M

For normal values of the collector voltage V_{CB} all the holes arriving at the collector side of the base are swept across the depletion layer to form the collector current. However, if V_{CB} is increased, avalanche multiplication will occur at some value V_B, the collector–base breakdown voltage. This gives rise to a multiplication factor, similar to the quantity M for the *p-n* diode (page 103), called the *collector efficiency*. For the transistor,

$$M = \frac{1}{1 - (V_{CB}/V_B)^n} \qquad (4.21)$$

where n has similar values to the *p-n* diode, and the full expression for α_0 becomes

$$\alpha_0 = \gamma\delta M = \frac{\gamma\delta}{1 - (V_{CB}/V_B)^n} \qquad (4.22)$$

Base Current Components

Under normal operating conditions when $V_{CB} \ll V_B$,

$$\alpha_0 = \gamma\delta \approx \left(1 - \frac{\sigma_n w_b}{\sigma_p L_{ne}}\right)\left(1 - \frac{w_b{}^2}{2L_p{}^2}\right) \qquad (4.23)$$

and for typical values of λ and δ given above, α_0 is 0·99.

The fraction of the emitter current reaching the collector depends, not only on the current I_{BF}, but also on a second component, I_{BE}, which is due to the electron component of the emitter current and is given by

$$I_{BE} = (1 - \gamma)I_E \tag{4.24}$$

Thus the total base current is

$$I_B = I_{BF} + I_{BE} - I_{CBO}$$

$$= I_E(\gamma - \gamma\delta + 1 - \gamma) - I_{CBO}$$

$$= (1 - \alpha_0)I_E - I_{CBO} \tag{4.25}$$

which is the same as eqn. (4.3). The various components of the base current are illustrated in Fig. 4.5.

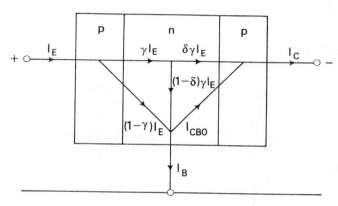

FIG. 4.5 Components of the base current

Variation of α_0 with Collector Current

The expression for α_0 (eqn. (4.23)) is independent of I_E and I_C, but in practice α_0 varies considerably with I_C as shown in Fig. 4.6. At low currents α_0 is also low, rising to a flat peak as I_C is increased and then falling again slowly. The letters (a), (b) and (c) in Fig. 4.6 correspond to Figs. 4.7(a), (b) and (c) respectively for a uniform-base transistor. Similar effects occur in a graded-base transistor.

The initial low value of α_0 is due to the loss of injected minority carriers by recombination at the *surface* of the base outside the edge of the emitter–base junction (Fig. 4.7(a)). The discontinuity of a crystal at its surface results in the formation of many trapping centres which are empty initially. As the current is increased injected carriers are lost by recombination at the surface traps, which effectively reduces the emitter efficiency γ.

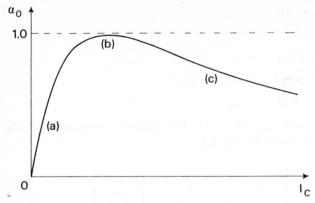

FIG. 4.6 Variation of α with collector current
The letters refer to Figs. 4.7(*a*), (*b*) and (*c*)

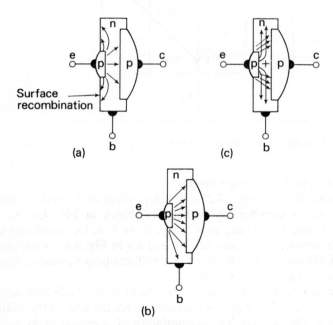

FIG. 4.7 Variation of current flow in a transistor

    (*a*) Surface recombination at low collector current
    (*b*) Normal current flow in a transistor
    (*c*) Internal biasing at high collector current

The traps are then filled progressively, which allows carriers to move into the base and so γ increases with the current to a more constant value (Fig. 4.7(*b*)).

At high currents the density of the injected carriers is large and requires a correspondingly large base current to replenish the majority carriers which have recombined in the base. The first consequence is that the conductivity of the base is reduced owing to conductivity modulation (page 63). It may be seen from eqn. (4.10) for γ in a *p-n-p* transistor that, if the base conductivity σ_n is reduced, γ also falls. A second consequence of

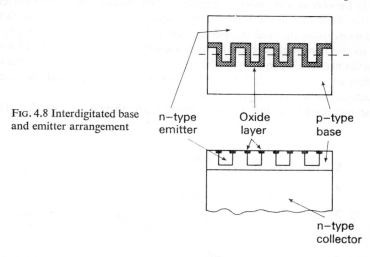

FIG. 4.8 Interdigitated base and emitter arrangement

n—type emitter   Oxide layer   p—type base

n—type collector

the large base current is that the central base area between emitter and collector becomes positively biased, so that only the edges of the emitter can inject carriers which will reach the collector (Fig. 4.7(*c*)). This reduces the effective cross-sectional area of the base, and both effects cause the fall in α_0 at high currents. They are counteracted by making the edge of the emitter junction as long as possible, as shown in Fig. 4.8 for a typical power transistor manufactured by means of the planar technology (page 183).

D.C. CHARACTERISTICS OF A TRANSISTOR

The currents flowing through a transistor are related to the voltages applied between the electrodes. For instance, I_E is an exponential function of V_{EB} because the emitter–base diode is forward biased. The arrangement so far discussed is called the *common-base* configuration, in which the emitter is the input electrode and the collector the output electrode, with the base as the electrode common to both the input and output circuits. Here I_E controls I_C and this process can be easily appreciated from a physical point of view.

However, this is not the only possible arrangement of the transistor. Greater efficiency is obtained if the base is used as the input electrode, so that the small base current controls the much larger collector current. This is known as the *common-emitter* configuration, since the emitter is now the electrode common to the input and output circuits. Although this is the most usual method of operation, it is less easy to appreciate physically, so the common-base characteristics will be discussed first. Two sets of curves are needed to give all the relationships between the voltages and currents of a transistor. These are known as the *input* and the *output* characteristics respectively.

The common-base input characteristics are shown in Fig. 4.9. Their

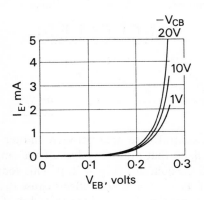

FIG. 4.9 Common-base input characteristics (*p-n-p* transistor)

general form follows the I/V characteristic of a p-n diode under forward bias (eqn. (3.4)), and the curves given correspond to $I_0 = 0.1\,\mu\text{A}$ when V_{CB} is low. However, it can be seen that, if V_{BE} is fixed, I_E rises as V_{CB} is made more negative. This is due to the effect of $V_{CB}^{\cdot}$ on the base width w_b which will now be considered in more detail. For a high emitter efficiency, I_E will be almost entirely due to holes and the hole distribution in the base region will be as shown in Fig. 4.10(a), with the density at the edge of the emitter–base depletion layer increasing with V_{EB} as in eqn.

132

(3.26) for the diode. The effective base width w_b is less than the actual distance W between the two metallurgical junctions, since both depletion

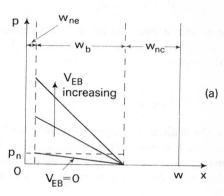

(a)

FIG. 4.10 Variation of excess base charge

(a) With emitter–base voltage
(b) With collector–base voltage
(c) In the saturated region

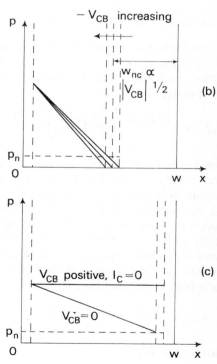

(b)

(c)

layers penetrate into the base. The distances penetrated are w_{ne} at the emitter and w_{nc} at the collector, so that

$$w_b = W - w_{ne} - w_{nc} \tag{4.26}$$

The general expression for w_n, the penetration of the depletion layer into the n-region, is given by eqn. (3.73). For a p-n-p transistor $N_a \gg N_d$, so the expression for w_{nc} becomes

$$w_{nc} \approx \left(\frac{2\epsilon}{eN_d}\right)^{1/2} (\psi - V)^{1/2} \tag{4.27}$$

For the emitter junction V is small and positive and for the collector junction, V is much larger and negative. Hence $w_{ne} \ll w_{nc}$, and is nearly constant over the allowed range of emitter voltages, while w_{nc} varies widely with the much larger changes in collector voltage. Thus we can write

$$w_b \approx W - w_{nc} \tag{4.28}$$

For a narrow n-region the emitter current may be obtained from eqn. (4.7). Putting $l_n = w_b = W - w_{nc}$, making S the area of the emitter junction, and neglecting the electron current term since γ is assumed almost unity,

$$I_E = \frac{eSD_pp_n}{(W - w_{nc})} \exp\left[\left(\frac{eV_{EB}}{kT}\right) - 1\right] \tag{4.29}$$

Since $w_{nc} \propto (\psi + V_{CB})^{1/2}$, $W - w_{nc}$ falls as V_{CB} is increased and I_E therefore rises, when V_{EB} is held constant. The effect on the charge distribution is illustrated in Fig. 4.10(*b*). At a sufficiently large value of V_{CB}, $w_{nc} = W$, so that $w_b \to 0$ and the depletion layer extends right through the base to the emitter junction. The emitter current then rises directly with V_{CB} and normal transistor action ceases. This condition is known as *punch-through*, which like Zener or avalanche breakdown is non-destructive if the current is limited to prevent excessive heat dissipation, It occurs at a collector–base voltage V_{PT}, obtained from eqn. (4.27):

$$V_{PT} \approx \frac{eN_dW^2}{2\epsilon} \tag{4.30}$$

putting $w_n = W$.

Junction Transistors with Uniform and Graded Bases

The common–base output characteristics of a small transistor (Fig. 4.11) show how I_C varies with V_{CB} for constant values of I_E, I_C being almost equal to I_E and practically independent of V_{CB}. When $I_E = 0$, $I_C = I_{CBO}$, which is the leakage current of the collector–base diode. This is less than a microampere and so cannot be seen on a milliampere scale. I_C is then due to a current $\alpha_0 I_E$ added directly to I_{CBO} as in eqn. (4.1). Both γ and δ are functions of w_0, so that α_0 and hence I_C increase slowly with V_{CB}. However, for high values of V_{CB} avalanche multiplication of

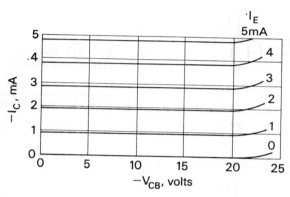

FIG. 4.11 Common-base output characteristics (*p-n-p* transistor)

I_{CBO} occurs, the factor M in α_0 becomes much greater than unity, and I_C rises rapidly with V_{CB}. I_C flows even when $V_{CB} = 0$, since the hole distribution in the base still has a gradient towards the collector when V_{EB} is positive. In fact a small forward bias on the collector is required to reduce I_C to zero, as shown in Fig. 4.10(*c*).

In common–emitter operation the base is used as the input electrode (Fig. 4.12(*b*)). The component of base current, $(1 - \alpha_0)I_E$, supplies the losses in emitter current occurring in the base region, so that if the base current I_B is increased both I_E and I_C also increase. I_B is in turn controlled by V_{BE}, but since the essential mechanism is due to I_B the transistor is still considered as a current-controlled device.

From eqn. (4.3),

$$I_E = \frac{1}{1 - \alpha_0}(I_B + I_{CBO}) \tag{4.31}$$

135

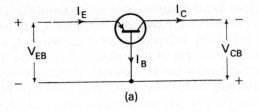

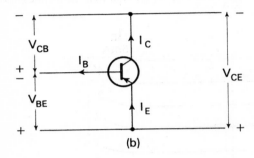

FIG. 4.12 Transistor
currents and voltages

(a) *p-n-p* common base
(b) *p-n-p* common emitter
(c) *n-p-n* common emitter

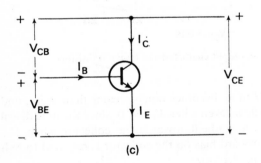

Substituting in eqn. (4.1),

$$I_C = \frac{\alpha_0}{1 - \alpha_0}(I_B + I_{CBO}) + I_{CBO}$$

$$= \frac{\alpha_0}{1 - \alpha_0}I_B + \frac{I_{CBO}}{1 - \alpha_0} \qquad (4.32)$$

We can then define

$$\beta_0 = \left.\frac{\partial I_C}{\partial I_B}\right|_{V_{CE}\text{ constant}} \qquad (4.33)$$

where $\beta_0 = \alpha_0/(1 - \alpha_0)$. We can also put

$$I_{CEO} = \frac{I_{CBO}}{1 - \alpha_0}$$

the collector current flowing when $I_B = 0$ (base open-circuited).

β_0 is the small-signal common-emitter current gain and is much greater than unity, as shown in Table 4.1, so that a small base current can control

Table 4.1 Dependence of β_0 and I_{CEO} on α_0

α_0	0·97	0·98	0·99
β_0	32	49	99
I_{CEO} (μA)	3·3	5·0	10

Values of I_{CEO} are given for $I_{CBO} = 0·1\,\mu$A.

a much larger emitter current. I_{CEO} is the collector leakage current for the common-emitter configuration and is much larger than I_{CBO}, as may be seen from the values given in the table. Both β_0 and I_{CEO} are very sensitive to changes in α_0 since the very small factor $(1 - \alpha_0)$ occurs in the denominator of each of them. The effect of a 2% change in α_0, from 0·97 to 0·99, causes more than a threefold increase in β_0 and I_{CEO}. Thus measurement of β_0 is a very sensitive method of detecting small changes in α_0.

Although I_{CEO} is still small compared with a collector current of about 1 mA, β_0 is not the large-signal current gain in a practical transistor owing to the variations introduced by small changes in α_0. The large-signal current gain is the h-parameter h_{FE}:

$$h_{FE} = \frac{I_C}{I_B} \tag{4.34}$$

An exact expression for I_C is

$$I_C = h_{FEL}I_B + I_{CEO} \tag{4.35}$$

where $h_{FEL} = h_{FE} - I_{CEO}/I_B$. Since $I_{CEO} \ll I_C$, $h_{FEL} \approx h_{FE}$, and there will be no further reference to h_{FEL}.

The common-emitter output characteristics are shown in Fig. 4.13. The collector–emitter voltage V_{CE} is developed across both junctions, so that

$$V_{CE} = V_{CB} + V_{BE} \tag{4.36}$$

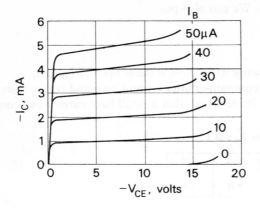

FIG. 4.13 Common-emitter output characteristics

In the *active* region of operation, where V_{CE} exceeds about 0·2 V, the emitter junction has a small forward bias, which is almost constant to maintain the base current constant, and the collector junction has a much larger reverse bias (Fig. 4.14(*a*)). In this region I_C rises slowly with

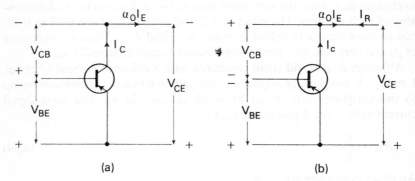

(a)   (b)

FIG. 4.14 Transistor junction biases

(*a*) In active region   (*b*) In saturated region

V_{CE} owing to the fall in base width as V_{CB} rises and the corresponding rise in α_0. This effect is much more noticeable than in the common-base characteristics owing to the factor $(1 - \alpha_0)$. For the same reason the increase in I_C due to avalanche breakdown becomes apparent at a lower collector voltage since the change in M and hence in α_0 is magnified in its effect on I_C. The breakdown voltage defines the upper voltage limit of the active region into which the transistor is biased when it is used as an amplifier, as described on page 154.

For values of V_{CE} below about 0·2 V, I_C falls very rapidly as V_{CE} is reduced, and the transistor is working in the *saturated* region. For a given base current the point at which the rapid fall in current begins is called the "knee" of the characteristics, and the corresponding value of V_{CE} defines the boundary between the saturated and active regions of the transistor. As V_{CE} is reduced in the active region, V_{BE} remains almost constant so that V_{CB} must fall until it is about zero at the knee of the curve. However, holes continue to flow into the collector since the slope of the density gradient in the base is still towards the collector (Fig. 4.10(c)). As V_{CE} is reduced below this voltage, V_{CB} becomes a *forward* bias and current I_R flows from the collector in the reverse direction to the current from the emitter (Fig. 4.14(b)). Both junctions are then forward biased so that, in the saturation region,

$$V_{CE} = -V_{CB} + V_{BE}$$

and

$$I_C = \alpha_0 I_E - I_R \tag{4.37}$$

neglecting the leakage current. Thus, as V_{CE} is reduced towards zero, $-V_{CB}$ and I_R both increase and I_C also tends to zero. A detailed analysis (Ref. 4.2) shows that, when $I_C = 0$, V_{CE} is not zero but has a small value called the *offset* voltage which is normally only a few millivolts. The transistor operates in the saturation region when it is used as a switch (page 142). It is particularly suitable for this application owing to the low value of V_{CE} when it is conducting heavily.

The input characteristics (Fig. 4.15(a)) depend on eqn. (4.29), which shows that I_E increases exponentially with V_{EB}. I_B therefore increases exponentially with V_{BE}, and is also a function of collector voltage since I_E depends on V_{CB} through the base-width term of eqn. (4.29). When the base current is zero, $I_E = I_{CBO}/(1 - \alpha_0)$, from eqn. (4.3), and a finite value of V_{BE} is required to maintain this value of I_E. The value of V_{BE} for $I_B = 0$ increases with V_{CE}, since the corresponding value of I_E increases with V_{CB}. Thus all the input characteristics cut the voltage axis, and finally, when $V_{BE} = 0$, $I_E = 0$ and $I_B = -I_{CBO}$. Input characteristics at the same value of V_{CE} are compared in Fig. 4.15(b) for germanium and silicon transistors having base-to-emitter leakage currents of 100 nA and 5 nA respectively.

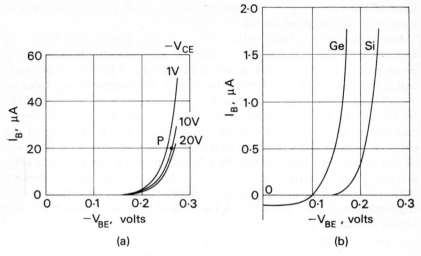

FIG. 4.15 Common-emitter input characteristics

(*a*) Germanium transistor
(*b*) Comparison of germanium and silicon transistors at low base currents
(leakage currents 100 nA and 5 nA respectively)

THE OPERATION OF A TRANSISTOR

Consider the circuit of Fig. 4.16 in which V_{CC} is a fixed direct voltage obtained from a power supply, and R_B and R_L are resistors connected

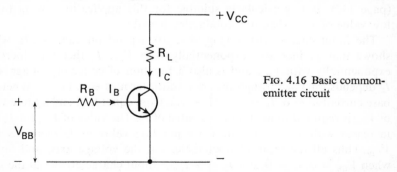

FIG. 4.16 Basic common-emitter circuit

in series with the base and collector respectively. The voltage V_{BB} controls I_B, which in turn controls I_C and hence the *output* voltage V_{CE}. A *change* in the magnitude of any of these direct quantities is known as a *signal*.

140

Such a circuit is the basis of many practical applications of transistors, which fall into two main classes. These involve large and small signals respectively, both of which depend closely on the physical principles which have been described so far.

An *n-p-n* transistor has been chosen in this section for convenience, making both I_B and I_C conventionally positive since they are due to negative charges leaving their respective electrodes. The basic principles, however, apply equally to *p-n-p* transistors. Then, applying Kirchhoff's law to the base circuit, $V_{BB} = I_B R_B + V_{BE}$, or

$$I_B = \frac{V_{BB} - V_{BE}}{R_B} \qquad (4.38)$$

If V_{BB} is a direct voltage which can be varied, I_B can be set at any desired value, which can be predicted easily when $V_{BB} \gg V_{BE}$, since this makes $I_B \approx V_{BB}/R_B$. Now, I_B is associated with I_C and V_{CE} through the common-emitter output characteristics, but the actual values of collector current and voltage obtained will also depend on the load resistance, R_L. This may be incorporated with the characteristics by means of a *load line*, whose equation is obtained by applying Kirchhoff's law to the collector circuit. This gives $V_{CC} = V_{CE} + I_C R_L$, or

$$I_C = -\frac{1}{R_L} V_{CE} + \frac{V_{CC}}{R_L} \qquad (4.39)$$

If I_C is measured along the *y*-axis and V_{CE} along the *x*-axis, eqn. (4.39) represents a straight line across the characteristics having a slope of $-1/R_L$ (Fig. 4.17), which is the *load line*. It cuts the current axis where $V_{CE} = 0$ and $I_C = V_{CC}/R_L$, and it cuts the voltage axis where $I_C = 0$

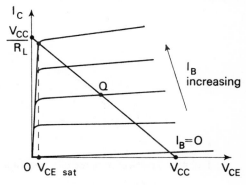

Fig. 4.17 Construction of load line

141

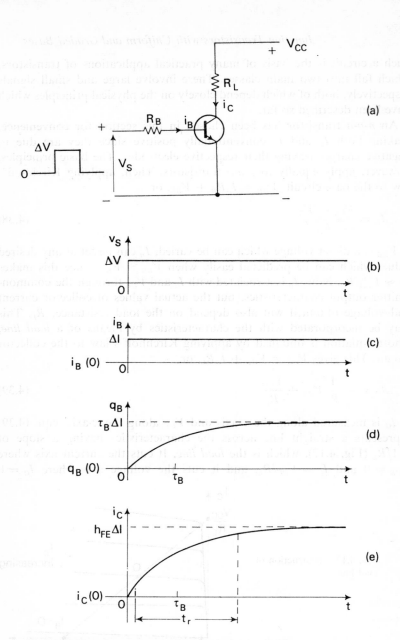

FIG. 4.18 Transient response of a transistor to switching on

(a) Circuit with signals   (d) Excess base charge
(b) Input-voltage step function   (e) Collector current
(c) Base-current step function

and $V_{CC} = V_{CE}$, as may easily be seen from eqn. (4.39). The operating values of I_C and V_{CE} are obtained where the load line cuts the characteristic corresponding to a given value of I_B. This occurs at the quiescent point Q, so-called because it refers to direct values without any alternating signals superimposed on them.

LARGE-SIGNAL OPERATION

In applications involving large signals, V_{BB} is replaced by the input signal voltage v_S, which commonly has a rectangular or pulsed waveform, and the transistor may be considered to act as a switch (Fig. 4.18(a)). v_S is made sufficiently positive for I_B to be large enough to take the transistor into saturation, where the resulting value of V_{CE} is known as $V_{CE\ sat}$. This is normally much less than V_{CC} so that the "switch" is closed with the transistor conducting heavily. When v_S has reached its maximum negative value, I_B is reduced to zero and may be reversed if v_S is large enough. I_C is then reduced to a leakage current which will be I_{CEO} when $I_B = 0$ (eqn. (4.33)), and can go down to I_{CBO} when I_B is reversed, which will occur when both junctions have become reverse biased. The switch is now opened with the transistor cut off.

In this switching operation the active region is rapidly traversed between the "on" and "off" conditions. The time taken to change from one of these steady states to the other is of vital importance in many high-speed switching applications, such as the logic circuits in a digital computer. This switching time is controlled by the rate at which the excess charge in the base can be established and dispersed, as described below. Here the pulsed waveform is considered as the sum of two step functions, and the switching times are obtained by analysing the response of the transistor to each step function in turn.

Charge Control of a Transistor

The steady-state operation of a junction transistor depends on the maintenance of an excess charge, q_B, in the base region as described on page 127. The emitter current is then given by eqn. (4.17) and

$$I_E = \frac{q_B}{\tau_C} \tag{4.40}$$

where τ_C is the transit time of the minority carriers across the base. A base current I_{BF} is also necessary to make good the loss of charge due

143

to recombination. Since $I_{BF} \approx I_B$ (page 127), we can write from eqn. (4.18) that approximately

$$I_B = \frac{q_B}{\tau_B} \tag{4.41}$$

where τ_B is the lifetime of the minority carriers in the base. Then from eqns. (4.2) and (4.34), since α_0 is almost unity and I_{CBO} is very small, we can write approximately that

$$I_E = I_C = h_{FE}I_B \tag{4.42}$$

which gives

$$I_C = \frac{q_B}{\tau_C} \tag{4.43}$$

and

$$\frac{I_C}{I_B} = h_{FE} = \frac{\tau_B}{\tau_C} \tag{4.44}$$

These equations refer to an ideal transistor which can be realized only approximately in practice, but nevertheless they give a useful guide to practical transistor operation. τ_C and τ_B are called *charge control parameters*.

If a signal is applied to the base a varying current is superimposed on the steady current I_B. Thus the charge distribution will change with time so that at any instant the total base current, i_B, may be written as the sum of the steady-state and varying currents:

$$i_B = \frac{q_B}{\tau_B} + \frac{dq_B}{dt} \tag{4.45}$$

a *charge control equation*.

TRANSIENT RESPONSE

Eqns. (4.44) and (4.45) are particularly useful when considering the response of a transistor to a transient or step function. In the simple common-emitter circuit of Fig. 4.18(a) it is assumed that R_B is much larger than the effective input impedance of the transistor, so that a sudden change or step of input voltage ΔV results in a sudden change of base current $\Delta I \approx \Delta V / R_B$. Suppose in the first case that ΔI is small enough to ensure that the transistor remains in the active region of operation. If the initial values of base charge, base current and collector current are $q_B(0)$, $i_B(0)$ and $i_C(0)$ respectively, the step input will increase the base charge by q and the base current by ΔI. Then initially,

$$i_B(0) = \frac{q_B(0)}{\tau_B} \tag{4.46}$$

and after the input has been applied the *change* in base current is

$$\Delta I = \frac{q}{\tau_B} + \frac{dq}{dt} \tag{4.47}$$

Then $\tau_B \, dq/dt = \tau_B \Delta I - q$, or

$$\frac{dq}{\tau_B \Delta I - q} = \frac{dt}{\tau_B} \tag{4.48}$$

so that, integrating,

$$-\log_e (\tau_B \Delta I - q) = \frac{t}{\tau_B} + C_1 \tag{4.49}$$

Putting $t = 0$, $q = 0$ initially, $C_1 = -\log_e \tau_B \Delta I$, so that

$$\log_e \left(1 - \frac{q}{\tau_B \Delta I}\right) = -\frac{t}{\tau_B} \tag{4.50}$$

and

$$q = \tau_B \Delta I \left[1 - \exp\left(-\frac{t}{\tau_B}\right)\right] \tag{4.51}$$

Hence the total base charge as a function of time is given by

$$q_B = q_B(0) + \tau_B \Delta I \left[1 - \exp\left(-\frac{t}{\tau_B}\right)\right] \tag{4.52}$$

From eqn. (4.43), $q_B = \tau_C i_C$, and from eqn. (4.44), $\tau_B = h_{FE} \tau_C$, so that, substituting into eqn. (4.52),

$$i_C = i_C(0) + h_{FE} \Delta I \left[1 - \exp\left(-\frac{t}{\tau_B}\right)\right] \tag{4.53}$$

for the collector current as a function of time. Eqns. (4.52) and (4.53) are illustrated in Figs. 4.18(d) and (e). The rise time t_r of the collector current is measured between the times corresponding to 10% and 90% of the total change $h_{FE}\Delta I$, since the time to reach 100% of $h_{FE}\Delta I$ is not well defined. It may be easily found from eqn. (4.53) that this gives

$$t_r = 2 \cdot 2\tau_B \tag{4.54}$$

Typical values for an alloy junction transistor are $\tau_C = 10^{-8}$ s, $\tau_B = 10^{-6}$ s and $h_{FE} = 100$, which makes $t_r = 2 \cdot 2\,\mu s$. This relatively long rise time may be unacceptable for fast switching applications, but it can be shortened by increasing the base current step so that the transistor saturates. Under saturation conditions, from Fig. 4.16,

$$V_{CC} = I_C R_L + V_{CE\,sat} \approx I_C R_L \tag{4.55}$$

so that

$$I_C \approx \frac{V_{CC}}{R_L} \tag{4.56}$$

and the collector current is determined by the external circuit conditions. Saturation is achieved when $h_{FE}\Delta I > V_{CC}/R_L$, so that when ΔI is large enough for this to occur the collector current is clamped at V_{CC}/R_L before the transient described by eqn. (4.53) is completed (Fig. 4.20(b)). It is normal to switch the transistor between zero and saturated currents, so that $q_B(0)$, $i_B(0)$ and $i_C(0) = 0$. It is evident from Fig. 4.20(b) that the rise time from 10% to 90% of the current V_{CC}/R_L can be made much less than $2 \cdot 2t_B$ as ΔI is increased. Indeed, the rise time can usually be measured over the whole range (0 to 100%) of V_{CC}/R_L since the point at which i_C reaches V_{CC}/R_L is better defined than before.

However, under saturation conditions both the emitter and collector junctions are forward biased and the base charge is supplied through both electrodes. In an alloy junction transistor, the total charge in the base is then the result of two triangular distributions as in Fig. 4.19(a), which may be more conveniently redrawn as shown at (b). Here q_{BO} represents the total base charge needed to take the transistor to the point of saturation and q_S represents the extra base charge needed to hold the transistor firmly in saturation, so that the total charge is

$$q_B = q_{BO} + q_S \tag{4.57}$$

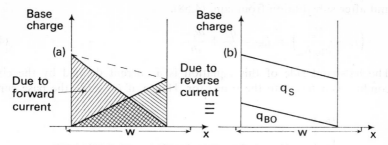

FIG. 4.19 Base charge distribution during saturation

(*a*) Excess base charges due to forward and reverse currents
(*b*) Saturation base charges

The component of base current associated with q_{BO} is i_{BO}, which is just sufficient to saturate the transistor. Hence

$$i_{BO} = \frac{q_{BO}}{\tau_B} = \frac{V_{CC}}{h_{FE}R_L} \tag{4.58}$$

Similarly the component of current associated with q_S is i_{BS}, which is the part of the base current needed to hold the transistor firmly in saturation, and in the steady state

$$i_{BS} = \frac{q_S}{\tau_S} \tag{4.59}$$

where τ_S represents the lifetime of the extra stored charge. Then, again in the steady state, the total base current is

$$i_B = i_{BO} + i_{BS} = \frac{q_{BO}}{\tau_B} + \frac{q_S}{\tau_S} \tag{4.60}$$

τ_S and τ_B are not equal since τ_S refers to minority carriers originating both in the emitter and the collector, while τ_B refers only to minority carriers from the emitter. It may be shown (Ref. 4.2) that $\tau_S = 0.44\,\mu s$ for the transistor whose charge control parameters are quoted on page 146.

For any transient change in the saturation region, q_{BO} may be regarded as constant while only q_S changes, so that the base current is given by

$$i_B = \frac{q_{BO}}{\tau_B} + \frac{q_S}{\tau_S} + \frac{dq_S}{dt} \tag{4.61}$$

147

and after substitution from eqn. (4.58),

$$\left(i_B - \frac{V_{CC}}{h_{FE}R_L}\right) = i_{BS} = \frac{q_S}{\tau_S} + \frac{dq_S}{dt} \tag{4.62}$$

The left-hand side of this equation is a current defined by the circuit conditions external to the transistor. Hence, if $i_B = I_{B1}$, the base current

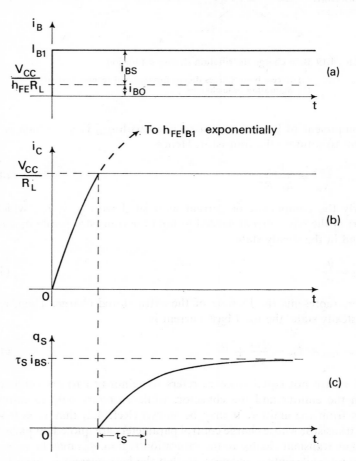

FIG. 4.20 Transient response with overdrive for switching on
(*a*) Base-current step function
(*b*) Collector current
(*c*) Saturation base charge

used to switch on the transistor, the time for q_s to reach its steady state value can be determined (Fig. 4.20(c)).

The transistor is switched off by reversing the input current step (Fig. 4.21(a)), but collector current continues to flow and remains constant at V_{CC}/R_L while the charge q_S is removed from the base. If it is assumed that

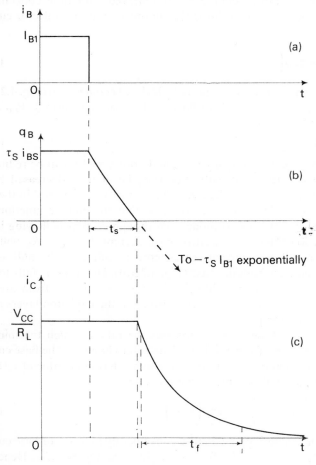

FIG. 4.21 Transient response to switching off

   (a) Base-current step function
   (b) Saturation base charge
   (c) Collector current

q_S has reached a steady value $\tau_S i_{BS}$ before turn-off, solution of eqn. (4.62) shows that it falls exponentially and attempts to reach a terminal value of $-\tau_S I_{B1}$. However, this exponential fall cannot be completed since after a time t_s the point $q_S = 0$ is reached and the transistor then enters the active region (Fig. 4.21(b)). t_s is the *storage time* after which the collector current begins to fall as charge q_{BO} is removed with time constant τ_B. Solution of eqn. (4.53) then shows that in this region the collector current is given by

$$i_C = \frac{V_{CC}}{R_L} \exp\left(-\frac{t}{\tau_B}\right) \tag{4.63}$$

where t is measured from the instant at which q_S becomes zero (Fig. 4.21(c)). The *fall time*, t_f, measured from 90% to 10% of the current V_{CC}/R_L is given by

$$t_f = 2{\cdot}2\tau_B \tag{4.64}$$

Again t_f may be unacceptably long, and both t_s and t_f can be reduced if the transistor is turned off with overdrive; i.e. if v_S is decreased below zero to some value V_2 (Fig. 4.22(a)). The minority carriers still in the base are then removed through the emitter, so that the emitter junction will behave like a forward-biased diode even though current is flowing in the reverse direction. Hence a negative base current $I_{B2} \approx V_2/R_B$ will also flow until the collector current ceases, corresponding to the bulk of the charge being cleared from the base (Fig. 4.22(b)). However, i_B falls to zero only gradually owing to the diffusion of minority carriers from areas of the base remote from the emitter and also from the depletion layers at the emitter and collector junctions.

This reversal of base current increases the rate at which the minority carriers are removed from the base. The total change in the base current is $I_{B1} - I_{B2}$, so that q_S will attempt to reach a terminal value of $\tau_S(I_{B1} - I_{B2})$. However, q_S will become zero after time t_s so that

$$q_S = \tau_S(I_{B1} - I_{B2})\left[1 - \exp\left(-\frac{t_s}{\tau_S}\right)\right] \tag{4.65}$$

This equation can be expressed more conveniently in terms of base current since $i_{BS} = I_{B1} - V_{CC}/h_{FE}R_L$, from eqn. (4.62), and $q_S = \tau_S i_{BS}$. Hence the storage time t_s will be given by

$$I_{B1} - \frac{V_{CC}}{h_{FE}R_L} = (I_{B1} - I_{B2})\left[1 - \exp\left(\frac{t_s}{\tau_S}\right)\right] \tag{4.66}$$

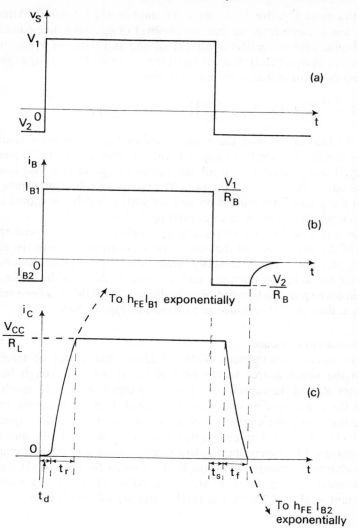

Fig. 4.22 Complete transient response with overdrive

(*a*) Input voltage pulse
(*b*) Base current
(*c*) Collector current

and the more negative I_{B2} is made the smaller will t_s become. After time t_s, q_S has become zero, so that the removal of q_{BO} with time-constant τ_B can begin, with the collector current aiming at a value of $h_{FE}I_{B2}$ instead of zero as in eqn. (4.63). Since it starts from a value V_{CC}/R_L the expression for the decay of collector current becomes

$$i_C = \frac{V_{CC}}{R_L} + \left(h_{FE}I_{B2} - \frac{V_{CC}}{R_L}\right)\left[1 - \exp\left(-\frac{t}{\tau_B}\right)\right] \tag{4.67}$$

and the fall time t_f over the whole range of V_{CC}/R_L becomes smaller as I_{B2} is increased, since it corresponds to a smaller part of the exponential decay. Finally when $i_C = 0$ all the excess charge in the base has been removed, so that $i_B = 0$ and $t = t_f$. The complete collector current waveform for a transistor turned on and off with overdrive is shown in Fig. 4.22(c), which corresponds to normal operation.

For a practical transistor there is also a *delay time* t_d between application of the step input and the time when the current begins to rise to 10% of V_{CC}/R_L. This is due to minority carriers filling the depletion layers in the emitter and collector junctions and so reducing their widths to the values which correspond to forward bias conditions. Until this has been achieved current flow cannot commence, as shown in Fig. 4.22(c).

Measurement of τ_B and τ_C

In the circuit arrangement of Fig. 4.23(a), charge will be transferred from the signal source to the base of the transistor through both the resistor R_B and the capacitor C. For a step input voltage ΔV much larger than the corresponding change in V_{BE}, which in practice means an input of a few volts, the charge transferred initially through the capacitor is $C\Delta V$ (Fig. 4.23(b)). This charge then leaks away exponentially through the transistor with time-constant τ_B to give a component of the base charge q_1. The charge q_2 transferred through R_B is given by eqn. (4.52) and rises exponentially to a value $\tau_B\Delta I$, where $\Delta I = \Delta V/R_B$, again with time-constant τ_B (Fig. 4.23(c)). The total charge transferred is then

$$q_B = q_1 + q_2 \tag{4.68}$$

so that the corresponding change in collector current (Fig. 4.23(d)) is given by

$$i_C = \frac{q_B}{\tau_C} = h_{FE}\frac{q_1 + q_2}{\tau_B} \tag{4.69}$$

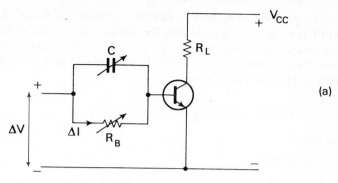

(a)

(b)

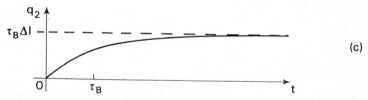

(c)

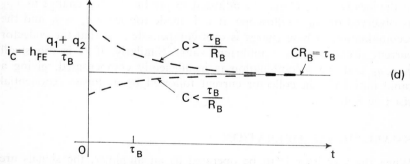

(d)

FIG. 4.23 Measurement of charge control parameters
(a) Measuring circuit
(b) Variation of charge q_1 introduced through capacitor C
(c) Variation of charge q_2 introduced through resistor R
(d) Collector current waveforms

153

The parameters τ_B and τ_C can be determined if C and R are adjusted until $q_1 = q_2$. In this condition the fall in q_1 due to recombination in the base is exactly compensated by the rise in q_2, so that q_B rises instantaneously and then remains constant with time, while i_C undergoes a similar step change (Fig. 4.23(d)). Then

$$C\Delta V = \tau_B \Delta I = \tau_B \frac{\Delta V}{R_B} \tag{4.70}$$

so that

$$\tau_B = CR_B \tag{4.71}$$

If ΔV is made just large enough to switch the transistor from zero collector current to saturation, the required change in base current is

$$\frac{V_{CC}}{h_{FE}R_L} = \frac{\Delta V}{R_B} \tag{4.72}$$

Then

$$h_{FE} = \frac{R_B}{R_L}\frac{V_{CC}}{\Delta V} = \frac{\tau_B}{\tau_F} \tag{4.73}$$

and

$$\tau_C = \frac{\Delta V R_L}{V_{CC}R_B}\tau_B = \frac{\Delta V}{V_{CC}}CR_L \tag{4.74}$$

If R_B is held constant at the value at which the transistor is just saturated in the steady state, C may be adjusted to produce a step change in V_{CE} as observed on an oscilloscope. If C is made too large $q_1 > q_2$ and the recombination of base charge is over-compensated, so that the collector current overshoots its equilibrium value. Similarly, if C is too small $q_1 < q_2$ and the recombination is insufficiently compensated, giving a rapid initial rise in collector current followed by a slower exponential rise (see Ref. 4.2).

SMALL-SIGNAL OPERATION

When the transistor is to be operated as an amplifier, the signals are normally small and the transistor is first biased to a particular operating point such as Q in Fig. 4.17. The required value of base current may be obtained by connecting the resistor R_B directly to the d.c. supply, as shown in the simple amplifier circuit of Fig. 4.24(a), so that $I_B \approx V_{CC}/R_B$ since $V_{BE} \ll V_{CC}$. Corresponding steady values of I_C and V_{CE} are obtained by the intersection of the load line and the characteristics at the point Q.

The input signal to the amplifier is obtained from a generator of voltage V_S and internal resistance R_S, which is connected to the base through a capacitor C. When the waveform of V_S is sinusoidal at frequency f the reactance of C is $1/2\pi f C$, which is made smaller than R_S by choosing a large value for C, say $5\,\mu$F, for operating frequencies within the audible range of 20 Hz to 20 kHz. The purpose of the capacitor is to block the direct voltage V_{BE} from the input generator, which it achieves since its reactance at zero frequency is infinite. The signal waveforms corresponding to sinusoidal changes are illustrated in Fig. 4.24, which shows that a rise in base current causes a rise in collector current and hence a fall in collector voltage. The opposite changes occur for a fall in base current, the changes always occurring with respect to a direct quantity.

In a typical amplifier let us suppose that $V_{CC} = -6\,$V, $R_L = 2\,$kΩ, $V_{CE} = -3\,$V, so that $I_C = -1.5\,$mA. Then if $h_{FE} = 100$, $I_B = -15\,\mu$A and $R_B = 400\,$kΩ. The peak values of the signals might be $I_{bm} = 1\,\mu$A, $I_{cm} = 100\,\mu$A and $V_{2m} = 0.2\,$V. These are clearly much less than the corresponding direct values and would be very difficult to measure accurately from the characteristics. Hence the performance of a small-signal amplifier is obtained from a circuit which is equivalent to the transistor operating at a particular point Q on its characteristics. This equivalent circuit is composed of resistive and capacitive elements to simulate the effect of high frequencies and also contains one or more generators. All the elements are related to the physical processes described previously so that their values are functions of the operating voltage, current and temperature.

One of these processes, which sets a limit to the highest frequency at which the transistor will operate, is the transit time of carriers across the base, τ_C (page 142). The instantaneous signal current i_b will cause corresponding currents i_e and i_c to flow through the emitter and collector respectively, and owing to the finite value of τ_C there will be a time delay between carriers leaving the emitter and reaching the collector in response to the input signal (Fig. 4.25). This will cause the collector current to lag behind the emitter current, so that if the instantaneous change in emitter current is given by

$$i_e = I_{em} \sin \omega t \tag{4.75}$$

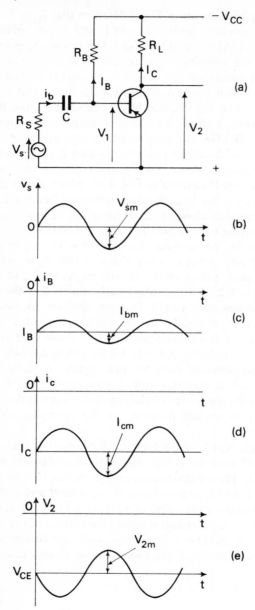

FIG. 4.24 Small-signal operation

(a) Amplifier circuit
(b) Input voltage to base
(c) Base current
(d) Collector current
(e) Collector–emitter voltage

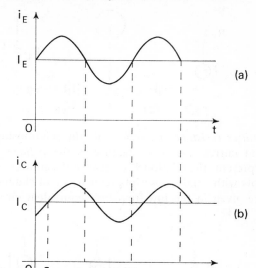

FIG. 4.25 Phase lag and fall in current gain at high frequencies

(*a*) Emitter current
(*b*) Collector current

(a)

(b)

the instantaneous change in collector current is given by

$$i_c = \alpha I_{em} \sin \omega(t - \tau_C) \qquad (4.76)$$

At low frequencies τ_C is negligible compared with the period T of the signal, so that the currents may be considered to be in phase, but at frequencies where $\tau_C > 0.2T$ the phase lag will become appreciable. In addition, it is shown on page 169 that α falls as the frequency rises, so that these two effects will cause the performance of the transistor to deteriorate.

It is convenient to derive initially a common-base equivalent circuit since in this arrangement the emitter and collector voltages are developed only across their respective junctions, which leads to a simple physical model. A common-emitter equivalent circuit can then be easily obtained by rearranging the elements of the common-base circuit. Consider the common-base amplifier of Fig. 4.26, whose input characteristic will depend on the I_E/V_{EB} characteristic of Fig. 4.9. If V_{CB} is about -3 V a characteristic between the -10 V and -1 V curves is needed, and for $I_E = 1$ mA, V_{EB} will be about 0.24 V. Small changes in V_{EB} about this mean value will occur due to the input signal, leading to corresponding changes in I_E and V_{CB} about their mean values. One feature of the equivalent circuit, the

157

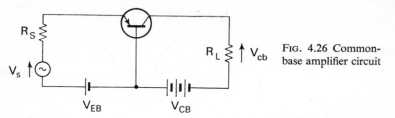

Fig. 4.26 Common-base amplifier circuit

emitter resistance r_ε, represents the relationship between emitter voltage and current. Another feature is the *voltage feedback factor* μ, which represents the relationship between the emitter and collector voltages. Symbols with small subscripts refer to small changes in the quantity, such as are given in the low-frequency equivalent circuit of Fig. 4.27, in which

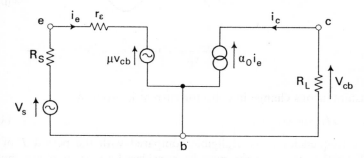

Fig. 4.27 Low-frequency equivalent circuit of a common-base amplifier

transit-time effects and internal capacitances are neglected. The circuit applies only to small changes of voltage and current, for which the values of the circuit elements may be considered constant. Also the d.c. supplies are supposed to be short-circuited since the internal impedance of a power supply is very low.

On the output characteristic of the transistor, Fig. 4.11, the quiescent point Q will be at $I_C = -0.99\,\text{mA}$ and $V_{CB} = -3\,\text{V}$ if $\alpha_0 = 0.99$. To a good approximation I_C is independent of V_{CB}, so the only element required to represent the effect of I_E at the output is a current generator $\alpha_0 i_e$. It is shown on page 169 that the parameter α is a function of frequency; α_0 refers to its value at low frequencies. The definition of α_0 is

$$\alpha_0 = \frac{\partial I_C}{\partial I_E}\bigg|_{V_{CB}\,\text{constant}} \qquad\qquad \text{(eqn. (4.1))}$$

or

$$\alpha_0 = \frac{i_c}{i_e}\bigg|_{V_{cb}=0} \tag{4.77}$$

which is close to the value obtained by considering the ratio of the quiescent currents at normal operating values and given by eqn. (4.23) in terms of the physical properties of the transistor.

Common-base Input Resistance, r_ε

I_E and V_{EB} are related by an expression of the form of eqn. (3.5) for a *p-n* junction, where I_0 represents the emitter–base leakage current with the collector open-circuited. This leads to

$$I + I_0 = I_0 \exp\left(\frac{eV}{kT}\right) \tag{4.78}$$

so that the gradient of the curve at any point is

$$\frac{dI}{dV} = I_0 \frac{e}{kT} \exp\left(\frac{eV}{kT}\right) \tag{4.79}$$

$$= \frac{e}{kT}(I + I_0) \tag{4.80}$$

Thus, for signals sufficiently small that their ratio equals the gradient of the curve,

$$\frac{dI}{dV} = \frac{i_e}{v_e} = \frac{eI_E}{kT} \tag{4.81}$$

since I_0 is normally negligible compared with the quiescent emitter current I_E. Then the resistance r_ε is given by

$$r_\varepsilon = \frac{v_e}{i_e} = \frac{kT}{eI_E} \tag{4.82}$$

its value being inversely proportional to the emitter current. At room temperature kT/e is about $25\,\mathrm{mV}$ (page 35), so that, if I_E isex pressed in milliamperes,

$$r_\varepsilon = \frac{25}{I_E} \tag{4.83}$$

159

Thus for $I_E = 1\,\text{mA}$, $r_\varepsilon = 25\,\Omega$, but for $I_E = 2\cdot5\,\text{mA}$, r_ε falls to $10\,\Omega$, such low values of input resistance being a feature of the common-base amplifier.

Voltage Generator, μv_{cb}

This represents the effect of changes in V_{CB} on the base width w_b, as described on page 134. For I_E to remain constant as V_{CB} is increased, V_{EB} must be decreased (Fig. 4.9), so that the definition of the voltage feedback factor is

$$\mu = \left.\frac{\partial V_{EB}}{\partial V_{CB}}\right|_{I_E \text{ constant}} = \left.\frac{v_{eb}}{v_{cb}}\right|_{i_e = 0} \tag{4.84}$$

Thus μv_{cb} represents the change in the emitter voltage to maintain constant emitter current as the collector voltage is changed. In order to obtain an expression for μ the density of excess holes, P_e, at the emitter junction is involved (in a *p-n-p* transistor):

$$\frac{\partial V_{EB}}{\partial V_{CB}} = \frac{\partial V_{EB}}{\partial p_e}\frac{\partial p_e}{\partial w_b}\frac{\partial w_b}{\partial V_{CB}} \tag{4.85}$$

From eqn. (3.26) for a *p-n* junction,

$$p_e = p_n \exp\left(\frac{eV_{EB}}{kT}\right) \tag{4.86}$$

so that

$$\frac{\partial q_e}{\partial V_{EB}} = \frac{ep_e}{kT} \tag{4.87}$$

Also, from eqn. (4.7),

$$I_E \approx \frac{eD_p p_e S}{w_b} \tag{4.88}$$

so, for constant I_E,

$$\frac{\partial p_e}{\partial w_b} = \frac{p_e}{w_b} \tag{4.89}$$

Again, from eqn. (4.28),

$$w_b \approx W - w_{nc}$$

so that

$$\frac{\partial w_b}{\partial V_{CB}} = -\frac{\partial w_{nc}}{\partial V_{CB}} \tag{4.90}$$

Substituting into eqns. (4.84) and (4.85),

$$\mu = -\frac{kT}{ew_b}\frac{\partial w_{nc}}{\partial V_{CB}} \tag{4.91}$$

For an alloy transistor with a highly-doped collector, w_{nc} is given by eqn. (3.73) for w_n, and putting $V = -V_{CB}$, which is much larger than ψ,

$$w_{nc} \approx \left(\frac{2\epsilon}{eN_d}\right)^{1/2} V_{CB}^{1/2} \tag{4.92}$$

Thus

$$\frac{\partial w_{nc}}{\partial V_{CB}} = \frac{w_{nc}}{2V_{CB}} \tag{4.93}$$

and

$$\mu = -\frac{kTw_{nc}}{2eV_{CB}w_b} \tag{4.94}$$

Typically w_{nc} is about $1\,\mu$m and w_b about $10\,\mu$m, so at room temperature and with $V_{CB} = -3\,$V, $\mu = 4 \times 10^{-4}$, which is a very small ratio. This means that the voltage fed back from collector to emitter is only important in an amplifier with a high gain in which v_{cb} is greater than $1\,$V, say, making $\mu v_{cb} \geqslant 0.4\,$mV. However, μ is important in determining some of the elements of the equivalent circuit as shown in later sections.

HIGH-FREQUENCY EQUIVALENT CIRCUITS

The equivalent circuit of Fig. 4.27 is only approximate and is restricted to audio frequencies. At high frequencies extra elements must be added to account for the transit-time effect described above, the depletion layer capacitances and the effect of the input signal on the excess base charge,
The excess charge q_B for a constant current I_E is a function of p_e, which depends on V_{EB} (eqn. (4.86)). Thus a change in V_{EB} will cause a

corresponding change in p_e, and this effect may be represented on the equivalent circuit by a capacitor C_d, whose capacitance is given by

$$C_d = \frac{dq_B}{dV_{EB}} \qquad (4.95)$$

C_d is called the *diffusion capacitance* since q_B is maintained by diffusion of charge across the base and it links the small-signal equivalent circuit with the charge-control model. Then

$$C_d = \frac{dp_B}{dI_E} \frac{dI_E}{dV_{EB}} \qquad (4.96)$$

$$= \frac{\tau_C e I_E}{kT} \qquad (4.97)$$

$$= \frac{\tau_C}{r_\varepsilon} \qquad (4.98)$$

using eqns. (4.17) and (4.82). Thus the diffusion capacitance is proportional to the emitter current and for $I_E = 1\,\text{mA}$ and $\tau_C = 10^{-8}\,\text{s}$, $C_d = 400\,\text{pF}$ at room temperature. A similar capacitance exists in a p^+-n diode under forward bias, and this is much greater than the depletion-layer capacitance measured under reverse bias conditions.

The effect of transit time may be represented after considering the section of the base across which most of the carriers travel from emitter to collector, which is shown shaded in Fig. 4.28(a). This is analogous to a transmission line in that the output current lags behind the input current (eqn. (4.76)), and it can be shown that a line composed of resistive and capacitive elements is mathematically equivalent to the behaviour in a transistor at high frequencies (Ref. 4.3). An exact representation is obtained by using a large number of consecutive sections similar to Fig. 4.28(b), but for many purposes the whole line may be approximated by the equivalent symmetrical π-network shown at (c). Here C_1 represents the diffusion capacitance C_d in parallel with the depletion-layer capacitance of the emitter–base diode C_E, and C_2 is equal to C_1. Resistances R_1 and R_2 represent the loss of carriers due to recombination in the base and are also equal.

The alternating base current flows to its terminal through the *base spreading resistance* $r_{bb'}$, which represents the effective distributed resistance of the base material and has a value between about 50 and $200\,\Omega$, while

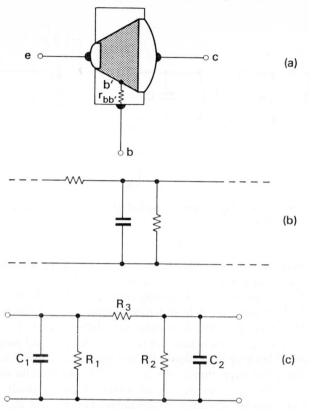

FIG. 4.28 Representation of a junction transistor at high frequencies
(*a*) Effective region of base for minority carrier transport
(*b*) Section of transmission line representing the base region
(*c*) π-equivalent circuit of the transmission line

b' is a point in the base on the edge of the region of maximum current density. The same general idea may also be applied to the graded-base transistor. The π-network, together with the collector depletion-layer capacitance C_c, is incorporated in the complete equivalent circuit of Fig. 4.29. The current i_2 represents minority carriers which have reached the edge of the collector depletion layer and are swept into the collector by the reverse bias. Thus i_2 reappears in the collector circuit as a current generator, so that at low frequencies,

$$i_2 = \alpha_0 i_1 \tag{4.99}$$

163

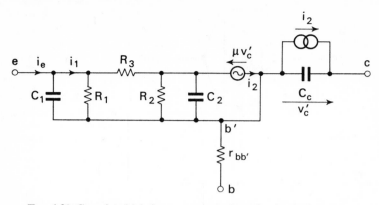

FIG. 4.29 Complete high-frequency equivalent circuit of a junction transistor

The signal voltage across the collector depletion layer is now v'_c, so that the generator accounting for the variation of base width is $\mu v'_c$.

Although this circuit is a good approximation to the behaviour of a transistor at high frequencies it is still too complicated for the purpose of analysis. Also it is necessary to express the elements in terms of the parameters r_ε, μ and α_0, which in turn are related to the physical properties of the transistor. This may be achieved by first considering low-frequency signals (for which the capacitances are negligible); the circuits of Figs. 4.29 and 4.27 are then equivalent to each other. Also for small signals the generator $\mu v'_c$ may be neglected, which means that C_2 and R_2, on the collector side, are virtually short-circuited. In physical terms this means that C_2 carries no charge, which is reasonable since there is no charge at the edge of the collector depletion layer. The currents flowing through R_1 and R_3 then lead to

$$(i_1 - i_2)R_1 = i_2 R_3 \tag{4.100}$$

and putting $i_2 = \alpha_0 i_1$,

$$(1 - \alpha_0)R_1 = \alpha_0 R_3 \tag{4.101}$$

Also, by comparing the circuits of Figs. 4.29 and 4.27,

$$\frac{R_1 R_3}{R_1 + R_3} = r_\varepsilon \tag{4.102}$$

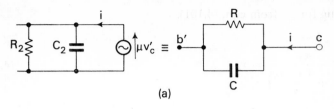

(a)

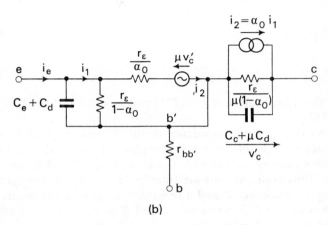

(b)

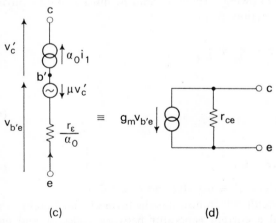

(c)   (d)

FIG. 4.30 Reduction of Fig. 4.29 to hybrid-π equivalent circuit

(a) Transfer of R_2 and C_2 to collector
(b) Simplified equivalent circuit
(c) Transformation of equivalent circuit between collector and emitter

Substituting for R_3 from eqn. (4.101),

$$\frac{R_1{}^2 \dfrac{1-\alpha_0}{\alpha_0}}{R_1\left(1+\dfrac{1-\alpha_0}{\alpha_0}\right)} = r_\varepsilon \tag{4.103}$$

so that

$$R_1 = \frac{r_\varepsilon}{1-\alpha_0} \tag{4.104}$$

$$R_3 = \frac{r_\varepsilon}{\alpha_0} \tag{4.105}$$

Thus, for the typical values of $r_\varepsilon = 25\,\Omega$ and $\alpha_0 = 0\cdot99$, $R_1 = 2500\,\Omega$ and $R_3 = 25\cdot3\,\Omega$.

The effect on the input impedance of the elements C_2 and R_2 on the collector side is negligible, owing to the effect of the generator $\mu v_c'$, but these elements are associated with i_2 and v_c', both of which are output quantities. Thus both elements are transferred to the collector circuit by postulating a capacitance C and a resistance R between the points c and b', and then finding C in terms of C_2, and R in terms of R_2. From Fig. 4.30(a), the current i flowing through C_2 and R_2 must also flow through C and R by transistor action, so that

$$i = \frac{\mu v_c'}{R_2} + \frac{\mu v_c'}{j\omega C_2} = \frac{v_c'}{R} + \frac{v_c'}{j\omega C} \tag{4.106}$$

Hence

$$R = \frac{R_2}{\mu} = \frac{r_\varepsilon}{\mu(1-\alpha_0)} \tag{4.107}$$

and

$$C = \mu C_2 = \mu(C_d + C_E) \tag{4.108}$$

Inserting typical values, $R = 6\cdot3\,\mathrm{M}\Omega$ and $C \approx 0\cdot2\,\mathrm{pF}$.

The equivalent circuit, which now has the form of Fig. 4.30(b), contains both a voltage and a current generator between collector and emitter, which are the output terminals in the common-emitter configuration. Before the circuit is rearranged to give this configuration the two generators are replaced by a single current generator in terms of the internal signal voltage $v_{b'e}$ (Fig. 4.30(d)). The new output circuit must be identical to the circuit between terminals c and e in Fig. 4.30(c), and the relationships between the elements of the two circuits are obtained by comparing first their short-circuit currents and then their open-circuit voltages. An a.c. short-circuit may be obtained by connecting a large capacitor between collector and emitter, which makes $v'_c = 0$. Then the current flowing is

$$i_{sc} = -i_2 = -\alpha_0 i_e \tag{4.109}$$

at low frequencies. The circuit of Fig. 4.30(c) is resistive, so at all frequencies

$$i_{sc} = g_m v_{b'e} = -\alpha_0 i_e \tag{4.110}$$

when the circuits are identical. But from eqn. (4.82), putting $v_{b'e} \approx -v_e$,

$$\frac{i_e}{v_{b'e}} = -\frac{eI_E}{kT} \tag{4.111}$$

so that

$$g_m = \frac{eI_C}{kT} = \frac{\alpha_0}{r_\varepsilon} \tag{4.112}$$

and at room temperature with I_C in milliamperes,

$$g_m = 40I_C \text{ millisiemens} \tag{4.113}$$

g_m is called the *mutual conductance*, or *transconductance*, of the transistor and is proportional to the collector current, having a value of 39·6 mS at $I_C = 0·99$ mA. Finally, the resistance r_{ce} must be identical to the resistance between the terminals c and e of Fig. 4.30(c). Comparing the open-circuit voltages of the two circuits.

$$v_{oc} = -g_m v_{b'e} = v'_c - \mu v'_c \approx v'_c \tag{4.114}$$

since $\mu \ll 1$. But, from Fig. 4.30(c),

$$v_{b'e} = -\mu v'_c$$

so that substituting into eqn. (4.114),

$$g_m \mu r_{ce} = 1 \tag{4.115}$$

and

$$r_{ce} = \frac{1}{\mu g_m} = \frac{r_\varepsilon}{\mu \alpha_0} \tag{4.116}$$

Physical Electronics

The complete high-frequency equivalent circuit of Fig. 4.31, known as the *hybrid-π* circuit, is obtained by combining Figs. 4.30(b) and (d) and rearranging the elements so that the emitter is common to the input and output. Its parameters, which are designated according to the points which they connect, and their equivalents are listed in Table 4.2. The

Table 4.2 Parameters of Junction Transistors

Hybrid-π parameters	Other common-emitter parameters	Common-base parameters	Typical values	
			Uniform-base transistor	Graded-base transistor
$r_{bb'}$			50–200 Ω	20–50 Ω
$r_{b'e}$	$\dfrac{\beta_0}{g_m}$	$\dfrac{r_\varepsilon}{1-\alpha_0}$	2·5 kΩ	2·5 kΩ
$r_{b'c}$	$\dfrac{\beta_0}{\mu g_m}$	$\dfrac{r_\varepsilon}{\mu(1-\alpha_0)}$	6·3 MΩ	6·3 MΩ
r_{ce}	$\dfrac{1}{\mu g_m}$	$\dfrac{r_\varepsilon}{\mu\alpha_0}$	63 kΩ	63 kΩ
$C_{b'e}$		$\dfrac{\tau_C}{r_\varepsilon}+C_E$	400 pF	11 pF
$C_{b'c}$			10 pF	1–2 pF
g_m	$\dfrac{eI_C}{k\tau}$	$\dfrac{\alpha_0}{r_\varepsilon}$	39·6 mS	39·6 mS
β_0	$g_m r_{b'e}$	$\dfrac{\alpha_0}{1-\alpha_0}$	100	100
f_τ	$\dfrac{g_m}{2\pi C_{b'e}}$	$\approx \dfrac{\alpha_0}{2\pi\tau_c}$	15 MHz	570 MHz
f_β	$\dfrac{1}{2\pi C_{b'e}r_{b'e}}$	$\approx \dfrac{1-\alpha_0}{2\pi\tau_c}$	155 kHz	5·7 MHz
r_ε		$\dfrac{kT}{I_E}$	25 Ω	25 Ω
τ_c			10^{-8} s	$2{\cdot}8\times 10^{-10}$ s
α_0			0·99	0·99
μ			4×10^{-4}	4×10^{-4}
I_C			1 mA	1 mA
T			293 K	293 K

numerical values of α_0 and I_C are appropriate to small high-frequency transistors. A transistor for power amplification, however, would operate at a much higher collector current, 0·1 A for example, and at a lower

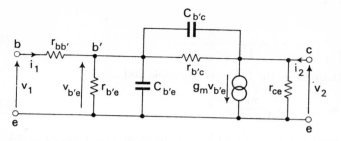

Fig. 4.31 Hybrid-π equivalent circuit of a junction transistor

value of α_0, say 0·98. These changes have a large effect on the parameters, causing g_m to rise to 4S and $r_{b'e}$ to fall to 12·5 Ω, for instance.

GAIN-BANDWIDTH PRODUCT f_τ

One of the most important small-signal parameters of a transistor is the common-emitter short-circuit current gain, h_{fe}, also given the symbol β. For sinusoidal signals,

$$h_{fe} = \frac{I_c}{I_b}\bigg|\, V_{ce} = 0 \tag{4.117}$$

and it is a function of frequency which can be obtained from the hybrid-π circuit of Fig. 4.31. It is related to the common-base short-circuit current gain α by the expression

$$h_{fe} = \frac{\alpha}{1 - \alpha} \tag{4.118}$$

so that α is also a function of frequency.

From Fig. 4.31, if the effects of $C_{b'c}$ and $r_{b'c}$ are neglected, and the collector and emitter are short-circuited, then

$$I_b = V_{b'e} \frac{(1 + j\omega C_{b'e} r_{b'e})}{r_{b'e}} \tag{4.119}$$

and

$$I_c = g_m v_{b'e} \tag{4.120}$$

169

so that

$$\left.\frac{I_c}{I_b}\right|_{V_{ce}=0} = h_{fe} = \frac{g_m r_{b'e}}{1 + j\omega C_{b'e} r_{b'e}} \tag{4.121}$$

At low frequencies the j term is negligible and

$$h_{fe} = g_m r_{b'e} = \frac{\alpha_0}{1 - \alpha_0} = \beta_0 \tag{4.122}$$

As the frequency is increased h_{fe} falls until, at a frequency f_β, $\omega C_{b'e} = 1$. Here the reactance of $C_{b'e}$ and the resistance $r_{b'e}$ are equal and

$$f_\beta = \frac{1}{2\pi C_{b'e} r_{b'e}} \tag{4.123}$$

Substitution into eqn. (4.121) then yields

$$h_{fe} = \frac{\beta_0}{1 + j\dfrac{f}{f_\beta}} \tag{4.124}$$

and f_β is the frequency at which $|h_{fe}|$ has fallen to $\beta_0/\sqrt{2}$, or 70·7% of its low-frequency value, β_0, which may be expressed as a fall of 3 dB from the low-frequency value.* At frequencies well above f_β, $1/\omega C_{b'e} \ll r_{b'e}$ and I_b may be considered to flow entirely through $C_{b'e}$ (Fig. 4.32(*a*)). Then

$$I_b = \frac{V_{b'e}}{\dfrac{1}{j\omega C_{b'e}}} = j\omega C_{b'e} V_{b'e} \tag{4.125}$$

and

$$h_{fe} = \frac{g_m}{j\omega C_{b'e}} \tag{4.126}$$

In a practical transistor $|h_{fe}|$ falls almost linearly as frequency rises in this region until, at a frequency f_1, $|h_{fe}| = 1$. In order to determine f_1 a lower frequency is chosen on the linearly falling part of the curve in Fig.

---

* For two powers, P_1 and P_2, the ratio in decibels (dB) is $10 \log_{10} (P_1/P_2)$. Where the powers are due to currents I_1 and I_2 flowing through the same value of resistance R the current ratio is $10 \log_{10} (I_1/I_2)^2 = 20 \log_{10} (I_1/I_2)$. Thus when $I_1/I_2 = 0·707$ this corresponds to a decibel ratio of 3·01 dB. Similarly for two voltages, V_1 and V_2, developed across the same resistance the voltage ratio is $20 \log_{10} (V_1/V_2)$.

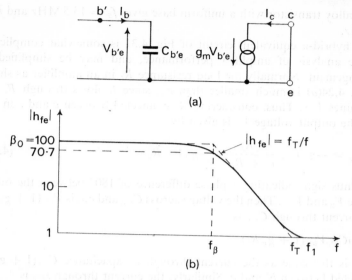

(a)

(b)

FIG. 4.32 High-frequency response of a junction transistor

(a) Equivalent circuit for short-circuit current gain
(b) Short-circuit current gain as a function of frequency

4.32(b) and $|h_{fe}|$ is measured. Then from eqn. (4.126) the product of current gain and frequency is

$$|h_{fe}| f = \frac{g_m}{2\pi C_{b'e}} = f_T \qquad (4.127)$$

f_T is called the *gain-bandwidth product* of the transistor. h_{fe} may be inversely proportional to frequency in practice, in which case $f_1 = f_T$, but this is not always so. The relationship between f_T and f_β may be obtained from eqn. (4.124), since when the j term is much larger than unity,

$$|h_{fe}| \frac{f}{f_\beta} = \beta_0 \qquad (4.128)$$

so that

$$f_T = \beta_0 f_\beta \qquad (4.129)$$

from eqn. (4.127). f_β then occurs at the intersection of the $|h_{fe}| = \beta_0$ and $|h_{fe}| = f_T/f$ parts of Fig. 4.32(b). Using typical values of g_m, $C_{b'e}$ and β_0

171

for an alloy transistor with a uniform base gives $f_T = 15\cdot5\,\text{MHz}$ and $f_\beta = 155\,\text{kHz}$.

The hybrid-π equivalent circuit of Fig. 4.31 is somewhat complicated for the analysis of amplifier performance, and may be simplified by rearrangement. Normally the load resistance R_L in an amplifier as shown in Fig. 4.24(a) is much smaller than r_{ce}, since I_C flows through R_L and determines V_{CE}. Thus, considering R_L connected between c and e in Fig. 4.31, the output voltage V_2 is given by

$$V_2 = -g_m V_{b'e} \frac{r_{ce}R_L}{r_{ce}+R_L} \approx -g_m V_{b'e} R_L \qquad (4.130)$$

the minus sign indicating a phase difference of 180° between the output voltage V_2 and $V_{b'e}$. Then the voltage across $C_{b'e}$ and $r_{b'c}$ is $V_{b'e}(1 + g_m R_L)$. The current through $C_{b'c}$ is

$$j\omega C_{b'c} V_{b'e}(1 + g_m R_L)$$

which is the same as the current through a capacitance $C_{b'c}(1 + g_m R_L)$ connected between b' and e. Similarly, the current through $r_{b'c}$ is

$$\frac{V_{b'e}(1 + g_m R_L)}{r_{b'c}}$$

which is the same as the current through a resistance $r_{b'c}/(1 + g_m R_L)$ connected between b' and e. Now, for values of $g_m R_L$ up to about 100 this new resistance will still be about $35\,\text{k}\Omega$, which is much larger than the value of $r_{b'e}$ (Table 4.2), so that it may be neglected. Then the approximate equivalent circuit becomes as shown in Fig. 4.33, where

$$C = C_{b'e} + C_{b'c}(1 + g_m R_L) \qquad (4.131)$$

Thus the frequency response is determined by the input circuit, which depends on R_L and so on the gain of the amplifier.

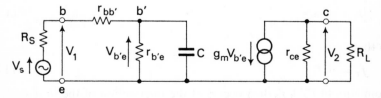

FIG. 4.33 Simplified hybrid-π equivalent circuit

Using the circuit of Fig. 4.33 to obtain $V_{b'e}$ in terms of V_S,

$$V_{b'e} = \cfrac{V_S\dfrac{r_{b'e}}{1 + j\omega Cr_{b'e}}}{R_S + r_{bb'} + \dfrac{r_{b'e}}{1 + j\omega Cr_{b'e}}}$$

$$= \cfrac{V_S\dfrac{r_{b'e}}{R_S + r_{bb'} + r_{b'e}}}{1 + \dfrac{j\omega Cr_{b'e}(R_S + r_{bb'})}{R_S + r_{bb'} + r_{b'e}}} \qquad (4.132)$$

The output voltage V_2 is equal to $-g_m V_{b'e} R_L$, so that, using eqn. (4.122) and taking the modulus of eqn. (4.132), leads to the modulus of the voltage gain:

$$|A| = \left|\frac{V_2}{V_S}\right| = \cfrac{\dfrac{\beta_0 R_L}{R_S + r_{bb'} + r_{b'e}}}{\left[1 + \omega^2 C^2 r^2_{b'e}\left(\dfrac{R_S + r_{bb'}}{R_S + r_{bb'} + r_{b'e}}\right)^2\right]^{1/2}} \qquad (4.133)$$

At low frequencies this equation reduces to

$$A_0 = \frac{\beta_0 R_L}{R_S + r_{bb'} + r_{b'e}} \qquad (4.134)$$

and at a high frequency given by

$$f_h = \cfrac{1}{2\pi Cr_{b'e}\dfrac{R_S + r_{bb'}}{R_S + r_{bb'} + r_{b'e}}} \qquad (4.135)$$

the gain has fallen to 70·7% of its low-frequency value; f_h is the upper 3 dB frequency which is normally taken as the upper limit of the frequency response of the amplifier.

Both the low-frequency and the high-frequency response depend on the value of the source resistance R_S, as well as on the load resistance R_L. Although A_0 is proportional to R_L, C also rises with R_L, from eqn. (4.131), so that a higher low-frequency gain is compensated by a reduced high-frequency response. When $R_S \gg r_{bb'} + r_{b'e}$,

$$f_h \approx \frac{1}{2\pi C r_{b'e}} \qquad (4.136)$$

and for smaller values of R_S, f_h is increased, so that the best frequency response is obtained with the lowest value of R_S.

It is common practice among transistor manufacturers to give a summary of the properties of a transistor in the form of numerical values of I_C, h_{fe} (at low frequencies) and f_T. It may be noted that three of the elements of the circuit of Fig. 4.31 may be obtained from this information, since g_m is a function of f_T, as shown in Table 4.2. These are the three most important parameters, since $r_{bb'}$ may well be much less than R_S, and the effect of $C_{b'c}$ will be negligible for small values of R_L. In addition, f_T will give an estimate of the charge control parameter τ_C, because $C_{b'e} \approx \tau_C/r_e$ when C_e is negligible so that $f_T \approx \alpha_0/2\pi\tau_C$. While the full data of a transistor should be used wherever possible, the summarized data will nevertheless give some basis for comparing the properties of different types of transistor.

At low frequencies, up to about 10 kHz for a uniform-base transistor and a few hundred kilohertz for a graded-base transistor, a commonly used equivalent circuit is based on the four *hybrid parameters*, which are purely resistive in this range of frequencies. The transistor is considered as a "black box" and its internal operation is expressed entirely in terms of the small-signal input and output voltages and currents (Fig. 4.34).

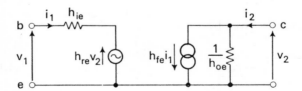

FIG. 4.34 Low-frequency hybrid equivalent circuit of a junction transistor

The h-parameters can also be defined in terms of the d.c. characteristics, and they are given below in terms of both small signals and small changes in the direct voltages and currents of a transistor in the common-emitter configuration.

Input Resistance

$$h_{ie} = \frac{v_1}{i_1}\bigg|_{v_2=0} = \frac{\partial V_{BE}}{\partial I_B}\bigg|_{V_{CE} \text{ const}} \tag{4.137}$$

Reverse Voltage Transfer Ratio

$$h_{re} = \frac{v_1}{v_2}\bigg|_{i_1=0} = \frac{\partial V_{BE}}{\partial_{CE}}\bigg|_{I_B \text{ const}} \tag{4.138}$$

Both of these parameters would be measured with the transistor operating at a point such as P on the input characteristic (Fig. 4.15(a)).

The condition $v_2 = 0$ implies that the output terminals are short-circuited to alternating current; and the condition $i_1 = 0$, that the input terminals are open-circuited to alternating current.

Forward Current Transfer Ratio

$$h_{fe} = \frac{i_2}{i_1}\bigg|_{v_2=0} = \frac{\partial I_C}{\partial I_B}\bigg|_{V_{CE} \text{ const}} \tag{4.139}$$

Output Conductance

$$h_{0e} = \frac{i_2}{v_2}\bigg|_{i_1=0} = \frac{\partial I_C}{\partial V_{CE}}\bigg|_{I_B \text{ const}} \tag{4.140}$$

In practice the parameters could be measured by biasing the transistor to the relevant operating points and applying a 1 kHz signal to the input terminals with the output short-circuited, or to the output terminals with the input open-circuited.

The parameters are called *hybrid* since two of them are dimensionless and the dimensions of the other two are different. The notation of the subscripts is explained by the quantities to which they refer: thus *i* represents input, *r* reverse (output to input), *f* forward (input to output) and *o* output; *e* refers to the terminal common to input and output, in this case the emitter. Since a circuit similar to Fig. 4.34 can be drawn for the common base or the common collector configuration, the corresponding hybrid parameters would have a subscript ending with *b* or *c* respectively, e.g. h_{fb}, h_{fc} and so on.

The *h*-parameters may be expressed in terms of the physical properties of the transistor by relating them to those *hybrid-π parameters* which are important at low frequencies. This leads to the low-frequency hybrid-π equivalent circuit of Fig. 4.35, which may be compared with Fig. 4.34.

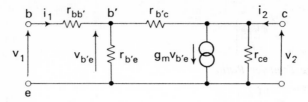

FIG. 4.35 Low-frequency hybrid-π equivalent circuit

Then with the output terminals short-circuited the input resistance is

$$\frac{v_1}{i_1}\bigg|_{v_2=0} = r_{bb'} + \frac{r_{b'e}r_{b'c}}{r_{b'e} + r_{b'c}} \approx r_{bb'} + r_{b'e} \tag{4.141}$$

or

$$h_{ie} \approx r_{bb'} + \frac{r_\varepsilon}{1 - \alpha_0} \tag{4.142}$$

Also the forward current transfer ratio is

$$\frac{i_2}{i_1}\bigg|_{v_2=0} = \frac{g_m v_{b'e}}{v_{b'e}/r_{b'e}} = g_m r_{b'e} \tag{4.143}$$

or

$$h_{fe} = \frac{\alpha_0}{1 - \alpha_0} \tag{4.144}$$

at low frequencies. If the input terminals are open-circuited and a voltage v_2 is applied to the output terminals the reverse voltage ratio is

$$\frac{v_1}{v_2}\bigg|_{i_1=0} = \frac{r_{b'e}}{r_{b'e} + r_{b'c}} = \frac{1}{1 + 1/\mu} \approx \mu \tag{4.145}$$

or

$$h_{re} \approx \mu \tag{4.146}$$

The output impedance is obtained from

$$i_2 = \frac{v_2}{r_{ce}} + g_m v_{b'e} \tag{4.147}$$

$$\approx v_2(1/r_{ce} + \mu g_m) \tag{4.148}$$

176

since $v_{b'e} = v_1 \approx \mu v_2$ with the input open-circuited. From eqn. (4.116), $\mu g_m = 1/r_{ce}$, which leads to

$$\left. \frac{i_2}{v_2} \right|_{i_1=0} = h_{oe} \approx \frac{2}{r_{ce}} \tag{4.149}$$

Since the hybrid parameters are expressed in terms of r_e, α_0 and μ their dependence on the transistor properties is obtained from eqns. (4.82), (4.23) and (4.94). Using the typical values given in Table 4.2, $h_{ie} \approx 2\,600\,\Omega$, $h_{fe} = 100$, $h_{re} \approx 4 \times 10^{-4}$ and $h_{oe} \approx 32\,\mu S$ with h_{ie} inversely proportional to I_C and h_{oe} proportional to I_C. It may be noted that, although measurement of h_{ie} gives a method for finding $r_{bb'}$, using eqn. (4.142), $r_{bb'}$ appears as the difference between two large quantities.

It may be seen from Fig. 4.34 that the following equations can be written down, which completely describe the operation of a transistor amplifier under low-frequency small-signal conditions.

$$v_1 = h_{ie}i_1 + h_{re}v_2 \tag{4.150}$$

$$i_2 = h_{fe}i_1 + h_{oe}v_2 \tag{4.151}$$

$$v_1 = v_S + i_1 R_S \tag{4.152}$$

$$v_2 = -i_2 R_L \tag{4.153}$$

Simultaneous solution of these equations leads to the current and voltage gains and the input and output resistances of the amplifier in terms of the transistor determinant

$$\Delta_e = h_{ie}h_{oe} - h_{re}h_{fe} \tag{4.154}$$

which is always a positive quantity, so that

Current gain

$$\frac{i_2}{i_1} = \frac{h_{fe}}{1 + h_{oe}R_L} \tag{4.155}$$

Voltage gain

$$\frac{v_2}{v_1} = -\frac{h_{fe}R_L}{h_{ie} + \Delta_e R_L} \tag{4.156}$$

Input resistance

$$\frac{v_1}{i_1} = \frac{h_{ie} + \Delta_e R_L}{1 + h_{oe}R_L} \tag{4.157}$$

Output resistance

$$\frac{v_2}{i_2} = \frac{h_{ie} + R_s}{\Delta_e + h_{oe}R_s} \tag{4.158}$$

In the common case of $R_L \ll 1/h_{oe}$, the first three equations reduce to the following useful approximations:

$$\frac{i_2}{i_1} \approx h_{fe} \tag{4.159}$$

$$\frac{v_2}{v_1} \approx -h_{fe}\frac{R_L}{h_{ie}} \tag{4.160}$$

$$\frac{v_1}{i_1} \approx h_{ie} \tag{4.161}$$

while for a very large value of source resistance, such that $R_s \gg h_{ie}$,

$$\frac{v_2}{i_2} \approx \frac{1}{h_{oe}} \tag{4.162}$$

THERMAL STABILIZATION OF QUIESCENT POINT OF A TRANSISTOR

The collector current, I_C, of a transistor operating in the common-emitter configuration is given by eqn. (4.35). It is sensitive to changes in I_{CEO}, h_{FE} and V_{BE}, all of which are functions of temperature, so that a change in temperature may lead to an unacceptably large change in I_C. It is desirable that I_C should remain constant for two reasons: firstly, the majority of the small-signal parameters for the hybrid-π and hybrid-parameter equivalent circuits are dependent on I_C, so that the properties of an amplifier will change with it; secondly, every transistor has a maximum collector dissipation which if exceeded will lead to destruction of the device by over-heating. Thus it is very important that the collector current does not change with temperature even though I_{CEO}, h_{FE} and V_{BE} do change, which is achieved by using a special circuit to provide *thermal stabilization.*

Comparing eqns. (4.32) and (4.35) we can write

$$I_C = h_{FE}I_B + (h_{FE} + 1)I_{CBO} \tag{4.163}$$

The effect of changes in one parameter may be considered by keeping the two other parameters constant at their room temperature values. Thus we define a *stability factor S* for changes in leakage current such that

$$S = \frac{\partial I_C}{\partial I_{CBO}} \qquad (4.164)$$

From eqn. (4.163), keeping h_{FE} constant,

$$S = h_{FE} + 1 + h_{FE}\frac{\partial I_B}{\partial I_{CBO}}$$

or

$$1 = \frac{h_{FE} + 1}{S} + h_{FE}\frac{\partial I_B}{\partial I_C} \qquad (4.165)$$

which leads to

$$S = \frac{h_{FE} + 1}{1 - h_{FE}\dfrac{\partial I_B}{\partial I_C}} \qquad (4.166)$$

For the simple amplifier circuit of Fig. 4.24(a), $I_B \approx V_{CC}/R_B$, since $V_{CC} \gg V_{BE}$. Thus I_B is held constant, $\partial I_B/\partial I_C = 0$ and $S = h_{FE} + 1$. This circuit is said to be unstabilized since S assumes its largest possible value. In a stabilized circuit, such as the one shown in Fig. 4.36(a), $\partial I_B/\partial I_C$ is negative and $S < (h_{FE} + 1)$. Here R_1 and R_2 form a potential divider taking a current from the supply which is much larger than I_B. The circuit is redrawn in Fig. 4.36(b) after applying Thévenin's theorem to the potential divider, so that

$$V_{BB} = V_{CC}\frac{R_2}{R_1 + R_2} \qquad (4.167)$$

and

$$R_B = \frac{R_1 R_2}{R_1 + R_2} \qquad (4.168)$$

Then, putting $I_E = I_B + I_C$,

$$V_{BB} = I_B R_B + V_{BE} + (I_B + I_C)R_E \qquad (4.169)$$

179

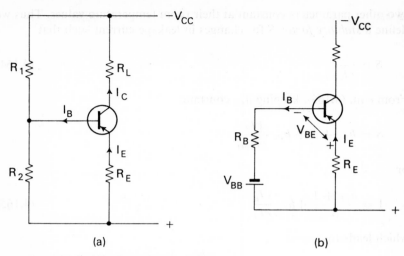

(a)   (b)

FIG. 4.36 Stabilized transistor amplifier

(a) Actual circuit
(b) Equivalent circuit

and

$$\frac{\partial I_B}{\partial I_C} = -\frac{R_E}{R_B + R_E} \tag{4.170}$$

which leads to

$$S = \frac{h_{FE} + 1}{1 + h_{FE}\dfrac{R_E}{R_B + R_E}} \tag{4.171}$$

The effect of changes in V_{BE} and h_{FE} may now be considered in terms of S. Putting

$$I_B = \frac{I_C - (h_{FE} + 1)I_{CBO}}{h_{FE}}$$

in eqn. (4.169) and rearranging,

$$V_{BE} = V_{BB} + \frac{(R_E + R_B)(h_{FE} + 1)I_{CBO}}{h_{FE}}$$
$$-I_C\left[\frac{R_B + (h_{FE} + 1)R_E}{h_{FE}}\right] \tag{4.172}$$

180

so that

$$\frac{\partial I_C}{\partial V_{BE}} = \frac{-h_{FE}}{R_B + R_E(h_{FE} + 1)} = \frac{Sh_{FE}}{(R_B + R_E)(h_{FE} + 1)} \qquad (4.173)$$

Thus the effect of V_{BE} is also reduced when S is small.

In order to consider the effect of h_{FE} eqn. (4.169) may be rewritten in the form

$$V_{BE} = V_{BB} + V' - I_C \frac{R_B + (h_{FE} + 1)R_E}{h_{FE}} \qquad (4.174)$$

where

$$V' = \frac{(R_E + R_B)(h_{FE} + 1)I_{CBO}}{h_{FE}} \qquad (4.175)$$

$$\approx (R_E + R_B)I_{CBO} \qquad (4.176)$$

so that V' is virtually independent of h_{FE}. Then, from eqn. (4.174),

$$\frac{\partial I_C}{\partial h_{FE}} = \frac{I_C}{h_{FE}} - \frac{I_C R_E}{R_B + R_E(h_{FE} + 1)}$$

$$= \frac{I_C}{h_{FE}} \frac{R_B + R_E}{R_B + R_E(h_{FE} + 1)} \qquad (4.177)$$

or

$$\frac{\partial I_C}{\partial h_{FE}} = \frac{SI_C}{h_{FE}(h_{FE} + 1)} \qquad (4.178)$$

Again, the effect of h_{FE} is reduced when S is small. Thus the total change in I_C is given by

$$\Delta I_C = \frac{\partial I_C}{\partial I_{CBO}} \Delta I_{CBO} + \frac{\partial I_C}{\partial V_{BE}} \Delta V_{BE} + \frac{\partial I_C}{\partial h_{FE}} \triangle h_{FE} \qquad (4.179)$$

the changes in each parameter being calculated from the change in temperature. As shown on page 102, the leakage current of a diode doubles for each $9.2°C$ rise in temperature for germanium and each $6.6°C$ rise for silicon and corresponds to I_{CBO}. Also the value of V_{BE} to maintain constant emitter current *falls* linearly at the rate of about $2\,mV$ per degree Celsius rise in temperature (page 103). Finally h_{FE} rises nearly linearly with temperature at about 1% per degree Celsius rise.

The smaller the value of S the smaller the derivatives become in eqn. (4.179), so that ΔI_C is reduced and I_C is maintained more constant for changes in I_{CBO}, h_{FE} and V_{BE}. In general, R_B should be as small as possible to reduce S, consistent with limiting the current drain from the V_{CC} supply and loading the previous stage which feeds the signal into the amplifier. For a transistor with $I_C = 1$ mA, $h_{FE} = 60$ and $I_{CBO} = 10$ nA at 20°C, typical resistance values are $R_E = 2$ kΩ, $R_1 = 20$ kΩ and $R_2 = 10$ kΩ, giving $R_B = 6.7$ kΩ (eqn. (4.168)). Then, from eqn. (4.175),

$$\frac{\partial I_C}{\partial I_{CBO}} = \frac{61}{1 + \dfrac{120}{8.7}} = 4.12 \tag{4.180}$$

and, substituting into eqns. (4.173) and (4.178),

$$\frac{\partial I_C}{\partial V_{BE}} = -\frac{4.12 \times 60}{61 \times 8.7 \times 10^3} = -0.467 \text{ mS} \tag{4.181}$$

and

$$\frac{\partial I_C}{\partial h_{FE}} = \frac{4.12 \times 10^{-3}}{60 \times 61} = 1.12 \,\mu\text{A} \tag{4.182}$$

Suppose now that the temperature rises from 20 to 40°C, making $I_{CBO} = 40$ nA and $h_{FE} = 72$, so that $\Delta I_{CBO} = 30$ nA, $\Delta h_{FE} = 12$ and $\Delta V_{BE} = -40$ mV. Then the change in collector current is

$$\Delta I_C = (4.12 \times 0.03) + (0.467 \times 40) + (1.12 \times 12)$$

$$= 0.1 + 18.7 + 13.4$$

$$= 32 \,\mu\text{A} \tag{4.183}$$

It may be noted that for this rise in temperature the effect of I_{CBO} is very small compared to the effects of h_{FE} and V_{BE} on the collector current, and that the total current rise is only 3%.

The maximum operating temperature of the collector junction in a transistor is normally set by the manufacturer at 85°C for germanium and 150°C for silicon. If these temperatures are exceeded a very rapid rise in current will occur which destroys the device by overheating, and is due to the density of intrinsic electron-hole pairs exceeding the majority carrier density (page 48). Adequate cooling of the collector must be ensured by means of a heat sink which increases the effective area for the radiation of thermal energy to the surroundings. The performance of a transistor will also be impaired if the operating temperature is very low, since almost all its characteristics are a function of temperature, and these limitations are an inherent feature of solid-state devices. They apply to junction transistors, both with uniform bases and with graded bases, which are described in the next section.

THE PLANAR GRADED-BASE TRANSISTOR

The upper frequency limit (f_T) of a uniform-base transistor manufactured by alloying emitter and collector to the base material, is about 20 MHz. This frequency has now been increased by two modifications to the base, namely reduction of its width and the introduction of an accelerating field across it. Both modifications, which reduce τ_C and hence increase f_T (page 168), are incorporated in the *graded-base transistor*. This is manufactured by a process which involves first diffusing the base into the collector and then diffusing the emitter into the base. The technique produces a transistor which is *planar* in form (Fig. 4.37) and it has

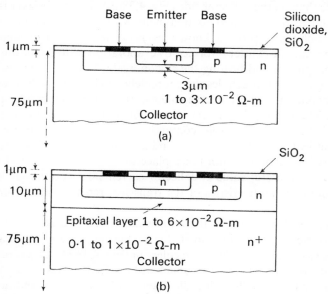

FIG. 4.37 Sections through transistors with graded bases

(*a*) Planar transistor   (*b*) Epitaxial planar transistor superseded other production methods to make available transistors with f_T up to about 500 MHz.

A cross-section through a planar transistor is shown in Fig. 4.37(a). The starting point of manufacture is a wafer of n-type silicon which forms the collector material. This is first exposed to steam while at a temperature of about 1 100°C and a layer of silicon dioxide is formed upon it. The oxide layer is chemically inert and only reacts with an etching agent such as hydrofluoric acid. It is then coated with a solution of *photoresist*, which is a material hardened by exposure to ultraviolet light; a transparent mask with an opaque area (*shadowmask*), which accurately defines the base, is placed over the photoresist; and the whole is illuminated by an ultraviolet source. This is followed by washing, which removes the photoresist only over the base area where it has not been exposed to the light, and etching, which thus can remove the oxide only within the base area. After removal of the remainder of the photoresist a p-type impurity such as boron is evaporated through the hole in the layer and the silicon is again heated to 1 100°C. The boron diffuses through it to form a graded collector junction and a base having more acceptor impurities on the emitter side than on the collector side (Fig. 4.38(a)). The oxide layer is then restored and again subjected to treatment with photoresist, ultraviolet light and etching, a process known as *photoetching*. This defines the emitter area, and the emitter is formed by evaporation of an n-type impurity such as phosphorus and heating the silicon again. Finally, after restoring the oxide layer once more, holes are photoetched in it to allow the evaporation of aluminium ohmic contacts to each of the three electrodes of the transistor. The oxide plays a very important part in protecting the electrodes and their junctions from contamination by the atmosphere. This ensures that leakage currents are low and also that there are fewer surface traps. Thus less recombination takes place at low emitter currents (page 129) and α has a useful value at collector currents down to about 10 μA. Furthermore the characteristics of the transistor remain stable over long periods.

By means of this technique the base width is reduced to about 3 μm, and an electric field is built into it due to the impurity gradient which accelerates electrons from emitter to collector (Fig. 4.38(b)). $C_{b'c}$ is reduced to 1 − 2 pF, which means that feedback at high frequencies is very small. The transistor is so small, with a diameter of about 100 μm, that a large number can be manufactured at the same time on one wafer of silicon, typically 2 500 on a silicon slice with a diameter of 4 cm. Thus the

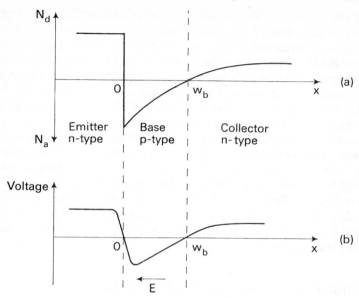

FIG. 4·38 Transistor with graded base
(*a*) Distribution of impurities
(*b*) Potential through transistor with zero applied bias

manufacturing process is very economical and uniform and also lends itself to the production of integrated circuits (page 208).

A consequence of the narrow base region is that the collector has to be of high resistivity ($1–3 \times 10^{-2}\,\Omega$-m) compared to the base. This is because the depletion layer at the collector–base junction penetrates further into the high-resistivity region than into the low-resistivity region (page 91), and if the resistivity of the base were higher than that of the collector the depletion layer would extend to the emitter, leading to breakdown at a low collector voltage. Furthermore, the wafer must have a thickness not less than about $75\,\mu$m in order to have sufficient mechanical strength, so that as a result the collector resistance is high. However, if the collector has a high resistance the charge stored in the base will take a relatively long time to disperse when the collector current is required to be switched off, say $0·5\,\mu$s. Furthermore, the saturation voltage when the transistor is conducting heavily will also be high, 1 V at $I_C = 20\,$mA, for example. These disadvantages have been overcome in the *epitaxial planar transistor*, (Fig. 4.37(*b*)).

185

Epitaxial means "with the same crystal orientation": a thin layer of silicon can be deposited on a silicon wafer epitaxially so that the atoms in the layer are arranged in the same way as in the wafer. This is achieved by exposing the wafer at a temperature of 1 200°C to a gaseous mixture of hydrogen containing 1% of silicon tetrachloride, which decomposes so that silicon is deposited. The wafer has a low resistivity (0·1–1·0 × 10^{-2} Ω-m) and again is not less than 75 μm thick, but the expitaxial layer has a high resistivity (1–6 × 10^{-2} Ω-m) and a thickness of only a few micrometres. The high resistivity ensures a high breakdown voltage, but the thin layer ensures that the whole collector has a much lower resistance than in the simple planar transistor. Consequently the storage time is reduced to about 0·1 μs and the saturation voltage to about 0·2 V at 20 mA. The result is that the silicon expitaxial planar transistor is a major step towards a transistor of high reliability, suitable for very-high-frequency amplification and high-speed switching applications, together with stable characteristics for d.c. operation.

DISTRIBUTION OF EXCESS BASE CHARGE

The mobility of the impurity atoms in a doped semiconductor depends on the temperature, and below 300°C they are quite immobile. Thus at all normal operating temperatures the junction between p- and n-materials is maintained. However, at temperatures over 1 000°C impurities introduced on the surface of a crystal are sufficiently mobile to diffuse into it with the depth of penetration depending on the time during which a high temperature is maintained. The rate of diffusion is approximately an exponential function of temperature, doubling approximately with every 10°C rise; thus 1 min at 1 000°C corresponds to about 2^{80} min at 200°C, which is about 2^{18} years, so that the rate of diffusion at normal operating temperatures is completely negligible.

When the material is cooled from 1 000°C to room temperature the distribution of impurities is fixed, with the greatest concentration at the surface, and in the base of a planar transistor the distribution may be taken as exponential (Fig. 4.38(a)). In an n-p-n transistor, if the acceptor density at the emitter end of the base region is N_{ao}, the distribution across the base is given by

$$N_a = N_{ao}\exp\left(-\frac{x}{b}\right) \tag{4.184}$$

where $x = 0$ at the emitter and the length b defines the rate of change of the impurity density across the base. Since at room temperature we can assume that all the acceptors are ionized, there will be a higher density of fixed negative charges at the emitter than at the collector side of the base. Thus a field E is set up which assists the flow of minority carriers (electrons) from emitter to collector. The density of majority carriers (holes) at any point is

$$p = N_{ao}\exp\left(-\frac{x}{b}\right) \tag{4.185}$$

and the field E will assist the flow of holes from collector to emitter, which is due to currents from collector to base and from base to emitter. Then the hole current is a combination of drift and diffusion currents, so that the hole current density is

$$J_p = e\mu_p pE - eD_p \frac{dp}{dx} \tag{4.186}$$

$$= e\mu_p E N_{ao}\exp\left(-\frac{x}{b}\right)$$

$$+ \frac{eD_p N_{ao}}{b}\exp\left(-\frac{x}{b}\right) \tag{4.187}$$

But at equilibrium $J_p = 0$ and it is still almost zero for low-level injection (of electrons), so from eqn. (4.187),

$$E = -\frac{D_p}{\mu_p b} = -\frac{D_n}{\mu_n b} = -\frac{kT}{eb} \tag{4.188}$$

using eqns. (2.107) and (2.108). The minus sign indicates that E opposes hole flow in the positive x-direction, from emitter to collector, and eqn. (4.188) shows that E is constant for an exponential acceptor distribution across the base.

In order to determine the effect of the built-in field on the distribution of excess charge, q_B, in the base we must consider the electron current density J_n. The equilibrium electron concentration in the base is n_p, which is increased to a new value n_e owing to the injection of excess electrons from the emitter. Then, for a steady current flow, the excess electron density is

$$\Delta n = n_e - n_p \tag{4.189}$$

187

and the electron current density is

$$J_n = e\mu_n \Delta n E + e D_n \frac{d\Delta n}{dx} \tag{4.190}$$

where the field E is unaffected by low-level electron injection and recombination is neglected. Substituting for E from eqn. (4.188) leads to

$$\frac{J_n}{eD_n} = -\frac{\Delta n}{b} + \frac{d\Delta n}{dx} \tag{4.191}$$

which, after rearranging and integrating, gives

$$\log_e\left(\frac{J_n b}{eD_n} + \Delta n\right) = \frac{x}{b} + C \tag{4.192}$$

Then, assuming that $\Delta n = 0$ at the collector where $x = w_b$,

$$C = \log_e \frac{J_n b}{eD_n} - \frac{w_b}{b} \tag{4.193}$$

so that the excess electron density is

$$\Delta n = -\frac{J_n b}{eD_n}\left[1 - \exp\left(\frac{x - w_b}{b}\right)\right] \tag{4.194}$$

the minus sign indicating negative excess charge. Typically $b \approx w_b/8$, in which case for values of x up to about $0 \cdot 7 w_b$ the exponential term is much less than unity and Δn is constant. This gives the distribution of excess electrons shown in Fig. 4.39, which is compared with the linear distribution for a uniform-base transistor of the same base width carrying the same current.

The total excess charge in the base is obtained by integrating $-e\Delta n$ between the limits of $x = 0$ and $x = w_b$, so that, from eqn. (4.194),

$$-\int_0^{w_b} e\Delta n\, dx = \frac{J_n b w_b}{D_n}\left[1 - \frac{b}{w_b}\left(1 - \exp\left(-\frac{w_b}{b}\right)\right)\right] \approx \frac{J_n b w_b}{D_n} \tag{4.195}$$

The approximation follows since b is always appreciably less than w_b in order to make the charge distribution constant over as much of the base as possible, which corresponds to the existence of the built-in field E. Multiplying eqn. (4.195) by the area of the base and putting $b = w_b/8$ leads to an expression for the excess base charge q_B:

$$q_B = \frac{w_b^2 I_E}{8 D_n} \tag{4.196}$$

which may be compared with the expression for the excess base charge in a uniform-base transistor obtained from eqn. (4.16) with $\gamma = 1$:

$$q_B\dot{} = \frac{w_b^2 I_E}{2 D_p} \tag{4.197}$$

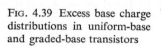

Fig. 4.39 Excess base charge distributions in uniform-base and graded-base transistors

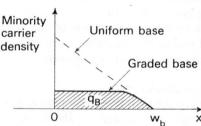

Frequency Response

For two transistors having the same base width and carrying the same current, with both base materials either n- or p-type, the excess charge in the graded base is only 25% of that in the uniform base. This alone accounts for a fourfold improvement in transient and frequency response. However, the true comparison is between a germanium p-n-p transistor with a uniform base of width $10 \, \mu$m and a silicon n-p-n transistor with a graded base of width $3 \, \mu$m. Since $\tau_C = q_B/I_E$ from eqn. (4.40), we have

$$\tau_C = \frac{w_b^2}{2 D p} \tag{4.198}$$

for the p-n-p transistor, and

$$\tau_C = \frac{w_b^2}{8 D_n} \tag{4.199}$$

for the n-p-n transistor with $b = w_b/8$. In each case

$$f_T \approx \frac{1}{2\pi\tau_C} \tag{4.200}$$

so that, using the appropriate values of D_n and D_p, derived from Table 2.3 and eqns. (2.107) and (2.108), $f_T \approx 15\,\text{MHz}$ and $f_T \approx 570\,\text{MHz}$ for transistors with uniform and graded bases respectively.

A further advantage is that the width of the collector depletion layer is increased owing to the low doping level at the collector side of the base, as may be seen from eqn. (4.27). This leads to a reduction in the value of $C_{b'c}$ to about $2\,\text{pF}$, so that the capacitance C in the equivalent circuit of Fig. 4.33 is also reduced and the response improved.

Another advantage is due to the high doping level at the emitter side of the base, which allows a base contact attached near the emitter to have access to the active region of the base through a low resistance. The result is that the resistance $r_{bb'}$ is reduced to about 20–$50\,\Omega$, again giving an improved frequency response.

Voltage Breakdown

Owing to the high level of doping near the emitter, the collector depletion layer will only extend to the emitter at very high collector voltages. Consequently the punch-through condition described on page 134 rarely occurs, and the graded-base transistor may be operated at higher collector voltages than the uniform-base transistor.

Current Gain

A disadvantage of the high doping level is that it tends to reduce the emitter efficiency, γ. If eqn. (4.10) is adapted for an *n-p-n* transistor,

$$\gamma = 1 - \frac{\sigma_p w_b}{\sigma_n L_{ne}} \tag{4.201}$$

so that, in order to make γ approach unity, σ_p must be much less than σ_n. But σ_p is high owing to the large value of N_a near the emitter, so that σ_n must be made even larger. This means that the impurity concentration on each side of the emitter–base junction may become so large that the emitter–base breakdown voltage is reduced and capacitance C_E is increased, which would tend to degrade the frequency response. Such considerations therefore set a limit on the maximum impurity concentration at the emitter side of the base.

REFERENCES

4.1 EBERS, J. J., and MOLL, J. L., "Large signal behaviour of junction transistors", *Inst. Radio Engrs.*, **42**, Pt. II, p. 1761 (Dec., 1954).

Junction Transistors with Uniform and Graded Bases

4.2 SEARLE, C. L., et al., Elementary Circuit Properties of Transistors, (Wiley, 1964).

4.3 LINDMAYER, J. and WRIGLEY, C. T., Fundamentals of Semiconductor Devices (Van Nostrand, 1965).

FURTHER READING

SPARKES, J. J., Junction Transistors (Pergamon, 1966).

GRAY, P. E., et al., Physical Electronics and Circuit Models of Transistors (Wiley, 1964).

PHILLIPS, A. B., Transistor Engineering (McGraw Hill, 1962).

FITCHEN, F. C., Transistor Circuit Analysis and Design, 2nd ed. (Van Nostrand, 1966).

MILLMAN, J., and TAUB, H., Pulse, Digital and Switching Waveforms (McGraw-Hill, 1965).

MILLMAN, J., and HALKIAS, C. C., Electronic Devices and Circuits (McGraw-Hill, 1967).

PROBLEMS

4.1 Explain the mechanism of current flow across the base of a p-n-p transistor operating under normal bias conditions. Sketch energy-level diagrams in illustration.

A certain transistor has the following properties: emitter efficiency, 99%; base transmission factor, 99·5%; collector-current multiplication factor, 100%. Calculate the collector current if the base current is $20\,\mu$A and the collector–base leakage current with open-circuited emitter is $1\,\mu$A. (L.U., B.Sc. (Eng.), 1966) (Ans. 1·38 mA)

4.2 Describe briefly the processes occurring during the transport of minority carriers across the base of a transistor in normal forward operation. Explain the terms *emitter efficiency*, *base transport factor* and *collector efficiency*.

An n-p-n transistor with equal junction areas of $1\,mm^2$ has an excess electron density of 10^{20} per m^3 maintained at the emitter–base junction. If the effective base width is $2 \times 10^{-5}\,m$ and the electron mobility is $0·39\,m^2/Vs$ at room temperature (300 K), sketch the approximate distribution of electrons in the base region and estimate the collector current. State clearly any assumptions you make. (L.U. B.Sc. (Eng.), 1968) (Ans. 7·8 mA)

4.3 The area of both the collector and emitter junctions of a p-n-p transistor is $10^{-2}\,cm^2$. The hole diffusion constant in the base is $50\,cm^2/s$. With a collector-to-base voltage V_{cb} of 1 V, a graph of the hole density in the base region is shown in Fig. 4.40.

Calculate the emitter current due to holes, neglecting leakage currents.

A direct voltage of 9 V is applied between the collector and the base of the transistor. Calculate the slope resistance between these two terminals if the width, W, is $(1 + \sqrt{V_{cb}}) \times 10^{-4}\,cm$. Assume that the base–emitter conditions remain constant and the same as in the diagram. (L.U., B.Sc. (Eng.), 1965) (Ans. 2 mA, 19·2 kΩ)

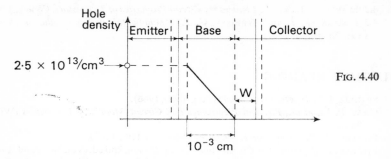

FIG. 4.40

4.4 The base of a silicon *p-n-p* transistor has a uniform donor density of $3 \times 10^{22}\,\mathrm{m^{-3}}$ and at room temperature the diffusion potential of the emitter–base diode is $0.7\,\mathrm{V}$. If it is operated with $V_{EB} = 0.5\,\mathrm{V}$ and $V_{CB} = -10\,\mathrm{V}$, estimate the distances penetrated by the depletion layers into the base at the emitter and the collector. If the metallurgical base width is $10\,\mu\mathrm{m}$ what is the value of the punch-through voltage?

(*Ans.* $0.094\,\mu\mathrm{m}$, $0.69\,\mu\mathrm{m}$, $2.35\,\mathrm{kV}$)

4.5 The base of an *n-p-n* silicon planar transistor is fabricated by diffusing boron into *n*-type silicon which has a resistivity of $10^{-2}\,\Omega$-m. The density of boron atoms at the emitter side of the base is $2 \times 10^{24}\,\mathrm{m^{-3}}$, and the base width is $2\,\mu\mathrm{m}$. Determine the field within the base and the transit time of electrons across it.

(*Ans.* $7.6 \times 10^{4}\,\mathrm{V/m}$, $0.2\,\mathrm{ns}$)

4.6 Discuss the factors that limit the speed at which transistors can be switched from the cut-off state to saturation and back again. In Fig. 4.41 the voltage at A is

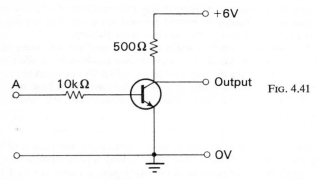

FIG. 4.41

suddenly switched from $+6\,\mathrm{V}$ to $-4\,\mathrm{V}$. Calculate the storage time and the fall time of the output voltage when the important parameters of the transistor are the current gain $\beta = 50$, the collector time factor $\tau_C = 0.02\,\mu\mathrm{s}$ and the saturation time factor $\tau_s = 0.8\,\mu\mathrm{s}$. (Assume that junction transition-region capacitances are negligible and that $V_{BE} = 0$ throughout the switching process.) (*L.U., B.Sc. (Eng.)*, 1968)

Junction Transistors with Uniform and Graded Bases

4.7 The transistor in Fig. 4.42 has current gain $h_{FE} = 80$, base time-constant $\tau_B = 0.8\,\mu s$ and storage time-constant $\tau_s = 0.4\,\mu s$. Making the same assumptions as in Problem 4.6, calculate the value of R_B which will ensure that V_{CE} changes from $10\,V$ to zero in $0.1\,\mu s$. Determine also the corresponding storage time. (*Ans.* $4.7\,k\Omega$, $230\,ns$)

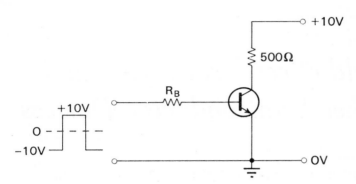

Fig. 4.42

4.8 A transistor has the following small-signal parameters: $r_{b'e} = 1\,k\Omega$, $r_{bb'} = 50\,\Omega$, $C_{b'e} = 1\,000\,pF$, $C_{b'c} = 10\,pF$, $r_{b'c} = 2\,M\Omega$, $g_m = 49\,mA/V$, $r_{ce} = 40\,k\Omega$. The transistor is connected between a signal source of $1\,mV$ and internal resistance $2\,k\Omega$ and a load of resistance $2\,k\Omega$ by means of capacitors whose reactances are negligible at all relevant frequencies. The emitter is common to input and output circuits, and suitable bias is provided by high-impedance circuits. Draw the equivalent circuit of the amplifier and calculate the output voltage and input impedance (*a*) at $5\,kHz$ and (*b*) at $500\,kHz$. (*L.U.*, *B.Sc.* (*Eng.*), 1967) (*Ans.* (*a*) $29.7\,mV$, $1.005\,k\Omega$, (*b*) $7.2\,mV$, $(77 - j157)\,\Omega$)

4.9 A certain transistor has $f_T = 100\,MHz$, low-frequency current gain $h_{fe} = 90$, $r_{bb'} = 50\,\Omega$ and $r_{b'e} = 1.2\,k\Omega$. It is used as an amplifier with a collector load of $500\,\Omega$ and is supplied from a source of $1\,mV$ e.m.f. with internal resistance $500\,\Omega$. Determine the output voltage at low frequencies and the bandwidth of the amplifier. The effects of r_{ce}, $C_{b'c}$, $r_{b'c}$ and the coupling components may be neglected. (*Ans.* $26\,mV$, $3.5\,MHz$)

4.10 Verify eqns. (4.155)–(4.158). Obtain the common-emitter hybrid parameters of a transistor operating at room temperature with $I_C = 2\,mA$, $\beta_0 = 70$, $\mu = 4 \times 10^{-4}$ and $r_{bb'} = 50\,\Omega$. (*Ans.* $h_{ie} = 925\,\Omega$, $h_{re} = 4 \times 10^{-4}$, $h_{fe} = 70$, $h_{oe} = 64\,\mu S$)

4.11 The transistor in Problem 4.10 is connected as a common-emitter amplifier between a source of e.m.f. $1\,mV$ and resistance $600\,\Omega$ and a collector load of $3\,k\Omega$. Determine (*a*) the input resistance, (*b*) the output current, (*c*) the output voltage, and (*d*) the output resistance. (*Ans.* (*a*) $856\,\Omega$, (*b*) $40\,mA$, (*c*) $0.12\,V$, (*d*) $27\,k\Omega$)

193

5

Field-effect Transistors and Other Semiconductor Devices

THE JUNCTION-GATE FIELD-EFFECT TRANSISTOR

The field-effect transistor, or FET, was proposed by Shockley in 1952, but it was not possible to manufacture it in large numbers until semiconductor techniques were sufficiently advanced. It has been commercially available since 1960 and has unique properties, some of which are complementary to those of a conventional (p-n-p or n-p-n) transistor, such as a very high input impedance.

One form of the device, known as a junction-gate field-effect transistor (JUGEET), normally consists of a bar of n-type silicon with an ohmic contact at each end known as the *source* and the *drain* respectively. The bar is enclosed for part of its length by heavily doped p-type silicon, known as the *gate*, so that a p-n junction is formed along it (Fig. 5.1(a)). If the voltage between gate and source, V_{GS}, is initially zero and the drain-to-source voltage, V_{DS}, is made positive an electron current I_D will flow from source to drain, whose magnitude depends on the effective resistance of the bar (Fig. 5.2(a)). The junction is subjected to a reverse bias which increases from zero at the source to a maximum at the drain, owing to the voltage drop along the bar. Consequently depletion layers whose width increases with the reverse bias extend into the n-region, since its resistivity is higher than that of the p-region (page 91) and I_D flows through the tapering channel between the depletion layers. This is called an *n-channel* device; the opposite arrangement with a p-type bar and an n-type gate would be a *p-channel* device, as symbolized in Fig. 5.1(b). A practical

194

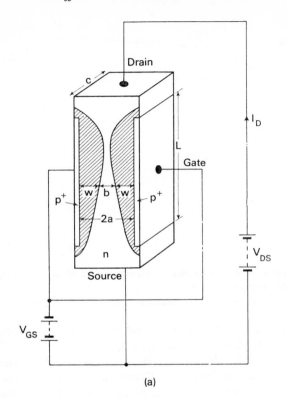

(a)

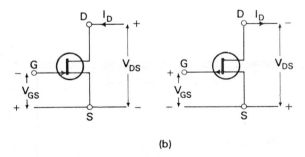

(b)

FIG. 5.1 The junction-gate field-effect transistor (JUGFET)

(*a*) Diagram and applied voltages   (*b*) Graphical symbols

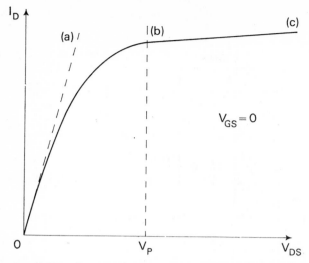

FIG. 5.2 Drain-current/voltage characteristic of a field-effect transistor, with zero gate voltage

(*a*) Ohmic region   (*b*) Pinch-off   (*c*) Pinch-off region

form is shown in Fig. 5.3, where the n^+- and p^+-regions are interdigitated in a manner similar to that shown in Fig. 4.8.

As V_{DS} is increased the channel becomes narrower and the resistance between source and drain increases. Eventually, when V_{DS} reaches a critical value V_P known as the *pinch-off voltage*, the two depletion layers

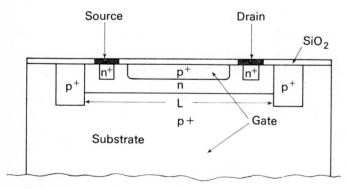

FIG. 5.3 Section through a planar JUGFET

almost meet at the drain and the channel is said to be *pinched off*. Thus I_D increases non-linearly with V_{DS} until pinch-off has been reached (Fig. 5.2(*b*)). At all values of V_{DS} the value of I_D at any point in the channel may be obtained by combining eqns. (2.49) and (2.60), so that

$$I_D = N_d e \mu E_D S \tag{5.1}$$

where E_D is the field accelerating electrons from source to drain. At pinch-off the cross-sectional area of the channel, S, is very small near the drain, so that E_D becomes very large to maintain the flow of current and thus prevents the depletion layers quite meeting each other. The current *density* is very high at this point and the electrons approach their maximum drift velocity (page 52) as they shoot through the very narrow gap. Thus for values of V_{DS} above pinch-off the drain current is almost independent of voltage (Fig. 5.2(*c*)).

Suppose now that V_{GS} is made negative. This will increase the reverse bias and cause a general widening of the depletion layers, so that pinch-off occurs at a lower value of drain voltage. Thus the electric field in the depletion layers due to V_{GS} may be considered to control their width and hence the current flowing in the conducting channel, which is the "field effect" in the name of the device. The JUGFET will then have an I_D/V_{DS} characteristic of the type shown in Fig. 5.4 with V_{GS} as a parameter. The drain current is completely cut off at any drain voltage for a negative value of gate voltage, V_P. This is related to the gate and drain voltages by the general expression

$$V_{DS} = V_{GS} - V_P \tag{5.2}$$

which is shown on the characteristics as a dotted line. When biased negatively the device is said to be working in the *depletion mode*, since increasing the bias depletes the channel of charge carriers. The input impedance is then virtually that of a reverse-biased silicon *p-n* junction, which is about $10^{10}\,\Omega$ at room temperature and far higher than for a conventional transistor. If the bias were made positive, the junction would be forward biased and the flow of carriers enhanced since the channel would become wider. The input impedance in this *enhancement mode* becomes that of a forward-biased *p-n* junction, r_e, which is very low (page 159), so that operation with positive values of V_{GS} is not normally recommended.

The current through a field-effect transistor consists only of majority carriers and so it is a *unipolar* device, while in junction transistors

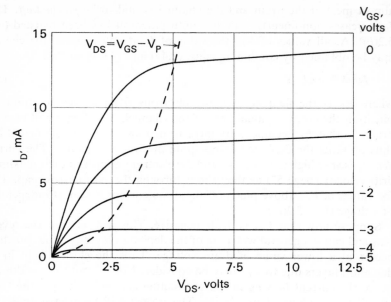

FIG. 5.4 Output characteristics of a JUGFET

minority carriers diffuse across the base and their space charge is neutralized by an equal number of majority carriers. Both types of carrier are involved, so they are called *bipolar* devices. Also the leakage-current effects present in conventional transistors are not met in field-effect transistors.

Theoretical Characteristics of a JUGFET

An approximate expression for the characteristics shown in Fig. 5.4 may be obtained by considering the device shown in Fig. 5.1(*a*). The width w of each depletion layer is given approximately by eqn. (3.70) for a p^+n diode, so that

$$w = \left(\frac{2\epsilon V_R}{eN_d}\right)^{1/2} \tag{5.3}$$

V_R is the reverse bias voltage, $V_{GS} + V_{DS}$, which is a function of distance x along the bar. Under pinch-off conditions $w \approx a$, so that, from eqn. (5.3),

$$V_P = \frac{eN_d a^2}{2\epsilon} \tag{5.4}$$

and

$$V_R = V_P \frac{w^2}{a^2} \tag{5.5}$$

Now, from eqn. (5.1),

$$I_D = \sigma E_D S = \sigma bc \frac{dV_R}{dx} \tag{5.6}$$

where σ is the conductivity of the bar.

Over the distance L (Fig. 5.1(a)), V_R will change from V_{GS} at the source to $V_{GS} + V_{DS}$ at the drain, which gives

$$\int_0^L I_D \, dx = \sigma c \int_{V_{GS}}^{V_{GS}+V_{DS}} b \, dV_R \tag{5.7}$$

But

$$b = 2a - 2w = 2a\left[1 - \left(\frac{V_R}{V_P}\right)^{1/2}\right] \tag{5.8}$$

using eqn. (5.5), which leads to

$$I_D = \frac{2ac\sigma}{L}\left[V_R - \tfrac{2}{3}\frac{V_R^{3/2}}{V_P^{1/2}} + C\right]_{V_{GS}}^{V_{GS}+V_{DS}} \tag{5.9}$$

after integration. When $V_R = 0$, $I_D = 0$, so that $C = 0$ and

$$I_D = \frac{2ac\sigma}{L}\left[V_{DS} - \tfrac{2}{3}\frac{(V_{GS}+V_{DS})^{3/2}}{V_P^{1/2}} + \tfrac{2}{3}\frac{V_{GS}^{3/2}}{V_P^{1/2}}\right] \tag{5.10}$$

which is an approximate expression for the characteristics *below* pinch-off, called the *triode* region.

In amplifier applications the JUGFET is normally operated in the saturation region *above* pinch-off, called the *pinch-off* region. An expression for the saturation current, $I_{D\,sat}$, is obtained by putting $V_{GS} + V_{DS} = V_P$ in eqn. (5.10), which gives

$$I_{D\,sat} = \frac{2ac\sigma}{L}\left[\frac{V_P}{3} - V_{GS} + \tfrac{2}{3}\frac{V_{GS}^{3/2}}{V_P^{1/2}}\right] \tag{5.11}$$

This expression is independent of V_{DS}, and when $V_{GS} = V_P$, $I_{D\,sat} = 0$, confirming that the drain current is cut off for a gate voltage more negative than the pinch-off voltage. A useful relationship between saturation current and gate voltage is obtained by considering the value of I_D for $V_{GS} = 0$. This is the current I_{DSS}, given by

$$I_{DSS} = \frac{2ac\sigma}{L}\frac{V_P}{3} \tag{5.12}$$

obtained from eqn. (5.11), and is a constant since V_P depends on the electrical properties of the bar. Substitution into eqn. (5.11) gives

$$I_{D\,sat} = I_{DSS}\left[1 - 3\frac{V_{GS}}{V_P} + 2\left(\frac{V_{GS}}{V_P}\right)^{3/2}\right] \tag{5.13}$$

The $I_{D\,sat} = V_{GS}$ curve is the transfer characteristic of the device, relating input voltage and output current for a device with alloyed junctions. If diffused junctions are formed, as in the planar JUGFET illustrated in Fig. 5.3, it is found experimentally (Ref. 5.1) that the transfer characteristic obeys the equation

$$I_{D\,sat} = I_{DSS}\left(1 - \frac{V_{GS}}{V_P}\right)^2 \tag{5.14}$$

the square-law relationship being very convenient in circuit design. Eqns. (5.13) and (5.14) are compared in Fig. 5.5, the average difference being about 10%, which suggests that a square law is quite a good approximation to eqn. (5.13) also. Typical values of I_{DSS} and V_P are 13 mA and -5 V respectively, with V_{DS} at 15 V. The drain current at a fixed value of V_{GS} is a function of temperature, typically falling at about 25 μA per degree Celsius temperature rise from a value of 8 mA at 20°C. This is due to the fall in conductivity of the bar, the rise in junction leakage current having no direct effect on $I_{D\,sat}$. Electrical breakdown of the junctions will occur at a sufficiently high value of V_{DS}.

The FET is used as an amplifier by connecting a load resistance in series with the drain, just as the conventional transistor amplifier has a collector load resistance. A similar load-line construction determines the operating point, while the input parameter is V_{GS} and not input current. For small-signal operation an equivalent circuit is required, and again the mutual conductance, g_m, is defined from the slope of the input characteristic, eqn. (5.14). Then

$$\frac{\partial I_{D\,sat}}{\partial V_{GS}}\bigg|_{V_{DS}\text{const}} = g_m = \frac{2I_{DSS}}{V_P}\left(1 - \frac{V_{GS}}{V_P}\right) \tag{5.15}$$

or

$$g_m = \frac{2}{V_P}(I_{D\,sat}I_{DSS})^{1/2} \tag{5.16}$$

g_m is not constant, but rises linearly with V_{GS} and is proportional to $I_{D\,sat}^{1/2}$, I_{DSS} and V_P being constants depending on the physical properties

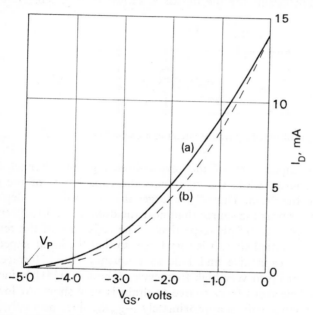

FIG. 5.5 Input characteristics of JUGFETS

(*a*) Diffused junction, square law   (*b*) Alloyed junction, eqn. (5.13)

of the device (eqns. (5.12) and (5.4)). It may be noted that the value of g_m at $V_{GS} = 0$ is

$$g_{mo} = \frac{2I_{DSS}}{V_P} = \tfrac{4}{3}\frac{ac\sigma}{L} \tag{5.17}$$

using eqn. (5.12), so that g_{mo} has a value $1\frac{1}{3}$ times the conductance of the bar, known as the *open-channel* conductance. Theoretically the output characteristics are horizontal straight lines, since eqn. (5.11) is independent of V_{DS}, but in practice $I_{D\,sat}$ rises slowly with V_{DS} and the output conductance g_{os} is given by

$$\left.\frac{\partial I_{D\,sat}}{\partial V_{DS}}\right|_{V_{GS}\text{const}} = g_{os} \tag{5.18}$$

Practical values of g_m and g_{os} are 5 mS and 50 μS respectively, so that the output resistance is about 20 kΩ and falls as $I_{D\,sat}$ is increased (Fig. 5.4).

An equivalent circuit for the device is shown in Fig. 5.6. Here C_{gc}

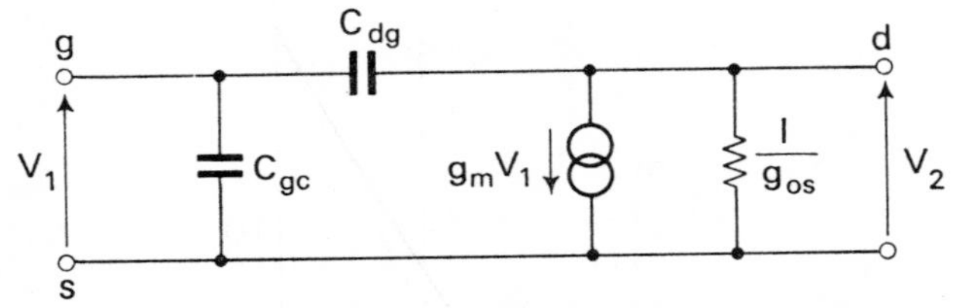

FIG. 5.6 Small-signal equivalent circuit of a field-effect transistor

represents the capacitance of the reverse-biased gate-to-channel diode close to the source, while C_{dg} represents the capacitance of the same diode close to the drain. Thus C_{gc} is larger than C_{dg} since the depletion layer is narrower near the source than near the drain (Fig. 5.1(a)), typical values being 3 pF and 0·5 pF respectively. The resistance of the reverse-biased diode is omitted since it is very large, so that the input impedance of the device is capacitive and falls as frequency rises. However, g_m remains constant up to very high frequencies (about 900 MHz) since the transit time of electrons from source to drain is very short. At low frequencies the voltage gain is approximately $-g_m R_L$, since normally $R_L \ll 1/g_{os}$.

The equivalent circuit of Fig. 5.6 is very similar to the hybrid-π equivalent circuit of a conventional transistor (Fig. 4.31). Its analysis is simplified by the transformation of C_{dg} to a capacitance $C_{dg}(1 + A)$ in parallel with C_{gc} by a method similar to that considered on page 172. Also a gain-bandwidth product may be defined, similar to eqn. (4.127), since at very high frequencies the short-circuit current gain of an FET will also be inversely proportional to frequency. The gain-bandwidth product of an FET is then $g_m/2\pi C_{gc}$ and insertion of the typical values given above yields a figure of 270 MHz.

The gain-bandwidth product may be related only approximately to the physical properties of the device, since C_{gc} is a capacitance distributed along a non-uniform channel. If we assume that the average width of the channel is a, then

$$C_{gc} \approx \frac{\epsilon_r \epsilon_0 cL}{a} \tag{5.19}$$

and if we take a value of g_m given by the open-channel conductance,

$$g_m \approx \frac{ac\sigma}{L} = \frac{N_d e \mu ac}{L} \tag{5.20}$$

then

$$\frac{g_m}{2\pi C_{gc}} \approx \frac{N_d e \mu a^2}{2\pi \epsilon_r \epsilon_0 L^2} \tag{5.21}$$

and the gain-bandwidth product increases with $(a/L)^2$. The width a is about $1\,\mu m$, in order that pinch-off will be achieved, so that for $N_d \approx 10^{22}/m^3$, L is about $30\,\mu m$ for the typical value of gain-bandwidth product. Thus for the best frequency response L must be as short as possible.

In addition to its application as a high-impedance wide-band amplifier, the FET is very useful as a switch. Since it is basically a device whose resistance is controlled by the gate voltage, all the characteristics pass through the point $I_D = 0$, $V_{DS} = 0$, so that the offset voltage is negligible compared with $0 \cdot 1 – 2\,mV$ for the conventional transistor (page 139). Another unique property is that, for positive or negative values of V_{DS} less than about $0 \cdot 1\,V$, I_D varies linearly with V_{DS}. Thus the device can operate as a resistance whose value increases with negative values of gate voltage and may be used to control the gain of another transistor amplifier.

THE INSULATED-GATE FIELD-EFFECT TRANSISTOR

Another form of field-effect transistor is the insulated-gate FET (IGFET). Owing to the method of construction it is more commonly called the metal-oxide-semiconductor field-effect transistor (MOSFET) or the MOS transistor (MOST).

One form is illustrated in Fig. 5.7(a), where the substrate material is lightly doped p-type silicon, nearly intrinsic and therefore of high resistivity. This is coated with a layer of silicon dioxide, which is an insulator,

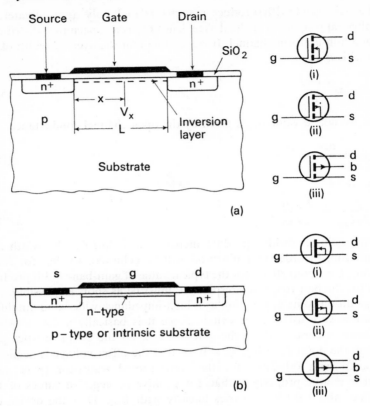

FIG. 5.7 Sections through MOS transistors and graphical symbols

(a) Enhancement type, single-gate. The symbols indicate
    (i) 3-terminal, *n*-channel
    (ii) 3-terminal, *p*-channel
    (iii) 4-terminal, *p*-channel, with substrate connection (b) brought out
(b) Depletion type, single-gate. The symbols indicate
    (i) 3-terminal, *n*-channel
    (ii) 3-terminal, *p*-channel
    (iii) 4-terminal, *p*-channel, with substrate connection (b) brought out

and then two heavily doped *n*-regions are diffused into the substrate through holes etched in the layer. Aluminium ohmic contacts are made to each of the *n*⁺-regions, again called the source and the drain respectively. Finally, a metal film, also of aluminium, is deposited on the oxide between source and drain to form the gate.

The gate and substrate now form a parallel-plate capacitor, with the silicon dioxide as the dielectric, so that a charge of one sign on the gate electrode induces a charge of the opposite sign on the substrate. Thus for a *negative* gate voltage holes will be attracted to the surface of the substrate, which becomes more strongly *p*-type. Since V_{DS} is positive as before, two reverse-biased *p-n* junctions are formed in series between source and drain and no drain current can flow. However, for a *positive* gate voltage, holes are repelled from and electrons attracted to the surface of the substrate, to form an *n*-type *inversion layer*. Thus a conducting channel is formed between the two n^+-regions and drain current flows, both the thickness of the layer and the current increasing with the gate voltage. Since the channel is conductive only when the gate voltage is positive, this is an *enhancement*-type MOST, with the characteristics shown in Figs. 5.8(*a*) and (*c*).

A *depletion*-type MOST is obtained if an *n*-type channel is diffused into the substrate to connect source and drain (Fig. 5.7(*b*)). In this case current flows when $V_{GS} = 0$ and a negative gate voltage is required to cut off the current. The negative gate repels electrons from the *n*-channel and so induces a positive charge in it, converting it to *p*-type material. The characteristics are shown in Figs. 5.8(*b*) and (*c*), and the advantage of this type of transistor is that it can be operated at zero bias, so that bias supplies and resistors are unnecessary.

Both enhancement and depletion types can be made with an *n*-type substrate, and the transistor can also be made in thin-film form. Here the electrodes are built up by evaporation onto an insulating substrate such as glass, with materials such as cadmium sulphide and tellurium being used for the semiconductor (Ref. 5.4). The input impedance of all types is very high, between 10^{11} and $10^{14}\,\Omega$, the particular value depending on the thickness of the oxide layer.

Theoretical Characteristics of a MOST

In an enhancement-mode device with $V_{GS} = 0$, the metal will be charged negatively with respect to the *p*-type semiconductor owing to the contact potential difference between them (page 81). Current will begin to flow only when $V_{GS} > 0$, so that the pinch-off voltage V_P is positive. Similarly in a depletion-mode device the metal will be positive with respect to the *n*-type semiconductor when $V_{GS} = 0$ so that V_P is negative. Thus V_P can be regarded as a voltage built into the device which depends on the mode of operation. An expression for the I_D/V_{DS} characteristics can then

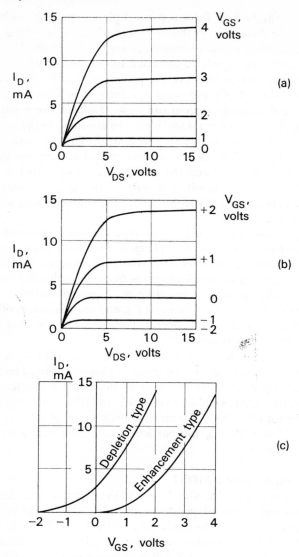

FIG. 5.8 Characteristics of MOS transistors

(*a*) Enhancement type, output characteristics
(*b*) Depletion type, output characteristics
(*c*) Input characteristics

be derived if we assume that the oxide is thicker than the conducting channel.

The electric field in the insulator is given by Gauss's law

$$E = \frac{D}{\epsilon_{ins}\epsilon_0} \tag{5.22}$$

where D is the charge per unit area on the gate electrode or in the channel. If the thickness of the insulator is d and the potential of the channel with respect to the source is V_x at a distance x from the source,

$$D = \epsilon_{ins}\epsilon_0 \frac{V_{GS} - V_x - V_P}{d} \tag{5.23}$$

which includes the effect of the contact potential.

Then the current in the channel is obtained from eqn. (5.6), so that

$$I_D = Dc\mu \frac{dV_x}{dx} \tag{5.24}$$

where c is the width of the device as before (Fig. 5.1).

Substituting from eqn. (5.23),

$$\frac{dV_x}{dx} = \frac{I_D d}{\epsilon_{ins}\epsilon_0 c\mu(V_{GS} - V_x - V_P)} \tag{5.25}$$

which leads to the integral

$$\int_0^{V_{DS}} (V_{GS} - V_x - V_P)\,dV_x = \int_0^L \frac{I_D d}{\epsilon_{ins}\epsilon_0 c\mu}\,dx \tag{5.26}$$

and, after integration,

$$I_D = \frac{\epsilon_{ins}\epsilon_0 c\mu}{Ld} V_{DS}\left(V_{GS} - \frac{V_{DS}}{2} - V_P\right) \tag{5.27}$$

the constant of integration being zero since $I_D = 0$ when $V_x = 0$. Eqn. (5.27) is an expression for the characteristics below saturation. In the MOST this occurs at the drain voltage for which $E = 0$ at the drain end of the gate. From eqns. (5.22) and (5.23) with $V_x = V_{DS}$, saturation occurs when $V_{DS} = V_{GS} - V_P$, and insertion of this condition into eqn. (5.27) leads to

$$I_{D\,sat} = \frac{\epsilon_{ins}\epsilon_0 c\mu}{Ld} \frac{(V_{GS} - V_P)^2}{2} \tag{5.28}$$

Then eqn. (5.28) applies only to drain currents for which $V_{DS} < V_P$ and again shows that $I_{D\,sat}$ rises with V_{GS} (compare eqn. (5.14)).

The mutual conductance in the saturated region is obtained from eqn. (5.28):

$$\left.\frac{\partial I_{D\,sat}}{\partial V_{GS}}\right|_{V_{DS}\text{const}} = g_m = \frac{\epsilon_{ins}\epsilon_0 c\mu}{Ld}(V_{GS} - V_P) \qquad (5.29)$$

so that once again g_m rises linearly with V_{GS}, with practical values similar to those of the junction-gate device. An equivalent circuit like that of Fig. 5.6 may be used, the various components having comparable values, and a gain-bandwidth product defined, $g_m/2\pi C_{gc}$. Here

$$C_{gc} = \frac{\epsilon_{ins}\epsilon_0 cL}{d} \qquad (5.30)$$

so that

$$\frac{g_m}{2\pi C_{gc}} = \frac{\mu}{2\pi L^2}(V_{GS} - V_P) \qquad (5.31)$$

which is still proportional to $1/L^2$, and rises with V_{GS} only because we cannot choose an open-channel conductance unless V_P is negative. For $V_{GS} - V_P = 1\cdot 5\,\text{V}$, $L = 30\,\mu\text{m}$ and $\mu_n = 0\cdot 14\,\text{m}^2/\text{Vs}$ for silicon the gain-bandwidth product is $330\,\text{MHz}$, again comparable to that of the junction-gate device.

The MOST is also very useful as a switch and as a variable resistance. Its high input impedance is retained for positive values of V_{GS}, unlike the junction-gate device. A practical consequence is that a comparatively low resistance should always be connected between the gate and the source electrodes, since excessive accumulation of charge on the gate can lead to breakdown of the dielectric.

INTEGRATED CIRCUITS

An *integrated circuit* (or *micro-circuit*) is a device in which a complete electronic circuit is manufactured in a single chip of semiconductor. Transistors, diodes, resistors and capacitors can be included as required, and the technology has progressed to the stage where at least 450 transistors can be fabricated in a wafer 3 mm square. Such a degree of miniaturization clearly presents great advantages in applications involving large numbers of components, such as occur in a digital computer. For instance, a particular computer using thermionic valves would occupy a volume of $8.4\,m^3$, which could be reduced to $0.67\,m^3$ using discrete transistors and components. Employment of integrated circuits could further reduce the volume to $0.02\,m^3$, equivalent to a cube of edge length 27 cm and only 3% of the volume of the transistorized version. An additional advantage is increased reliability, owing to a large reduction in the number of interconnections required, since only complete circuits are connected externally instead of individual components.

Transistors

The fabrication of integrated circuits has developed from the technology used in the manufacture of planar transistors (page 183). Thus successive masking, etching and diffusion processes are used to lay down components of the same type simultaneously over the whole chip, which leads to good uniformity of characteristics. Since all the components are formed in one silicon crystal the problem arises of maintaining adequate insulation between those components which are not required to be in electrical contact.

One method of achieving this is illustrated in Fig. 5.9(*a*), which shows an *n-p-n* planar transistor formed in a lightly doped *p*-type substrate. The substrate is maintained at a low negative voltage with respect to the collector *n*-region, so that the collector–substrate diode, or *isolation diode*, is reverse biased. Since the substrate has a higher resistivity than the collector, the depletion layer extends further into the substrate than into the collector (page 91). A very high resistance is thus formed between the transistor and the substrate and also between the transistor and any other components similarly isolated from the substrate. However, the complete circuit of the component is now as shown in Fig. 5.9(*b*), where C_I is the reverse-biased capacitance of the isolation diode and R_N the resistance between the collector and its contact, the other contact resistances being negligible. Thus while this method of isolation is satisfactory for low-frequency applications, the capacitance C_I will introduce an unwanted component at high frequencies or fast switching speeds, which will limit the response of the transistor.

The MOST is also in a very convenient form for incorporation in integrated circuits and is finding an increasing application in logic circuits.

Diodes

When diodes are used in an integrated circuit on the same chip of silicon as transistors it is often convenient to produce only transistors, in a set

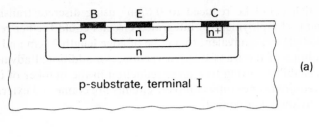

(a)

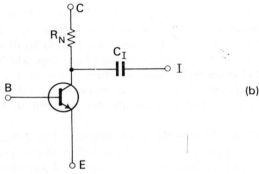

(b)

FIG. 5.9 Planar transistor in an integrated circuit

(a) Section (b) Equivalent circuit

of operations common to them all. Some of the transistors can then be connected for use as diodes, giving a choice of characteristics since the base–emitter junction has a smaller area than the base–collector junction, for example. An alternative arrangement for a single diode is shown in Fig. 5.10 with its equivalent circuit.

Resistors
An n-type strip diffused into a p-type substrate with metal contacts at each end will form a resistor (Fig. 5.11). The substrate potential is held more negative than any part of the strip in order to isolate it electrically. Suppose the strip has length l, width w and thickness t as shown in Figs. 5.11(a) and (b). Its resistance is

$$R = \frac{\rho l}{wt} \tag{5.32}$$

210

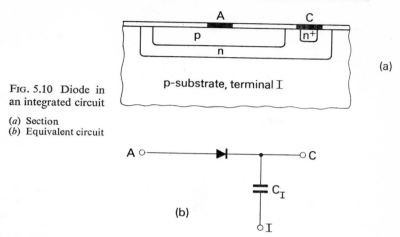

Fig. 5.10 Diode in an integrated circuit

(a) Section
(b) Equivalent circuit

while if it is square, $l = w$ and the resistance becomes

$$R_s = \frac{\rho}{t} = \rho_s \tag{5.33}$$

ρ_s is the *resistance per square* of the material and may be used instead of resistivity, so that in general

$$R = \rho_s \frac{l}{w} \tag{5.34}$$

This equation is very useful for comparing the resistances of strips or films of the same thickness such as occur in integrated circuits, and it shows that their resistance depends on their length-to-width ratio, l/w. Typical values of ρ_s lie between about 25 and 500 Ω per square, and l/w ratios up to about 20:1 can be achieved, with w constant at about 25 μm. Thus it is possible to produce diffused resistors with values between about 100 and 10,000 Ω, having a temperature coefficient of about $+10^{-3}$ per degree Celsius above room temperature (compare Fig. 2.19). The tolerance of single resistors is about $\pm 10\%$, but it is possible to make pairs of resistors whose resistance ratio may be held to $\pm 1\%$ by diffusing them simultaneously.

A feature of a diffused resistor is that it is purely resistive only at low frequencies. This is due to the capacitance of the reverse biased isolating diode, which is distributed along the resistor. The leakage current of the

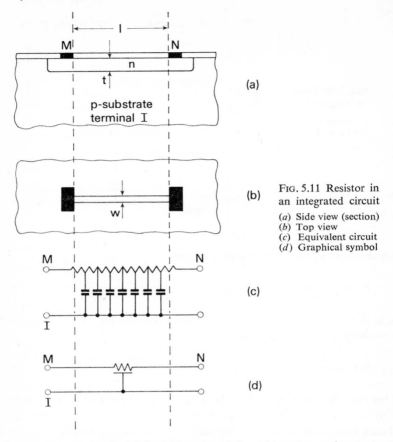

FIG. 5.11 Resistor in an integrated circuit

(*a*) Side view (section)
(*b*) Top view
(*c*) Equivalent circuit
(*d*) Graphical symbol

diode may be neglected and the resistance of the substrate may be assumed very low, which leads to the equivalent circuit shown in Fig. 5.11(*c*) together with a circuit symbol, Fig. 5.11(*d*). The diffused resistor may be analyzed in terms of transmission line theory (Ref. 5.2).

Capacitors

Capacitors can be formed either by using the capacitance of a reverse-biased diode (Fig. 5.10) or by producing two parallel plates. The diode capacitance is useful when low values are required. Values up to about $10\,\text{pF/mm}^2$ can be obtained with a reverse breakdown voltage of about $100\,\text{V}$, both capacitance and breakdown voltage falling as the doping densities are increased, and the capacitance being voltage dependent (page 110).

An improved component is obtained if a parallel-plate capacitor is formed with a silicon-dioxide dielectric. No bias voltage is required, the capacitance is independent of voltage and the leakage resistance is much higher than that of the diode. In Fig. 5.12(*a*) the metal layer forms one

Fɪɢ. 5.12 Capacitor in an integrated circuit

(*a*) Isolated with a *p-n* diode
(*b*) Isolated with a silicon-dioxide layer

(a)

(b)

plate and the *n*-type region the second plate, so that the capacitance per unit area is given by

$$C = \frac{\epsilon_{ins}\epsilon_0}{d} \tag{5.35}$$

where d is the thickness of the dielectric, which has relative permittivity ϵ_{ins}. A *p-n* junction is used to isolate this capacitor, a typical value being $200\,pF/mm^2$ with a breakdown voltage of 100 V using a dielectric thickness of $0{\cdot}2\,\mu m$. The same capacitor may also be isolated with a second silicon-dioxide layer (Fig. 5.12(*b*)), which removes the need for biasing the substrate and will lead to a reduced isolation capacitance if this layer is relatively thick.

In fact, silicon dioxide can also be used to isolate transistors, diodes and resistors. Each component is formed in a "pocket" lined with the oxide which has a very much smaller isolation capacitance and a much larger isolation resistance than a *p-n* junction. In whatever way the components are formed the final deposition is that of conducting layers in a pattern to form the required interconnections. Aluminium is normally used with ρ_s in the range $0 \cdot 1$ to $1 \cdot 0 \, \Omega$ per square.

At present integrated circuits are mainly used in logic applications such as the circuitry of a digital computer. Here the binary numbers 0 and 1 are used, with 0 corresponding to zero voltage and 1 corresponding to a positive voltage pulse in the *positive logic* system. An example of a logic circuit known as an inverter is shown in Fig. 4.18(*a*), where a positive input causes the transistor to conduct so that $V_{CE} = 0$. A negative input cuts off the transistor so that $V_{CE} = V_{CC}$ and a "1" input has then been inverted to a "0" output, and vice versa. Circuits of this type do not require resistors of closely controlled value, but they do require small transistors with fast switching times when pulses of short duration are used. As the technology has improved it is now possible to form amplifiers such as the type shown in Fig. 4.36(*a*) using small signals. Such a device is known as a *linear integrated circuit* and requires components with closely controlled values so that a specified gain and frequency response may be attained. When either digital or linear integrated circuits are formed on a single chip of silicon they are termed *monolithic*. However, in order to reduce the coupling between different parts of a complicated circuit, each section may be mounted on an individual chip and the chips interconnected by wires. This is known as a *multi-chip* integrated circuit, which normally has a better performance than the monolithic type, but may be a little less reliable owing to the interconnections.

THIN-FILM CIRCUITS

Finally, mention must be made of *thin film* integrated circuits, in which resistors and capacitors are formed by vacuum deposition on to a substrate such as borosilicate glass or a ceramic such as glazed alumina (Ref. 5.4). Resistors of nickel-chromium, for instance, can be deposited as films, with thicknesses between 100 nm and 10 nm, having values in the range 100 Ω to 1 MΩ and a tolerance of $0 \cdot 1 \%$. Capacitors are formed by depositing a metal film onto which a dielectric and a second metal film are deposited in succession to form a sandwich, using the same materials as in monolithic integrated-circuit capacitors. At present, it is difficult to produce transistors in thin-film form which have sufficiently stable and reproducible characteristics, but the MOST seems to offer the best hope of a solution to this problem. The thin-film circuit offers advantages over the monolithic circuit in terms of close component tolerances and reduced interaction, and hybrid devices are available using both types of integrated circuit. A problem common to both types is that it is difficult to produce an inductor with a value larger than about $1\,\mu\text{H}$, since a two-dimensional coil is required. Thus circuits are devised using R and C elements only which have the properties of circuits using coils.

THE THYRISTOR

The thyristor is a semiconductor device with *three p-n* junctions formed in the same material, which is usually silicon. This gives a *p-n-p-n* structure (Fig. 5.13(*a*)), with an ohmic contact called the *anode* at the end *p*-region and a second contact called the *cathode* at the end *n*-region. A third contact known as the *gate* is usually made to the inside *p*-region, and this is a control electrode for the two other electrodes through which the main current flow takes place. The thyristor is finding increasing application as a solid-state replacement of the thyratron (page 309), and is capable of switching currents up to several hundred amperes at voltages up to about 1·5 kV.

Operation of the device depends on the polarity of the voltage applied to the anode. When this is negative the two outer junctions, J_1 and J_3, are reverse biased while the inner junction, J_2, is forward biased (Fig. 5.13(*c*)). Most of the applied voltage thus appears across J_1 and J_3 and only the leakage current of these junctions can flow through the thyristor. If the voltage is increased sufficiently, avalanche breakdown will occur as in the junction diode (page 103) and the current will increase rapidly at the breakdown voltage V_{RA} (Fig. 5.14).

If the anode voltage is positive with respect to the cathode J_1 and J_3 are forward biased and only J_2 is reverse biased (Fig. 5.13(*d*)). However, most of the voltage will appear across J_2 and again only leakage current will flow; this is known as the *blocking condition*. When the breakdown voltage of this junction is reached the anode current increases rapidly with voltage, until at the *forward breakover voltage*, V_{BO}, and at current I_{BO} the device switches itself into a low-impedance state (Fig. 5.14). The voltage drop then remains at about 1 V up to high values of current, the I/V characteristic being similar to that of a silicon rectifier. When the thyristor is operating in this condition the anode voltage remains low until the anode current has been reduced below I_H, the *holding current*. In practice I_H is about 10 mA and V_{BO} may be up to about 1 kV.

So far, the gate has been assumed open-circuited, i.e. $I_G = 0$. If the gate is biased positively with respect to the cathode a small current flows

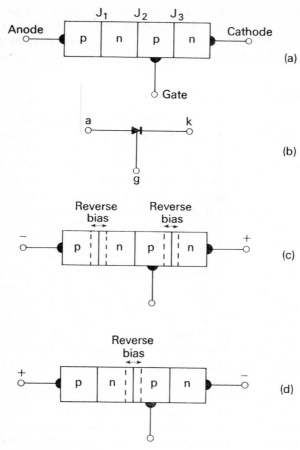

FIG. 5.13 The thyristor

(a) Arrangement of p-n junctions   (c) Application of reverse voltage
(b) Graphical symbol   (d) Application of forward voltage

between gate and cathode, a part of which is added to the leakage current across J_2. This means that breakdown of the junction can take place at a smaller value of V_{BO}, and the larger I_G is made the smaller the corresponding value of V_{BO}. In fact if I_G is large enough V_{BO} may be reduced so far that the blocking characteristic in the forward direction disappears and the device behaves as a low impedance for all values of anode voltage. It

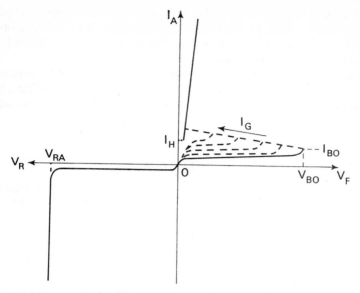

Fig. 5.14 Characteristics of thyristor

should be noted that the gate only controls the turning-on of the anode current and has no further effect once the anode voltage has fallen to its low value. The blocking condition is restored only by reducing the anode current below I_H, and in the reverse direction the gate has a negligible effect on the breakdown voltage, although it can cause an increase in reverse current.

Theory of Operation

Since J_2 is reverse biased during forward operation, the electrons and holes which are thermally generated within the depletion layer will give rise to a leakage current I_{CO}. Thus an electron current will flow to the anode, which is p-type, and a hole current will flow to the cathode, which is n-type. When the electrons cross J_1 and move into the anode they will cause holes to be emitted from it in order to maintain charge balance, and in a similar manner the arrival of holes at the cathode will cause electrons to be emitted. Thus the new holes from the anode will pass through J_2 causing further electrons to be released by collision, while the new electrons from the cathode will also pass through J_2 and similarly

217

cause further holes to be released. This constitutes an internal multiplication of current, similar in effect to that occurring during avalanche breakdown, so that the anode current will eventually reach a value limited only by the circuit conditions outside the junctions.

The forward operation of the thyristor may be analysed by supposing that it consists of a *p-n-p* transistor TR_1 and an *n-p-n* transistor TR_2, both sharing junction J_2 (Fig. 5.15). If $I_G = 0$, the two emitter currents and the

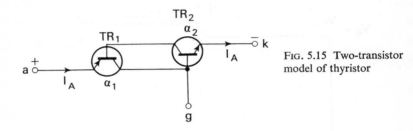

FIG. 5.15 Two-transistor model of thyristor

total current across J_2 are each equal to I_A. The hole current from the anode is then $\alpha_1 I_A$ and the electron current from the cathode is $\alpha_2 I_A$, since these correspond to the two collector currents. Hence the total current flowing across J_2, which includes the reverse leakage current of the junction, I_{CO}, is given by

$$I_A = \alpha_1 I_A + \alpha_2 I_A + I_{CO}$$

$$= \frac{I_{CO}}{1 - (\alpha_1 + \alpha_2)} \tag{5.36}$$

When the reverse voltage across J_2 approaches the breakdown value a factor M must be introduced, given by eqn. (4.21). Thus the leakage current becomes MI_{CO} and the two collector currents $M\alpha_1 I_A$ and $M\alpha_2 I_A$ respectively, so that near breakdown

$$I_A = \frac{MI_{CO}}{1 - M(\alpha_1 + \alpha_2)} \tag{5.37}$$

The thyristor is designed so that at low voltages and currents $\alpha_1 + \alpha_2 \ll 1$, and since $M \approx 1$ under these conditions,

$$I_A \approx I_{CO} \tag{5.38}$$

However, as the voltage is increased, avalanche multiplication occurs at J_2 making $M > 1$, and I_A increases until, at a certain value,

$$M(\alpha_1 + \alpha_2) = 1 \qquad (5.39)$$

Thus from eqn. (5.37) it can be seen that I_A would become infinite if it were not limited by the applied voltage and circuit resistance, and break-over has occurred corresponding to V_{BO} and I_{BO} in Fig. 5.14. As a result of the increase in current α_1 and α_2 will also have increased (page 129) until $\alpha_1 + \alpha_2 > 1$, and since the collector current of one transistor is the base current of the other, the base and collector currents in each transistor will now be of comparable magnitude. Thus the two transistors are in a saturated state (page 139), and in order to achieve this J_2 must now also be forward biased (Fig. 5.16), with its bias having opposite polarity to the

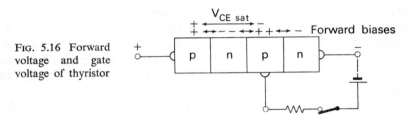

FIG. 5.16 Forward voltage and gate voltage of thyristor

biases on J_1 and J_3. Thus the total voltage across the thyristor is the forward bias of one emitter junction plus the saturated voltage of the other transistor, which accounts for the very low voltage across the device after breakover.

The thyristor remains switched on at a low voltage (even though $M = 1$) owing to the increase in $\alpha_1 + \alpha_2$, and it remains in the low-impedance state until the current falls below the value which makes $\alpha_1 + \alpha_2 = 1$. This is the holding current, I_H, below which the device switches off. Eqns. (5.36) and (5.37) apply to the operation *before* breakover, but are not applicable after breakover since the common collector is forward biased under these conditions.

Finally, if gate current I_G is allowed to flow by closing the switch in Fig. 5.16, a current $\alpha_2 I_G$ will be added to the currents flowing across J_2. Hence at low voltages

$$I_A = \frac{I_{CO} + \alpha_2 I_G}{1 - (\alpha_1 + \alpha_2)} \qquad (5.40)$$

and if I_G is large enough to make $\alpha_1 + \alpha_2 = 1$, the thyristor is switched on at a low value of V_{BO}. In order to trigger a 10 A thyristor into its low-impedance state, a gate current of about 100 mA at a minimum gate voltage of 3–4 V is required. An average gate dissipation up to 0·5 W is commonly allowed, and if a continuous supply is used to trigger the thyristor this dissipation may be exceeded, resulting in failure of the device. Hence the battery and switch of Fig. 5.16 are replaced by a pulse generator, so that the required values of I_G and V_G are obtained at the pulse peaks. Thus the average values are much less than the peak values, which ensures that the average dissipation is kept within a safe limit.

The duration of each pulse depends on the time required to *initiate* current multiplication within the thyristor, which is normally a few microseconds, although establishment of the full forward current takes an additional time owing to the spreading velocity, as explained below. Similarly the time needed to turn off the thyristor is determined by the rate at which stored charge can flow out of the device, which can take up to about 30 μs. Gate pulses are not normally applied when the anode is negative as the gate current adds to the leakage current, and this may result in a large current at a high reverse voltage which can easily cause over-heating of the device.

The construction of a thyristor is shown in Fig. 5.17. The anode is

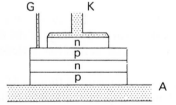

FIG. 5.17 Construction of thyristor

normally mounted on a heat sink to keep the temperature of the device below a maximum of about 150°C. It may be seen that the gate contact is made to a single point, which means that conduction due to internal multiplication is initiated near that point. Thus a high density of electrons and holes is set up in a small area of the cathode and these current carriers then diffuse sideways until the whole cathode is supporting the current. The *spreading velocity* with which this process occurs is constant for a given thyristor and may be about 0·1 mm/μs for a 10 A thyristor. Thus, if the cathode has a typical diameter of 3·5 mm, it will take about 35 μs for complete conduction to be established across the cathode. Hence the rate of rise of current dI/dT must be restricted by the external circuitry, so that a small cathode area will not be required to sustain the whole anode current, which could lead to overheating and failure of the device.

A second consideration is the rate of application of the forward blocking voltage dV/dt. This is due to C_j, the depletion-layer capacitance of J_2, through which a displacement current flows given by

$$i_D = C_j \frac{dV}{dt} \tag{5.41}$$

If i_D is larger than the holding current I_H, the electrons and holes forming i_D can cause sufficient emission from the cathode and anode respectively to turn on the transistor even though the forward voltage is below the required breakover value. For this reason dV/dt is limited to about $100\,\text{V}/\mu\text{s}$ to keep i_D down to a safe level.

PHOTODIODES AND PHOTOTRANSISTORS

Let us now consider in more detail the effect of illuminating a semiconductor which was introduced on page 62. The minimum frequency of radiation at which electron-hole pair generation can occur is given by

$$hf_0 = W_g \tag{5.42}$$

as shown in Fig. 5.18. In terms of wavelength,

$$\frac{hc}{\lambda_0} = W_g \tag{5.43}$$

and

$$\lambda_0 = \frac{hc}{W_g} = \frac{1\cdot24}{W_g}\ \text{micrometres} \tag{5.44}$$

where W_g is in electronvolts and λ_0 is the maximum or threshold *wavelength*. Thus for germanium with $W_g = 0\cdot72\,\text{eV}$, λ_0 is $1\cdot73\,\mu\text{m}$; while for silicon with $W_g = 1\cdot1\,\text{eV}$, λ_0 is $1\cdot14\,\mu\text{m}$. In both materials electron-hole pair generation occurs for wavelengths well into the infra-red region, but for wavelengths greater than λ_0 the effect cannot take place.

For radiation at the threshold wavelength the density of electron-hole pairs will be low since the density of states is small (Fig. 5.18 and Appendix 3), so that the conductivity of the semiconductor will show only a slight increase. As the wavelength is reduced, corresponding to an increase in

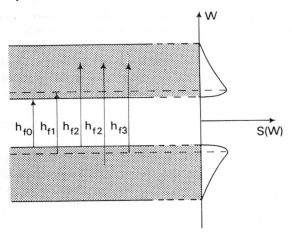

Fig. 5.18 Relationship between energy bands of a semiconductor
and radiation of various frequencies

frequency, the density of states and the conductivity will increase to
reach a maximum at λ_1. This corresponds to energy hf_1 and the maximum
density of available states for electron-hole pair generation. At a higher
frequency f_2 an electron may be excited from the top of the valence band
to the top of the conduction band, so that emission of the electron from
the semiconductor is possible and electron-hole pair generation is less
probable. Thus the conductivity falls as the frequency is raised to f_3, at
which emission is much more probable than the generation of electron-
hole pairs. A typical response curve is shown in Fig. 5.19 which applies
to a photodiode, whose operation is as follows.

Consider a *p-n* junction of the form shown in Fig. 5.20(*a*). The *p*-
region has a thickness of only a few micrometres so that radiation can
penetrate to the vicinity of the junction. Electrons formed near the junction will be swept into the *n*-region by the junction field and holes formed
near the junction will be similarly swept into the *p*-region (Fig. 5.20(*b*)).
Electron-hole pairs formed away from the junction may recombine
before they can be separated by the junction field. Thus the excess carriers
near the junction provide a photocurrent I_P which adds directly to the
current generated thermally, the leakage current I_0. In a reverse-biased

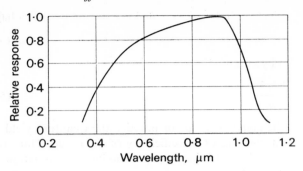

FIG. 5.19 Spectral response of a silicon photodiode

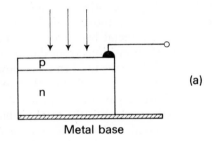

(a)

Metal base

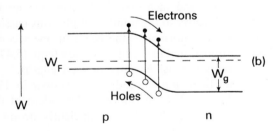

(b)

FIG. 5.20 Semiconductor photodiode

(*a*) Construction
(*b*) Energy diagram

photodiode with zero illumination, I_0 is known as the *dark current*, and after illumination the reverse current rises from I_0 to $I_0 + I_P$.

I_P is related to the power P of the incident radiation, since if there are n_P photons incident per second on the diode,

$$P = n_P hf = \frac{n_P hc}{\lambda} \tag{5.45}$$

223

and if there are n_{iP} electron-hole pairs created per second by the radiation

$$I_P = 2n_{iP}e \tag{5.46}$$

Dividing eqn. (5.46) by eqn. (5.45)

$$\frac{I_P}{P} = \frac{2e\lambda n_{iP}}{hcn_P} = \frac{2e\lambda}{hc}\zeta \tag{5.47}$$

where $\zeta = n_{iP}/n_P$, the *quantum yield*. In practice the quantum yield is a few per cent and is a measure of the efficiency of the diode. Thus, from eqn. (5.47), I_P is proportional to the power of the incident radiation at a particular wavelength.

In a light source such as a tungsten filament at a high temperature, photons will be emitted covering a continuous range of frequencies, rather than at one frequency as given above. This is due to the energy levels in the conduction band of a metal being continuous, which leads to a very large number of possible transitions from the top of the valence band. Hence the radiated power is here the total emissive power, which for a black body is equal to σT^4, where σ is the Stefan–Boltzmann constant.

We are concerned with the visible part of the power radiated by the source, known as the *luminous flux*, which is the total visible light energy emitted by a source in one second and is measured in *lumens* (lm)*. When a surface is illuminated only a part of the visible light energy radiated by the source is intercepted, so that the *illumination* of the surface is defined in terms of a flux density and measured in lumens per square metre or *lux*. This is also related linearly to the number of photons passing through one square metre of surface normal to the beam in one second. Thus the photocurrent is still proportional to the illumination (eqn. (5.47)), with λ replaced by a term covering the visible range of wavelengths.

The I/V characteristic of a silicon *p-n* photodiode is shown in Fig. 5.21(a). The reverse current, $I_R = I_0 + I_P$, is almost independent of the reverse voltage, since the field across the junction is sufficiently strong to

---

* The standard light source is a full radiator (black body) at the temperature of solidification of platinum, about 2 042 K. The unit of luminous intensity is the *candela* (cd) and 1 cm² of the surface of a full radiator at the above temperature has a luminous intensity of 60 cd normal to its surface. A source of light, considered as being at a point at the centre of a sphere, radiates uniformly in all directions. The total luminous flux from a point source of one candela is then 4π lumens (12·57 lm), where the solid angle subtended over the surface of the sphere is 4π steradians (sr).

It is found experimentally that a 1 W source of power radiating at the wavelength at which the eye is most sensitive (555 nm) gives a luminous flux of 680 lm. The luminous efficiency of a practical source radiating within the visible spectrum is much lower, being about 100 lm/W for the sun and about 40 lm/W for a daylight fluorescent tube.

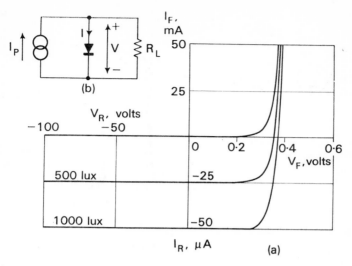

FIG. 5.21 Silicon photodiode

(*a*) Characteristics   (*b*) Equivalent circuit

extract all the current carriers created at the junction, even for low values of applied voltage. The reverse current is proportional to the illumination and can be measured directly; alternatively a resistor can be connected in series with the diode and supply and current changes observed as voltage changes across the resistor. Such a diode is useful as a detector of visible or infra-red radiation and greater sensitivity is obtained if the junction is the collector junction of a *phototransistor*. This can be a normal transistor in a transparent encapsulation, which is connected in the common-emitter configuration with the base open-circuited. In the absence of radiation the collector current is then $(h_{FE} + 1)I_{CBO}$ (eqn. (4.163)) or approximately $h_{FE}I_{CBO}$. When radiation falls on the collector–base junction (Fig. 5.22(*a*)) the leakage current I_{CBO} is increased by the photocurrent I_P, so that the collector current now becomes $I_C = h_{FE}(I_{CBO} + I_P)$. Thus, owing to the current gain of the transistor, the photocurrent is greater than the diode photocurrent and leads to the characteristics of Fig. 5.22(*b*). The base may be connected to the emitter through a resistor in order to reduce the dark current $h_{FE}I_{CBO}$ (Ref. 4.1), and hence to improve the ratio I_P/I_{CBO} at high temperatures.

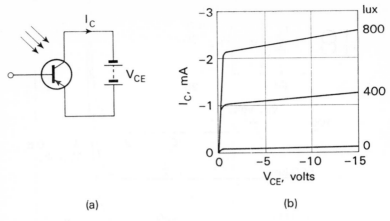

(a)

(b)

FIG. 5.22 The phototransistor

(a) Biasing arrangement   (b) Characteristics

THE SOLAR CELL

It may be seen from Fig. 5.21, not only that the reverse current has increased, but also that the characteristic passes through a forward voltage at zero current. Thus an open-circuited photodiode generates an e.m.f. when the junction is illuminated, since extra holes move into the *p*-region and extra electrons move into the *n*-region owing to the junction field. This means that the junction is converting light energy directly into electrical energy, and such a device is called a *solar cell*. A silicon solar cell can produce an open-circuit voltage up to about 0·5 V and a short-circuit photocurrent of about 0·3 μA per square millimetre of junction area. Higher voltages can be produced by connecting junctions in series and higher currents by connecting them in parallel, the main application being the generation of electrical power from sunlight, in particular for use in space satellites.

A photodiode may be represented by the circuit shown in Fig. 5.21(*b*), where the effect of illumination is represented by a current generator I_P across a normal *p-n* diode. R_L is a load resistance, and, in general, the diode current and voltage are related by eqn. (3.5), which leads to

$$\log_e (I + I_0) = \log_e I_0 + \frac{eV}{kT}$$

and

$$V = \frac{kT}{e} \log_e \left(1 + \frac{I}{I_0} \right)$$ (5.48)

When the diode is open-circuited by disconnecting R_L illumination of the diode makes $I = I_P$, so that the open-circuited voltage is

$$V_{oc} = \frac{kT}{e} \log_e \left(1 + \frac{I_P}{I_0} \right)$$ (5.49)

V_{oc} increases with the illumination as shown in Fig. 5.21(a).

With R_L connected across the diode,

$$I = I_P - I_L$$ (5.50)

and

$$V = \frac{kT}{e} \log_e \left(1 + \frac{I_P - I_L}{I_0} \right)$$ (5.51)

which is less than V_{oc}. Finally for $R_L = 0$ the diode behaves as a current generator with $I_L = I_P$ and $V = 0$.

Among the many other applications of photodevices a photodiode can be used in an exposure meter to generate a current, and this is read on an ammeter calibrated in units of illumination. Also a punched tape carrying data to be fed into a computer can be "read" by several photodiodes mounted next to each other across the tape, which is illuminated from the opposite side to the diodes. Thus, whenever a hole appears in the tape as it moves along, the appropriate diode generates a pulse, since there is one diode for each row of holes. Fire detection, automatic control of street lighting, and counting and control of articles on a production line may also be included among the uses of photodiodes and photo-transistors.

THE SEMICONDUCTOR LAMP (ELECTROLUMINESCENT DIODE)

The excess carriers produced by the absorption of light recombine at a rate determined by their lifetime. It might appear that recombination would occur through the direct transition of an electron from the conduction band to the valence band, with the emission of a photon of radiation having a maximum wavelength λ_0. However, this effect is not observed in germanium or silicon, but it does occur in some compound semiconductors produced by combining equal amounts of elements from Groups III and V of the Periodic Table. Important examples are gallium arsenide, which is also used in the manufacture of tunnel diodes (page 345), and gallium arsenide-phosphide. Their crystal structures are similar to those of germanium and silicon and the electrical properties of GaAs and GaP are compared in Table 5.1. Gallium arsenide can provide a light

Table 5.1

	W_g	μ_n	μ_p	ϵ_r
	eV	m²/Vs	m²/Vs	
Germanium	0·72	0·390	0·190	16
Silicon	1·09	0·140	0·048	12
Gallium arsenide	1·35	0·500	0·040	11
Gallium phosphide	2·25	0·010	0·002	8·5

source in the near infra-red at a wavelength of about $0·9\,\mu m$, which may be confirmed by using eqn. (5.44) for the energy gap of $1·35\,eV$. The energy gap of gallium arsenide-phosphide depends on the proportion of phosphorus and varies between 1·35 and 2·25 eV as the phosphorus content is changed from 0 to 100%. However, the material ceases to provide a direct transition when the proportion of phosphorus is greater than about 44%, and at lower phosphorus contents it provides a light source in the red region of the spectrum. Photon energy, wavelength and colour in the visible region are related in Fig. 5.23, energy and wavelength being related by eqn. (5.44).

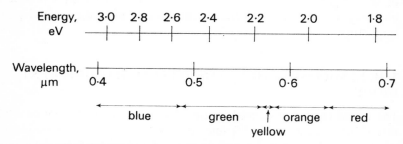

Fig. 5.23 Relationships between photon energies, wavelengths and colours for visible light

The suitability of a semiconductor as a light source depends on the variation of energy with momentum in the crystal. This is introduced in Appendix 2, where it is shown that the energy W rises with the square of the momentum vector k, except near a forbidden band. The situation is more complicated in a real crystal, and the dependence of the energy on the momentum vector is shown diagrammatically for germanium and silicon in Fig. 5.24(b), and for a material such as gallium arsenide in Fig. 5.24(a). The energy gap W_g corresponds to the difference between the

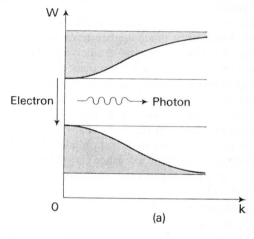

Fig. 5.24 Energy/momentum-vector diagrams

(a) Direct-gap semi-conductor
(b) Indirect-gap semi-conductor

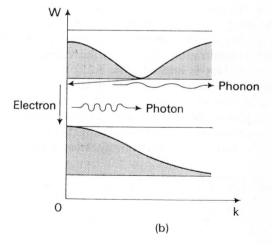

maximum energy in the valence band and the minimum energy in the conduction band. In gallium arsenide these energies both occur at the same value of k, which allows a direct transition with the emission of a photon, so that gallium arsenide is known as a *direct-gap* semiconductor. However, in germanium and silicon the maximum and minimum energies occur at different values of k, so that an electron must lose momentum in order to have the value of k corresponding to the maximum energy of the valence band, after which a transition back to the valence band can take place.

The momentum is lost through the release of a phonon (page 49), which has a very much higher momentum than a photon.* The probability of a phonon having the correct momentum being released at the same time as a photon of the correct energy is very small, so that this type of transition is very unlikely to happen. Germanium and silicon are known as *indirect-gap* semiconductors, and in these materials recombination takes place through traps (page 62), which give rise to one or more intermediate levels in the forbidden band. An electron can then fall from the conduction band into a trapping level and thence return to the valence band either directly or via other trapping levels to recombine with a hole.

Even in a direct-gap semiconductor significant light output is obtained only when large numbers of electrons and holes recombine in unit time. For a single crystal with no junctions a high temperature would be required to increase the density of electron-hole pairs (eqn. (2.30)), and doping introduces only one type of carrier at the expense of the other type. However, if a p-n junction is formed using degenerate semiconductors the energy bands are as shown in Fig. 5.25(a) at zero bias, with a high concentration of electrons in the conduction band of the n-region and a high concentration of holes in the valence band of the p-region. When forward bias is applied (Fig. 5.25(b)), the electrons at the edge of the conduction band occupy energy levels directly above the levels of holes at the edge of the valence band. Direct recombination takes place with one photon emitted for each electron transition, and since the carrier concentrations are high a useful light output is obtained.

The form of an electroluminescent diode, or semiconductor lamp, is

---

* The momentum of a quantum is $p = h/\lambda$ (eqn. (1.12)). The wavelength of a photon of visible light is around 10^{-6} m, while that of a phonon is about equal to the interatomic spacing of 10^{-10} m, so that its momentum is about 10 000 times that of a phonon. The energy of a quantum is hc/λ, where c is the velocity of light, 3×10^8 m/s, for a photon or the velocity of sound in a solid, around 3×10^3 m/s, for a phonon. Thus the energy of a phonon is only about one-tenth of that of a photon.

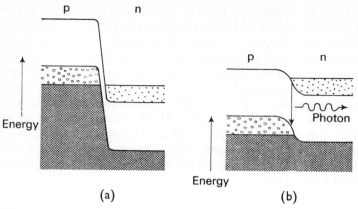

(a)　　　　　　　　　　　　　　(b)

FIG. 5.25 *p-n* junction formed from degenerate direct-gap semiconductors

(*a*) Zero bias　　　(*b*) Forward bias

shown in Fig. 5.26. Light is emitted normal to the junction, which has an area of about 1 mm^2, at forward currents between about 1 A and 30 A. The high current density means that the forward bias has to be applied in pulses lasting a few microseconds, the peak voltage being about

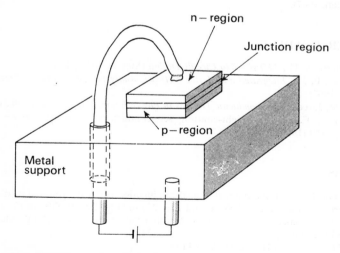

FIG. 5.26 The semiconductor lamp

Light is emitted from all the exposed sides of the crystal, but most intensely from the region of the junction.

1·6 V. For a mean electrical input of 500 mW the light output is typically in the milliwatt region, giving an efficiency around 1%.

An important property of the semiconductor lamp is that the output can be modulated at frequencies up to about 100 MHz by modulating the diode current. The lamp can thus be used as a transmitter of information, which can be received by a photodiode to form the basis of a short-range communication system. A further development is the semiconductor laser, described on page 396. Other III–V compounds such as indium arsenide, with $W_g = 0·33$ eV, and indium phosphide, with $W_g = 1·25$ eV can also be used in electroluminescent devices. Thus a range of wavelengths is available which is extended by the use of alloys, such as GaAs–InAs and GaAs–InP.

REFERENCES

5.1 Sevin, L. J., *Field Effect Transistors* (McGraw-Hill, 1965).

5.2 Lindmayer, J. and Wrigley, C. Y., *Fundamentals of Semiconductor Devices* (Van Nostrand, 1965).

5.3 Bergman, G. D., "The gate-triggered turn-on process in thyristors", *Solid State Electronics*, **8**, p. 757 (1965).

5.4 Coombe, R. A. (ed.), *The Electrical Properties and Applications of Thin Films* (Pitman, 1967).

FURTHER READING

Crawford, R. H., *MOSFET in Circuit Design* (McGraw-Hill, 1967).

Chirlian, P. M., *Integrated and Active Network Analysis and Synthesis* (Prentice-Hall, 1967).

Seymour, J. (ed.), *Semiconductor Devices in Power Engineering* (Pitman, 1968).

Morehead, F. F. Jr., "Light-emitting semiconductors", *Scientific American*, **216**, p. 109 (1967).

PROBLEMS

5.1 The donor density in the channel of a silicon junction FET is $10^{22}/m^3$ and its dimensions are $a = 1·0\,\mu m$, $c = 100\,\mu m$ and $L = 25\,\mu m$ (Fig. 5.1(*a*)). Obtain the open-circuit conductance and the gain-bandwidth product. If $V_p = -5·0$ V calculate I_{DSS} and the mutual conductance at $V_{GS} = -2·0$ V.
(*Ans.* 11·5 mS, 520 MHz, 2·9 mA, 6·9 mS)

5.2 A basic MOST device has an input capacitance C_{gc} of 3 pF, and the length of the channel is $10\,\mu m$. For an enhancement-mode device with $V_p = +1·5$ V, calculate

$I_{D\,sat}$ and g_m when $V_{GS} = +3\,\text{V}$. For a depletion-mode device with $V_p = -2\,\text{V}$ calculate $I_{D\,sat}$ and g_m for $V_{GS} = 0$.
(*Ans.* 4·6 mA, 6·1 mS, 8·1 mA, 8·1 mS)

5.3 An FET has the following small-signal parameters: $C_{gc} = 4\,\text{pF}$, $C_{gd} = 1\,\text{pF}$ and $g_m = 3\,\text{mS}$. It is used as an amplifier between a 600 Ω source and a drain load of 2 kΩ. Calculate (*a*) the gain at low frequencies, (*b*) the high frequency at which the gain has fallen 3 dB below the value in (*a*) (see the analysis of the hybrid-π circuit). (*Ans.* (*a*) −6, (*b*) 9·2 MHz)

5.4 A single solar cell has a short-circuit current of 70 μA at room temperature when the illumination is 1 000 lx. If the dark current is 10 nA, how many similar cells must be connected in series to give a minimum open-circuit voltage of 1 V at an illumination of 200 lx?
(*Ans.* six)

6

Electron Emission and Vacuum Devices

EMISSION OF ELECTRONS

Inside a solid an electron moves between the atoms in a random manner owing to thermal agitation. Near the surface it may have a component of velocity which will take it away from the atoms, but as soon as it escapes from the solid it leaves behind a positive charge. This exerts a force trying to prevent its escape, and this force may be obtained by the method of electrical images. If the electron, charge $-e$, is distant x outside the surface, the force F acting on it may be represented by a charge of $+e$ at a distance x below the surface (Fig. 6.1). Then, by Coulomb's law,

$$F = \frac{e^2}{4\pi\epsilon_0(2x)^2} = \frac{e^2}{16\pi\epsilon_0 x^2} \tag{6.1}$$

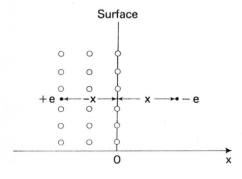

FIG. 6.1 Electron image near a surface

When x is large, $F \to 0$ and the electron has then escaped from the solid. However, when $x = 0$ the electron is just on the surface and eqn. (6.1) suggests that F then becomes infinite. In fact F must tend to zero, since within the solid the electron is surrounded by atoms and the net force on it is zero. Thus the image theory does not apply close to the solid, and in practice the image value of F is reached at about 1 nm from the surface (Fig. 6.2(a)). From this distance the row of atoms at the surface appears

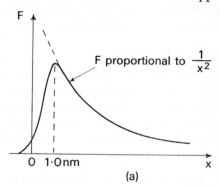

FIG. 6.2 Work function of a surface

(a) Force acting on an electron
(b) Energy barrier

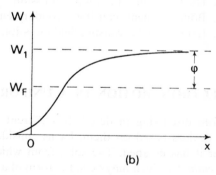

to be almost continuous, since their spacing is about 0·1 nm.

The work done in moving the electron a distance x from the surface is then

$$W = \int_0^x F\,dx \qquad (6.2)$$

which is the area under the F/x curve. The variation of W with x is shown in Fig. 6.2(b), where it will be seen that it reaches a constant value, W_1, asymptotically, in practice at about 10 nm from the surface. In order to escape, the electron must have potential energy greater than the energy barrier W_1. If its kinetic energy of emission is less than W_1 the electron will travel beyond the surface until it comes to rest and then returns to the solid. On the energy level diagram W_1 may be taken as the zero energy an electron possesses when it leaves the solid, so that its energy within the solid is negative with respect to W_1. The maximum energy it can possess at the absolute zero of temperature is W_F, the Fermi level. The difference between these two energies is given by

$$W_1 - W_F = \phi \tag{6.3}$$

where ϕ is the *work function*, usually measured in electronvolts, which corresponds to the minimum extra energy that must be given to the solid to allow an electron to escape. The values of ϕ range from 1 to 6 eV, depending on the atomic structure of the material, and at room temperature very few electrons have enough energy to escape. This energy can be supplied in four ways:

1. Radiation, causing photoemission
2. Heat, causing thermionic emission
3. Bombardment, causing secondary emission
4. External field, causing field emission

ELECTRON MOTION IN A UNIFORM ELECTRIC FIELD

Before discussing in detail the different types of electron emission and the devices based on them, let us consider what happens to an electron after it has escaped. The solid from which it has been emitted is called the *cathode*, which may be in the form of a flat plate, and it can be collected by a second plate called the *anode*. The plates are known as *electrodes* and many other forms are possible. They are mounted in a vacuum so that the motion of the electron is not interrupted by collision with air particles (see page 299).

Suppose the spacing between the plates is d and the anode is held at a potential V_A which is positive with respect to the cathode (Fig. 6.3). An electric field E_z is set up which is uniform away from the edges of the electrodes, leading to electron motion in the x-direction. It is assumed that

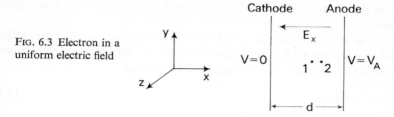

FIG. 6.3 Electron in a
uniform electric field

there is negligible motion in the y- and z-directions. The force on the
electron is given by

$$F_x = -eE_x = e\frac{dV}{d_x} \tag{6.4}$$

and is in the direction of increasing potential so that work will be done
on the electron in moving it towards the anode. In moving between the
points 1 and 2 the work done is

$$W_{12} = \int_1^2 F_x\,dx = e\int_1^2 dV \tag{6.5}$$

$$= e(V_2 - V_1) \tag{6.6}$$

where V_1 and V_2 are the potentials at 1 and 2 respectively. The force is
also given by the rate of change of momentum, so that, if u is the velocity
in the x-direction,

$$F_x = \frac{d}{dt}mu = m\frac{du}{dt} + u\frac{dm}{dt} \tag{6.7}$$

When relativistic effects are negligible (Appendix 4), $dm/dt = 0$ and

$$W_{12} = \int_1^2 F_x\,dx = \int_1^2 m\frac{du}{dt}dx = \int_1^2 mu\,du \tag{6.8}$$

$$= \tfrac{1}{2}m(u_2{}^2 - u_1{}^2) \tag{6.9}$$

Equating (6.6) and (6.9),

$$e(V_2 - V_1) = \tfrac{1}{2}m(u_2{}^2 - u_1{}^2) \tag{6.10}$$

where u_1 and u_2 are the velocities at 1 and 2 respectively. Thus the kinetic
energy of the electron has been increased because it has gained energy

from the electric field, and this is of fundamental importance in vacuum devices.

In the common case of an electron starting from rest at the cathode and moving through a potential difference V, both V_1 and u_1 are zero, so that the velocity u is obtained from the expression

$$eV = \tfrac{1}{2}mu^2 \tag{6.11}$$

whence

$$u = \sqrt{\frac{2eV}{m}} \tag{6.12}$$

The velocity will reach its greatest value at the anode, where it is given by

$$u_A = \sqrt{\frac{2eV_A}{m}} \tag{6.13}$$

Here the electron is brought to rest and gives up its energy to the anode, both in the form of heat and through the release of secondary electrons (page 257). For one electron and for a relatively low density of electrons emitted from the cathode, the voltage will rise linearly from cathode to anode, so that

$$E_x = -\frac{V_A}{d} \tag{6.14}$$

everywhere between anode and cathode. The acceleration of an electron from rest may be obtained by combining eqns. (6.4), (6.7) and (6.14) to give

$$\frac{du}{dt} = \frac{eV_A}{md} \tag{6.15}$$

and this is also uniform, so that u rises linearly with time as shown in Fig. 6.4. Then the average velocity is $\tfrac{1}{2}u_A$ and the transit time τ_A from cathode to anode is given by

$$\tau_A = \frac{2d}{u_A} = d\sqrt{\frac{2m}{eV_A}} \tag{6.16}$$

using eqn. (6.13).

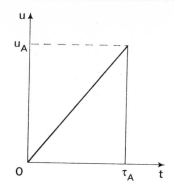

FIG. 6.4 Velocity/time
graph for an electron

CURRENT IN THE EXTERNAL CIRCUIT

Although the emitted electrons are moving in the space between the electrodes a measuring instrument in the external circuit will indicate a flow of current. This is because each electron repels a negative charge from the anode, which must equal $-e$, in the time, τ_A, it takes the electron to travel from cathode to anode. Thus a current i flows in the external circuit for each electron and energy is supplied to it from the external power supply (Fig. 6.5(a)). For an electron moving from rest through a potential V this energy is given by

$$\tfrac{1}{2}mu^2 = Ve = V\int_0^t i\,dt \qquad (6.17)$$

Differentiating,

$$mu\frac{du}{dt} = Vi \qquad (6.18)$$

and, substituting from eqns. (6.4) and (6.7),

$$mu\frac{du}{dt} = -eE_zu = e\frac{V}{d}u \qquad (6.19)$$

so that

$$i = \frac{eu}{d} \qquad (6.20)$$

Thus i increases linearly with time, since it is proportional to u (Fig. 6.5(b)).

239

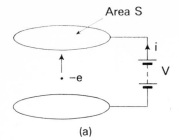

(a)

FIG. 6.5 Induced current in a diode

(a) Production of induced current
(b) Variation of induced current with time

(b)

The total current for a number of electrons will be the sum of the individual currents, provided the electron density is low enough for interaction to be negligible. If the electrode area is S and the electron density is n, the effective volume is Sd and the total number of electrons is nSd. Thus the total current flowing in the external circuit is

$$I = nSdi \qquad (6.21)$$

$$= neuS \qquad (6.22)$$

from eqn. (6.20). The current density between the electrodes is

$$J = neu \qquad (6.23)$$

since ne is the charge density. Thus, since $J = I/S$, the same current flows in the external circuit as between the electrodes.

PHOTOEMISSION

Let us now consider how electrons can be released from a cathode by the absorption of radiation. When radiation of frequency f falls on a solid each energy packet or photon will release one electron if $hf = e\phi$. The electron will have received just enough energy to surmount the energy barrier and so will be released with zero initial velocity. If $hf > e\phi$ the electron will have received more than enough energy and so will have an initial velocity u_e. The value of u_e is obtained by equating energies to give Einstein's photoelectric equation:

$$hf = e\phi + \tfrac{1}{2}mu_e{}^2 \tag{6.24}$$

Threshold Wavelengths λ_0

For a given surface the threshold wavelength λ_0 is the maximum wavelength that will allow an electron to be released with zero initial velocity. This is obtained from eqn. (6.24) with $u_e = 0$ and corresponds to the minimum frequency f_0, where $hf_0 = e\phi$. By comparison with eqn. (5.43) and with ϕ measured in electronvolts,

$$\lambda_0 = \frac{hc}{\phi} = \frac{1\cdot24}{\phi}\,\mu\mathrm{m} \tag{6.25}$$

Photoemission is not possible with light of wavelength longer than λ_0, and if visible light is taken to have an average wavelength of $0\cdot6\,\mu\mathrm{m}$, the work function of the cathode must be not greater than 2 eV for photoemission to occur (Fig. 5.23).

Maximum Emission Velocity u_e

u_e corresponds to the maximum velocity with which an electron can leave the surface. An electron can be released below the surface and be emitted with zero velocity due to collisions on its way through the material. Thus when $\lambda < \lambda_0$ electrons can be emitted with all values of velocity between 0 and u_e. If an anode is mounted near the cathode, photoelectrons will flow to it even when the anode-to-cathode voltage V_A is zero, owing to the initial energy with which they are emitted. If V_A is made increasingly negative the corresponding current will fall, since only those electrons with sufficient initial energy to overcome the potential barrier will be able to reach the anode. When the current is just zero only the most energetic electrons are reaching the anode, so that, if this occurs at an anode voltage V_0,

$$\tfrac{1}{2}mu_e{}^2 = eV_0 \tag{6.26}$$

and the maximum emission velocity u_e can be measured. Combining eqns. (6.24) and (6.26),

$$V_0 = \frac{hf}{e} - \phi \tag{6.27}$$

so that if the cathode is illuminated with light of various frequencies and the corresponding values of V_0 are measured, a plot of V_0 against f will be a straight line (Fig. 6.6). The slope of this line will give the value of

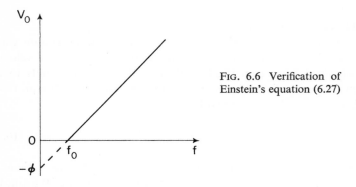

Fig. 6.6 Verification of Einstein's equation (6.27)

Planck's constant h, and the intercept on the V_0 axis will give the work function ϕ of the cathode in electronvolts. The line will cut the frequency axis at the threshold frequency f_0, so that an experiment of this type can be used to confirm the validity of the quantum theory.

The intensity of the light will not affect the initial velocity of emission, but the anode current is proportional to the intensity as may be seen by using an argument similar to that on page 224. The density of the photoelectrons is relatively low, so that the voltage rises linearly from cathode to anode and a small positive voltage is sufficient to draw all the photoelectrons to the anode. Thus a *vacuum photocell* has current/voltage characteristics as shown in Fig. 6.7, the current being nearly independent of voltage when $V_A > 15\,\text{V}$.

Photocathode Materials

In many applications photocells are illuminated by visible radiation and so are required to provide a response to wavelengths between about 0·4 and 0·7 μm. This means that the work function of the emitter must be less than 2 eV (eqn. (6.25)) for the threshold wavelength to be high enough. Of the pure metals only caesium, with $\phi \approx 1\cdot9\,\text{eV}$ satisfies this condition, so composite cathode materials with higher threshold wavelengths are

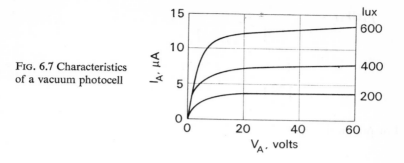

FIG. 6.7 Characteristics of a vacuum photocell

used which incorporate caesium. These include caesium-antimony (Cs-Sb), caesium-oxygen-silver (Cs-O-Ag) and caesium-oxygen-silver-bismuth (Cs-O-Ag-Bi). The corresponding values of λ_0 are 0·7, 1·2 and 0·85 μm respectively, and their response curves are shown in Fig. 6.8. The response of

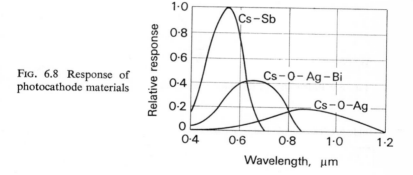

FIG. 6.8 Response of photocathode materials

Cs-O-Ag-Bi is similar to that of the human eye and is used in television camera tubes.

THERMIONIC EMISSION

As the temperature of a material in a vacuum is raised above room temperature the probability of an electron escaping is increased, according to the Fermi–Dirac function (Fig. 6.9). The probability of an electron having energy W_1 is given by eqn. (2.17):

$$p_F(W_1) = \cfrac{1}{1 + \exp\!\left(\cfrac{W_1 - W_F}{kT}\right)}$$

$$\approx \cfrac{1}{\exp\!\left(\cfrac{e\phi}{kT}\right)} = \exp\!\left(-\cfrac{e\phi}{kT}\right) \tag{6.28}$$

For practical values of ϕ and T see Table 6.1.

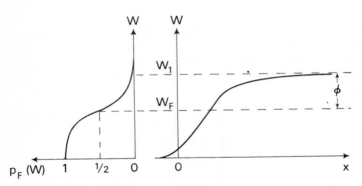

FIG. 6.9 Probability of thermionic emission

Table 6.1 Thermionic Cathode Materials

Cathode	Work function eV	Working temperature K	J_s A/cm²	Life h
Tungsten	4·5	2 500	0·25	3 000
Thoriated tungsten	2·6	1 900	1·5	10 000
BaO + SrO on nickel	1·0	1 000–1 100	1·0	20 000

The number of electrons available and hence J_s, the maximum current per unit area of cathode depends on $p_F(W)$ but is independent of the anode voltage (see page 249 for the effect of high anode voltages). J_s is known as the *saturated* or *temperature-limited* current density. The relationship is Richardson's law (1911):

$$J_s = AT^2 \exp\left(-\frac{e\phi}{kT}\right) \tag{6.29}$$

where A is a constant depending on the material of the cathode, varying from $60\,\text{A/cm}^2\text{-K}^2$ for tungsten to $0\cdot01$ for a mixture of barium and strontium oxides, the *oxide cathode* (Ref. 6.1). The exponential term has a much greater effect on J_s than the AT^2 term and may be rewritten in the form $\exp(-b/T)$, where $b = e\phi/k = 11\,600\,\phi$. Then, taking logarithms,

$$\log_e J_s = \log_e A + 2\log_e T - \frac{b}{T}$$

or

$$\log_e J_s + 2\log_e \frac{1}{T} = \log_e A - b\frac{1}{T} \tag{6.30}$$

Then, if $\log_e J_s + 2\log_e(1/T)$ is plotted against $1/T$, the result will be a straight line with slope $-b$ and intercept $\log_e A$, known as a *Richardson plot.*

This is illustrated in Fig. 6.10 for the three main cathode materials,

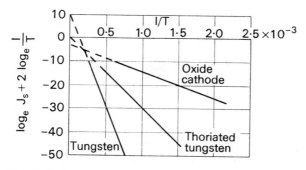

Fig. 6.10 Characteristics of thermionic cathode materials

tungsten, thoriated tungsten and oxide. Their main properties are given in Table 6.1, where it can be seen that a low work function allows a low working temperature and a long life, which results mainly from reduced evaporation of cathode material. The oxide cathode is considered to be an *n*-type semiconductor with excess barium atoms providing the donor impurity. First introduced by Wehnelt in 1905, it gives the greatest current density of the three types and is used in the great majority of thermionic valves. However, where the anode voltage is greater than about 2 kV, bombardment of the cathode by positive ions quickly destroys the thin ($\approx 50\,\mu$m) coating of barium and strontium oxides. For anode voltages up to 15 kV, which occur in transmitting valves, thoriated tungsten is used, and where the anode voltage reaches 20 kV, as in high-power transmitting valves and X-ray tubes a tungsten filament is used.

A thoriated-tungsten cathode consists of a tungsten filament coated with a layer of thorium about one atom thick (a monatomic layer). Since the work function of thorium is less than that of tungsten, valence electrons from the thorium mix with the free electrons in the tungsten leaving ionized thorium atoms on the surface, which is thus positively charged. Since the body of the cathode has acquired a negative charge a *dipole layer* or *electrical double layer* is formed on the surface (Fig. 6.11(*a*)).

(a)

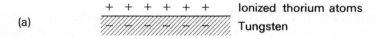

+ + + + + +   Ionized thorium atoms

Tungsten

(b)

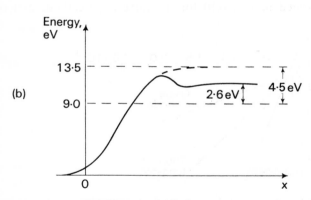

FIG. 6.11 Thoriated-tungsten cathode

(*a*) Dipole layer   (*b*) Energy barrier

This layer creates an electric field at the surface which assists electrons to escape, so that the overall work function is less than that of tungsten. The surface energy barrier then has a peak (Fig. 6.11(*b*)), which is thin enough to allow a finite probability that an electron can "tunnel" through it (Appendix 2).

THE SCHOTTKY EFFECT

When a sufficiently high voltage is applied to the anode all the electrons available for emission are collected so that the potential rises linearly from cathode to anode (Fig. 6.12(a)). The resulting electric field, $-V_A/d$,

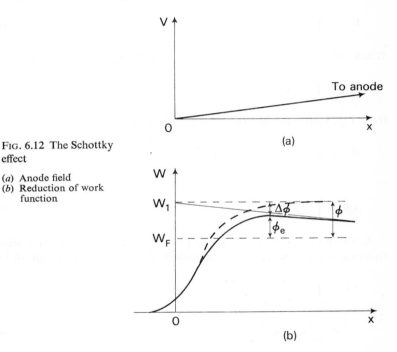

Fig. 6.12 The Schottky effect

(a) Anode field
(b) Reduction of work function

lowers the energy of the emitted electrons by an amount eV, where V is the potential at any point, so that the energy barrier is reduced by this amount at each point; the reduction may be significant for a high anode voltage. A peak then occurs in the new barrier which is too wide for an electron to penetrate and the work function is reduced from ϕ to ϕ_e (Fig. 6.12(b)). This means that the saturated current increases slowly with voltage instead of being independent of voltage, which occurs for fields up to about $10^5 \, \text{V/m}$.

247

9

Physical Electronics

The reduction in the potential barrier, $\Delta\phi$, may be obtained by considering the work done by an electron in escaping to a distance x from the surface, given by eqn. (6.2):

$$W = \int_0^x F\,dx$$

When the anode field is applied the force assisting the electron to escape is eV_A/d, so that the work done by the electron is reduced to

$$W_s = \int_0^x F\,dx - \frac{eV_Ax}{d} \tag{6.31}$$

When there is no external field the work done by the electron is

$$W_1 = \int_0^\infty F\,dx$$

from Fig. 6.2(b), so that the reduction in work done is

$$W_1 - W_s = \int_0^\infty F\,dx - \int_0^x F\,dx + \frac{eV_Ax}{d}$$

or

$$\Delta\phi = \int_x^\infty F\,dx + \frac{eV_Ax}{d} \tag{6.32}$$

Assuming that the peak in the potential barrier occurs at a distance from the cathode where image theory applies, F is given by eqn. (6.1) and

$$\Delta\phi = \int_x^\infty \frac{e^2}{16\pi\epsilon_0 x^2} + \frac{eV_Ax}{d}$$

$$= \frac{e^2}{16\pi\epsilon_0 x} + \frac{eV_Ax}{d} \tag{6.33}$$

Then

$$\frac{d\Delta\phi}{dx} = -\frac{e^2}{16\pi\epsilon_0 x^2} + \frac{eV_A}{d} \tag{6.34}$$

which is zero when

$$x = \left(\frac{ed}{16\pi\epsilon_0 V_A}\right)^{1/2} \tag{6.35}$$

248

Substituting eqn. (6.35) in eqn. (6.33),

$$\Delta\phi = \left(\frac{e^3 V_A}{4\pi\epsilon_0 d}\right)^{1/2} \tag{6.36}$$

so that the effective work function is

$$\phi_e = \phi - \Delta\phi = \phi - \left(\frac{e^3 V_A}{4\pi\epsilon_0 d}\right)^{1/2} \tag{6.37}$$

The current/voltage characteristic of a thermionic diode is shown in Fig. 6.13. The saturated current, I_s, at the knee of each curve increases

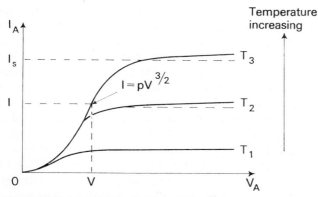

FIG. 6.13 Characteristics of a thermionic diode

with cathode temperature. I_s would be independent of V_A but for the Schottky effect, which causes the current to rise slowly with voltage. Combining eqns. (6.29) and (6.37), the current density in the saturated region is

$$J = J_s \exp\left(\frac{e^{3/2} V_A^{1/2}}{kTd^{1/2}}\right) \tag{6.38}$$

The Schottky effect is more pronounced in the oxide cathode, which has a porous surface, than in the tungsten cathode, which has a relatively smooth surface. This is due to the penetration of the pores of the coating by the field due to the anode, which causes electrons to be extracted from the inside surfaces of the coating. For voltages less than the saturation value the current rises with voltage. In this region the current is limited by the space charge as discussed below.

Although the thermionic diode has by now been replaced in most circuit applications by the semiconductor diode, it is still necessary to gain an understanding of its properties. This is because it is the basis of the electron gun (page 277), which is used extensively in cathode-ray tubes, electron microscopes and microwave valves. The diode is also fundamental to the operation of the triode, which again has been replaced almost entirely by the transistor but is still used in high-power amplifiers.

SPACE-CHARGE LIMITATION OF CURRENT

So many electrons are emitted from a thermionic cathode that only a fraction of them are collected by the anode when V_A has its working value. The remainder form a "cloud" of negative charges, known as a *space charge*, near the cathode. The space charge has the effect of causing a retarding field, which tends to prevent the emission of further electrons.

In order to understand how the space charge controls the anode current, suppose an anode voltage V is applied with the cathode at room temperature. The distribution of potential between cathode and anode will then be linear, as shown in Fig. 6.14(a), which ignores the effects of initial velocities and contact potential. If the cathode is heated to temperature T_1 the current will be temperature-limited as shown in Fig. 6.13, and the potential distribution will still be linear since all the available electrons are reaching the anode and no space charge is formed. The field at the cathode is then given by $-\tan \theta$, where θ is the angle the voltage curve makes with the horizontal at the cathode ($x = 0$). When the cathode temperature is raised to its working value, T_3 in Fig. 6.13, say, the resulting current I is less than its saturated value I_s, so that a space charge will be formed. The field at the cathode is then due to the combined effects of the accelerating field due to the anode and the retarding field due to the electrons of the space charge (Fig. 6.14(b)).

An equilibrium will be set up in which the two fields are equal and opposite, so that $\tan \theta = \theta = 0$. Thus, when the space charge loses an electron to the anode, θ becomes positive and its place is taken by another electron drawn from the cathode. If the space charge gains a surplus electron, θ becomes negative and an electron is returned to the cathode. The equilibrium condition, $\theta = 0$, corresponds to the same number of electrons entering the space charge as are leaving it for the anode, so that the space charge is controlling the anode current.

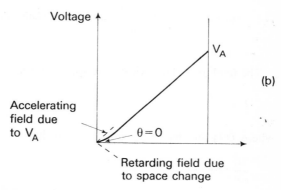

FIG. 6.14 Voltage distribution in a thermionic diode

(*a*) Temperature-limited current
(*b*) Space-charge-limited current

Calculation of Space-charge-limited Current

Consider two plane parallel electrodes sufficiently large and close that a uniform electric field can be set up between them (Fig. 6.15). The cathode at $x = 0$ is assumed to have an infinite supply of electrons, and the anode at $x = d$ is at a potential V_A. Electrons are assumed to leave the cathode with zero initial velocity, and the electric field at the cathode is zero under space-charge-limited conditions. Thus the electron velocity at any point is related to the potential V at that point by eqn. (6.12),

$$u = \sqrt{\frac{2eV}{m}}$$

The current density at any point is then

$$J = neu \tag{6.39}$$

where n is the density of electrons. The value of J is constant through the valve, so that the greatest density of space charge occurs near the cathode where the velocity is least.

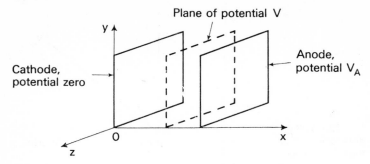

FIG. 6.15 Electrode arrangement for calculating the space-charge-limited current

The field at a plane within the space charge is

$$E = \frac{D}{\epsilon_0} \tag{6.40}$$

where D is the charge per unit area, so that

$$\frac{dE}{dx} = -\frac{ne}{\epsilon_0} \tag{6.41}$$

But $E = -dV/dx$, so that

$$\frac{-dE}{dx} = \frac{d^2V}{dx^2} = \frac{ne}{\epsilon_0} = \frac{J}{\epsilon_0 u} \tag{6.42}$$

This is a form of Poisson's equation in one dimension and may be used if it is assumed that the potential does not vary in the y- and z-directions. Combining eqns. (6.12) and (6.42),

$$\frac{d^2V}{dx^2} = \frac{J}{\epsilon_0}\sqrt{\frac{m}{2eV}} = AV^{-1/2} \tag{6.43}$$

where

$$A = \frac{J}{\epsilon_0}\sqrt{\frac{m}{2e}} \tag{6.44}$$

This equation must be integrated twice to obtain J in terms of V, so that, multiplying both sides by $2dV/dx$,

$$2 \frac{dV}{dx} \frac{d^2V}{dx^2} = 2AV^{-1/2} \frac{dV}{dx} \tag{6.45}$$

or

$$\frac{d}{dx}\left(\frac{dV}{dx}\right)^2 = 2AV^{-1/2} \frac{dV}{dx} \tag{6.46}$$

Integrating,

$$\left(\frac{dV}{dx}\right)^2 = 4AV^{1/2} + C_1 \tag{6.47}$$

But $dV/dx = 0$ when $V = 0$, so $C_1 = 0$. Therefore

$$\frac{dV}{dx} = 2A^{1/2}V^{1/4}$$

or

$$V^{-1/4} dV = 2A^{1/2} dx \tag{6.48}$$

Integrating again,

$$\tfrac{4}{3}V^{3/4} = 2A^{1/2}x + C_2 \tag{6.49}$$

But $V = 0$ when $x = 0$, so $C_2 = 0$. Thus

$$2V^{3/4} = 3A^{1/2}x \quad \text{or} \quad V^{3/4} = \tfrac{3}{2}A^{1/2}x \tag{6.50}$$

Squaring

$$V^{3/2} = \tfrac{9}{4}Ax^2 \tag{6.51}$$

$$= \tfrac{9}{4} \frac{J}{\epsilon_0} \left(\frac{m}{2e}\right)^{1/2} x^2 \tag{6.52}$$

substituting for A.

Hence

$$J = \frac{4\epsilon_0}{9} \left(\frac{2e}{m}\right)^{1/2} \frac{V^{3/2}}{x^2} \tag{6.53}$$

and inserting the numerical values of the constants,

$$J = \frac{2 \cdot 33}{10^6} \frac{V^{3/2}}{x^2} \quad \text{amperes/metre}^2 \tag{6.54}$$

at any point between the electrodes. At the anode the current density is

$$J = \frac{2 \cdot 33}{10^6} \frac{V_A^{3/2}}{d^2} \tag{6.55}$$

PROPERTIES OF THE THERMIONIC DIODE

Eqn. (6.55) is known as Child's law, and although it has been derived for a planar geometry the three-halves power relationship between current and voltage holds theoretically for *any* geometry. This is because the current density depends only on the density and velocity of the electrons (eqn. (6.39)). Suppose the anode voltage of a diode of arbitrary geometry is changed from V_A to V_A', so that

$$V_A' = aV_A \tag{6.56}$$

where a is any number greater than zero. Then, from eqn. (6.12), the new electron velocity is $a^{1/2}u$, and from eqn. (6.39) the new electron density is ane. The new current density is therefore

$$J_A' = a^{3/2}neu = a^{3/2}J_A \tag{6.57}$$

$$\frac{J_A'}{J_A} = a^{3/2} = \left(\frac{V_A'}{V_A}\right)^{3/2} \tag{6.58}$$

Eqn. (6.58) can be written in the form

$$I_A = PV_A^{3/2}$$

where P is a constant depending on the geometry and electrode area and is known as the *perveance*. For a planar diode

$$P = \frac{2 \cdot 33}{10^6} \frac{S}{d^2} \tag{6.59}$$

from eqn. (6.55), and in general P varies inversely as the square of some characteristic length. This is important in the design of an electron gun, which is essentially a thermionic diode with a hole in the anode through which a beam of electrons emerges (page 277). When a gun of the desired geometry is available, all the dimensions can be changed by the same factor without changing P, since the area S depends on the square of an electrode dimension. Thus the beam diameter can be controlled without changing the cathode current, provided that the current density does not exceed the saturated value.

The potential, the electron velocity and the density all vary with distance measured from the cathode. Thus, from eqns. (6.54) and (6.55),

$$V^{3/2} = V_A^{3/2}\left(\frac{x}{d}\right)^2 \quad \text{or} \quad V = V_A\left(\frac{x}{d}\right)^{4/3} \tag{6.60}$$

The velocity is proportional to $V^{1/2}$, so that

$$u = u_A\left(\frac{x}{d}\right)^{2/3} \tag{6.61}$$

where

$$u_A = \sqrt{(2eV_A/m)}$$

Putting ρ for the charge density ne,

$$J = \rho u \tag{6.62}$$

and at the anode

$$J_A = \rho_A u_A \tag{6.63}$$

Then, from eqn. (6.53),

$$\rho_A = \tfrac{4}{9}\epsilon_0 \frac{V_A}{d^2} \tag{6.64}$$

so that

$$\rho = \rho_A\left(\frac{x}{d}\right)^{-2/3} \tag{6.65}$$

This equation cannot hold very close to the cathode since it implies that ρ becomes infinite at $x = 0$.

The final property to be considered is the transit time of electrons from cathode to anode in the presence of space charge. Since their velocity increases with distance we must write

$$\tau_A = \int_0^d \frac{dx}{u} \tag{6.66}$$

$$= \frac{d^{2/3}}{u_A} \int_x^d x^{-2/3}\,dx \tag{6.67}$$

using eqn. (6.61). Hence

$$\tau_A = \frac{d^{2/3}}{u_A} \left[3x^{1/3} \right]_0^d \tag{6.68}$$

$$= \frac{3d}{u_A} \tag{6.69}$$

Thus the transit time has been increased by a factor of 3/2 compared with the value without space charge (eqn. (6.16)).

FIELD EMISSION

If a very high field is maintained between cathode and anode (up to 10^9 V/m), the effective potential barrier will be reduced further than in the Schottky effect (Fig. 6.16). ϕ_e will be smaller and the width of the barrier

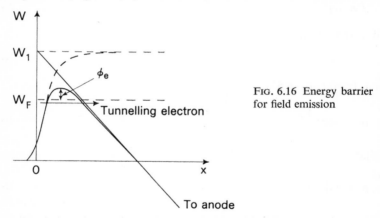

Fig. 6.16 Energy barrier for field emission

less than for the Schottky effect, with the result that there is a small but finite probability that electrons will be able to tunnel through the barrier (Appendix 2). Field emission can thus occur even at room temperature, and in practice the cathode is in the form of a point to increase the field intensity to the required value. Typically the effective cathode area is about 10^{-12} m² and the emission current density about 10^6 A/m², so that emission currents are of the order of microamperes. An application is in the *field emission microscope*, which provides such a large magnification that the positions of the atoms on the tip of the tungsten cathode are made visible by allowing the electrons to strike a fluorescent screen.

SECONDARY EMISSION

When the surface of a solid is bombarded by a stream of electrons (or other particles), other electrons may be removed from the surface. These are called *secondary* electrons, the incident beam consisting of *primary* electrons. They transfer their energy to the surface electrons, which can then surmount the energy barrier $W_1 - W_F$. The number of secondary electrons removed per second in this way is directly proportional to the primary current, and the constant of proportionality is the *secondary emission coefficient* δ:

$$\delta = \frac{\text{Secondary electron current}}{\text{Primary electron current}} = \frac{I_S}{I_P} \tag{6.70}$$

δ depends on the nature of the surface and the energy of the primary beam, which must exceed a few electronvolts before secondary emission can occur.

The variation of δ with primary energy is illustrated in Fig. 6.17(*a*) for three materials. It will be seen that δ at first rises with energy; this is due to the removal of electrons from the surface, since these have the greatest probability of escape. As the primary energy is increased, the bombarding electrons travel faster and so spend less time at the surface before penetrating below it. Thus the surface electrons receive less energy and excitation of electrons *below* the surface occurs. For example, for primary energies of a few hundred electronvolts, the primary electrons striking a platinum target can penetrate up to 30 atomic layers below the surface before all their energy is expended. The primary electrons only come to rest when their energy has been reduced to that of the conduction electrons in the target. Many of the electrons excited below the surface lose their energy before reaching it due to collisions with the lattice atoms.

Hence, as the primary energy is increased, δ rises more slowly than at first, reaches a maximum value δ_{max} and then falls. For pure metals δ shows a similar variation to the curve for nickel in Fig. 6.17(*a*) and δ_{max} is normally less than 1·8. A larger value of δ_{max} is obtained with compound surfaces such as barium oxide which is deposited on the surfaces of the electrodes of thermionic valves owing to evaporation from the cathode. Here δ_{max} is about 4·2, so considerable secondary emission occurs (page 292). Finally caesium oxide on silver has a value of δ_{max}

257

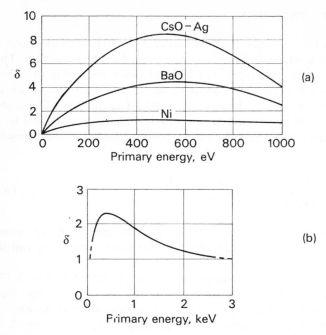

FIG. 6.17 Secondary-emission coefficient and target voltage

(*a*) Conductors and semiconductors   (*b*) Insulators

greater than 8 and is used as a secondary source of electrons in the photo-multiplier (page 260). A similar curve is obtained for insulators. Fig. 6.17(*b*) shows a plot of δ against primary energy for mica. For $\delta < 1$ the specimen will become negatively charged and primary electrons will be repelled, so that δ cannot be determined in this region.

The phenomena associated with secondary emission may be studied by fixing the primary energy and measuring the distribution of the energy of the secondary electrons. A curve similar to that of Fig. 6.18 is obtained for pure metals for a primary energy of about 150 eV. The majority of the electrons have an energy less than 50 eV and these are considered to be the true secondary electrons, with the most probable energy for metals lying between 1·3 and 6 eV. The narrow peak occurring at the primary energy is due to primary electrons which have been reflected *elastically* from the surface, losing very little energy in the process. The energies between 50 and 150 eV are due to electrons which have penetrated the

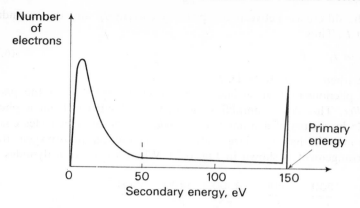

FIG. 6.18 Distribution of energy of secondary electrons

surface and have then been scattered out of it by the lattice vibrations. These are known as *inelastically* reflected primary electrons, since they have given up energy to the lattice without causing electrons to be removed from it.

An experimental arrangement similar to that shown in Fig. 6.19 may

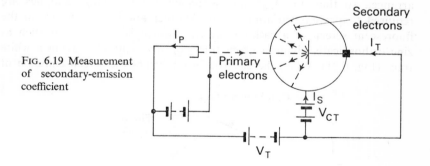

FIG. 6.19 Measurement of secondary-emission coefficient

be used to investigate secondary emission. The target is enclosed to ensure that all the secondary electrons are collected, and the primary electrons are obtained from an electron gun. The energy of the primary electrons is controlled by the target voltage V_T, and the density of the beam is controlled by the final anode potential. The target current I_T is then the difference between the primary current I_P and the secondary current I_S. Thus

$$I_T = I_P - I_S \qquad (6.71)$$

so that when $I_S > I_P$ ($\delta > 1$), I_T reverses.

The phenomenon of secondary emission is exploited in the *photomultiplier*. This device amplifies the very small current from a photocathode by means of a number of *dynodes*, which are electrodes coated with a material having a large value of δ such as caesium-oxygen-silver. The arrangement is shown in Fig. 6.20. Where there are n dynodes the

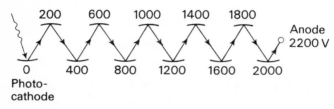

FIG. 6.20 Principle of photomultiplier

final current is δ^n times the original photocurrent, and current gains up to 10^6 can be obtained in this way. The dynode voltages are normally arranged so that $\delta = \delta_{max}$, and extremely stable voltage supplies are required to prevent variations in δ. Another application occurs in the fluorescent screen of a cathode-ray tube which is an insulator such as zinc orthosilicate, giving a green trace, or zinc sulphide, giving a white trace (Fig. 6.21). When the electron beam strikes the screen emission of

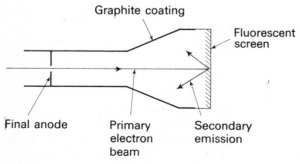

FIG. 6.21 Secondary emission in a cathode-ray tube

both light and secondary electrons occurs. Provided that $\delta > 1$, a condition which depends on the final anode potential relative to the cathode, more electrons are emitted from the screen than arrive at it so that the screen takes up a potential slightly more positive than the final anode. The secondary electrons are attracted to a graphite coating on the inside walls of the glass bulb of the tube, which is held at final anode potential. If this process did not take place the screen would charge up negatively and serious defocusing of the primary beam would occur. Similar considerations apply to the examination of insulants in an electron microscope (page 277). This is similar in principle to an optical microscope but the object is "illuminated" by means of a focused beam of electrons. The primary electron energy must always be large enough to ensure that $\delta > 1$, which avoids charging the specimen negatively and hence defocusing the primary beam.

ELECTRON MOTION IN STEADY ELECTRIC AND MAGNETIC FIELDS

The motion of an electron in a direction parallel to an electric field has already been discussed (pages 236 and 251). Its path is also influenced by a magnetic field so that the equations of motion are derived by applying the laws of electromagnetism in addition to Newton's laws of motion.

Owing to its very small mass it is quite easy to cause an electron to move at a velocity approaching that of light. Under these conditions its mass m is no longer constant but increases according to relativity theory. It is shown in Appendix 4 that

$$m = \frac{m_0}{(1 - u^2/c^2)^{1/2}} \tag{6.72}$$

where m_0 is the rest mass, i.e. the mass for low velocities such that $u^2/c^2 \ll 1$. For electrons accelerated through a potential difference of 5 kV, the increase in mass is just less than 1% and so may be ignored. Thus at voltages below 5 kV the velocity and voltage are related by eqn. (6.12), but above 5 kV the relativistic expression of eqn. (A.4.16) in Appendix 4 must be used. This leads to the curve relating u/c and V given in Fig. 6.22. The mass increase is important in high-power microwave valves and in electron microscopes, the latter using electron guns with anode voltages at and above 100 kV.

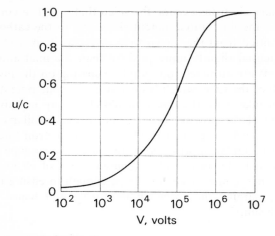

Fig. 6.22 Relativistic variation of speed with energy

Electrostatic Deflection

Consider a narrow beam of electrons which have been accelerated through a potential V_A. Their velocity in the x-direction (Fig. 6.23(a)) is given by

$$u_x = \sqrt{\frac{2eV_A}{m}} \tag{6.73}$$

Suppose the beam passes between two parallel deflecting plates, having separation s, and finally strikes a fluorescent screen as would occur in a cathode-ray oscilloscope. If there is no potential difference between the plates the beam is unaffected, but if a potential V exists between them, which sets up an electric field assumed to be uniform over the length l of the plates, the electrons will be deflected and strike the screen at a distance D from the undeflected position.

The electrons will have acquired a vertical velocity u_v as they leave the plates and will continue to the screen with no further change in direction. If their horizontal velocity after leaving the plates is u_h then their total energy is unchanged after passing through the plates, so that

$$\tfrac{1}{2}m(u_v{}^2 + u_h{}^2) = \tfrac{1}{2}mu_x{}^2 \tag{6.74}$$

Since $u_v{}^2$ and $u_x{}^2$ are proportional to V and V_A respectively then provided that $V \ll V_A$, which is normally the case in a cathode-ray tube, $u_h \approx u_x$.

The deflecting force is given by

$$F_y = \frac{m\,du_y}{dt} = eE = e\frac{V}{s} \tag{6.75}$$

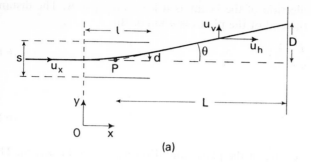

(a)

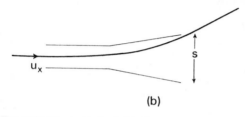

(b)

Fig. 6.23 Electrostatic deflection

(a) Flat deflecting plates (b) Flared deflecting plates

Hence

$$u_y = \frac{e}{m}\frac{V}{s}t \tag{6.76}$$

and, after integration,

$$y = \tfrac{1}{2}\frac{e}{m}\frac{V}{s}t^2 \tag{6.77}$$

But $t = x/u_x$, so that

$$y = \tfrac{1}{2}\frac{e}{m}\frac{V}{s}\frac{x^2}{u_x{}^2} \tag{6.78}$$

Thus $y \propto x^2$ and the path of the electrons between the plates is a parabola, taking $x = 0$ at the left-hand edge of the plates.

The slope of the parabola at $x = l$ is

$$\frac{dy}{dx} = \frac{e}{m}\frac{V}{s}\frac{l}{u_x{}^2} = \tan\theta \tag{6.79}$$

where d is the deflection of the beam as it leaves the plates. The distance of point P from the end of the plates is $d/\tan \theta$. But

$$d = \tfrac{1}{2} \frac{e}{m} \frac{V}{s} \frac{l^2}{u_x{}^2} \tag{6.80}$$

so that

$$\frac{d}{\tan \theta} = \frac{l}{2} \tag{6.81}$$

Hence P is at the centre of the plates and distant L from the screen. Then the final deflection is given by

$$D = L \tan \theta = \frac{eVlL}{msu_x{}^2} \tag{6.82}$$

using eqn. (6.79), so that

$$D = \tfrac{1}{2} \frac{V}{V_A} \frac{lL}{s} \tag{6.83}$$

using eqn. (6.73).

Thus $D \propto V$, and if V is sinusoidal D will correspond to the instantaneous value. The upper frequency is limited by the time an electron spends between the plates. If $l = 2\,$cm and $V_A = 1\,000\,$V, then $t = 1 \cdot 07 \times 10^{-9}\,$s so that in this case V may be taken as an instantaneous value for frequencies up to the megahertz region.

The deflection sensitivity is given by

$$\frac{D}{V} = \tfrac{1}{2} \frac{lL}{V_A s} \tag{6.84}$$

and is often quoted in millimetres per volt. It is proportional to the length of the plates and their distance from the screen, which are both limited by the required size of the tube. It is inversely proportional to V_A, but the electron velocity and brightness of the spot increase with V_A. The sensitivity is also increased by bringing the plates closer together, but the beam must not foul the edge of the plates. Consequently they are often flared to prevent this, the final separation being the same as for the parallel plates (Fig. 6.23(*b*)). The deflecting voltage V is normally applied to the plates through a phase-splitting amplifier, so that one plate is at a potential of $+V/2$ and the other at $-V/2$. This avoids distortion due to the position of zero voltage not being on the tube axis.

Motion in a Uniform Magnetic Field

The force F on a conducting element length ds carrying a current I in a magnetic field of density B making an angle θ with the conductor (Fig. 6.24(a)) is given by

$$F = BI\,ds\sin\theta \qquad\qquad (6.85)$$

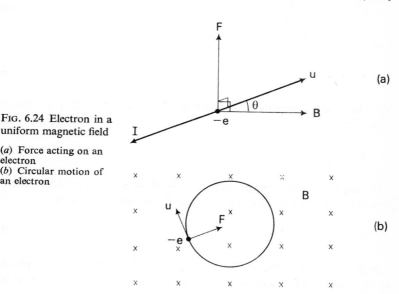

Fig. 6.24 Electron in a uniform magnetic field

(a) Force acting on an electron
(b) Circular motion of an electron

An electron moving with velocity u corresponds to a current moving in the opposite sense to a conventional current and also experiences a force in a magnetic field. From eqn. (6.85),

$$F = B\frac{dq}{dt}\,ds\sin\theta$$

$$= B\,dq\,\frac{ds}{dt}\sin\theta$$

$$= Beu\sin\theta \qquad\qquad (6.86)$$

265

The force acts at right angles to the directions of both B and u, and for electrons the relative directions are given by the right-hand rule. It should be noted that B can only alter the direction of u and not its magnitude, since the resulting force is perpendicular to the motion of the electron. The force on a stationary electron is zero and the maximum force occurs when $\theta = 90°$. Here $F = Beu$, and since it is always perpendicular to the direction of motion the electron moves in a circle (Fig. 6.24(b)). If r is the radius of the circle,

$$F = Beu = \frac{mu^2}{r} \tag{6.87}$$

and

$$r = \frac{mu}{Be} \tag{6.88}$$

The time for a complete orbit is $2\pi r/u$, so that the period of the motion is given by

$$T = \frac{2\pi m}{Be} \tag{6.89}$$

This depends only on the magnetic field B, since the radius increases linearly with the velocity. The angular frequency is given by

$$\omega_c = \frac{Be}{m} \tag{6.90}$$

which is known as the *cyclotron frequency*. There are many applications of the motion of an electron in a uniform magnetic field, some of which are discussed below.

Magnetic Deflection
An application of this principle occurs in the cathode-ray tube, magnetic deflection having been used extensively in television receivers. Coils are arranged outside the tube and current through them produces a magnetic field perpendicular to the electron beam. If a uniform field B is assumed over length l of the coils and zero field elsewhere, the electrons will move in a circular path of radius r while they are in the field and then continue without further deflection to the screen (Fig. 6.25). The length L is measured from the screen to the point of intersection of the tangent to the

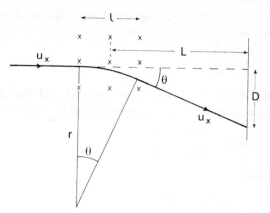

FIG. 6.25 Magnetic deflection

circle with the path of the undeflected beam. Then, for small angles of deflection θ,

$$\tan \theta = \frac{D}{L} = \frac{l}{r} \tag{6.91}$$

From eqn. (6.88), $r = mu_x/Be$ so that

$$D = \frac{Ll}{r} = \frac{LlBe}{mu_x} \tag{6.92}$$

If where V_A is the final anode voltage, $u_x = \sqrt{(2eV_A/m)}$, so that

$$D = \frac{BLl}{\sqrt{(2V_a)}} \sqrt{\frac{e}{m}} \tag{6.93}$$

The value of B depends on the coil design and is proportional to the current, so that the deflection sensitivity is normally quoted in millimetres per milliampere of coil current. The sensitivity depends on $\sqrt{V_A}$ and so a larger accelerating potential may be used than in electrostatic deflection, where the sensitivity depends on V_A. This results in a greater spot brilliance on the screen.

THOMSON'S DETERMINATION OF e/m

J. J. Thomson used combined electric and magnetic fields to determine the value of e/m for cathode rays (electrons) (see Fig. 1.1). The magnitudes of the two fields were adjusted until the net deflection of the beam was zero, so that $Beu_x = eE$, or

$$u_x = \frac{E}{B} \tag{6.94}$$

The electric field was then removed and the deflection D due to the magnetic field alone was measured. The angle of deflection was given by

$$\tan \theta = \frac{Bel}{mu_x} = \frac{D}{L} \tag{6.95}$$

Hence

$$\frac{e}{m} = \frac{Du_x}{BlL} = \frac{DE}{B^2lL} \tag{6.96}$$

using eqn. (6.94), or, since $E = V/s$,

$$\frac{e}{m} = \frac{DV}{B^2lLs} \tag{6.97}$$

THE MASS SPECTROMETER

The mass spectrometer is a device which can be used to measure e/m for positive ions (Fig. 6.26). Ions are accelerated from the source to the

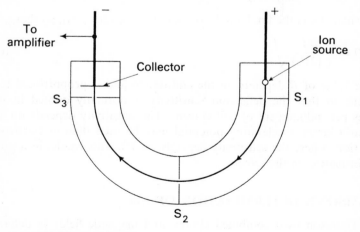

FIG. 6.26 Mass spectrometer

slit S_1, beyond which a magnetic field B exists. This causes the ions to move in circular paths whose radii depends on their value of e/m (eqn. (6.88)) Where the path corresponds with the positions of slits S_2 and S_3 the ions can pass through but others are rejected Those reaching the collector give rise to a current which can be amplified and measured, and the value of e/m is obtained by combining eqns. (6.12) and (6.88) to give

$$\frac{e}{m} = \frac{2V}{B^2 r^2}$$

(6.98)

The accelerating voltage V is adjusted for maximum collector current, and B and r are also known. The mass spectrometer can also be used to detect the presence of an ion having a known value of e/m in a vacuum system.

THE CYLINDRICAL MAGNETRON

A magnetron is a thermionic valve in which electrons move in combined electric and magnetic fields, which are usually perpendicular to each other. Its main application is as a generator of microwave oscillations (page 339), but in its cylindrical form it may be used to determine e/m for electrons.

Consider a magnetron with a filamentary cathode, having a radius much less than the radius of the cylindrical anode (Fig. 6.27(a)). If r_1 is the

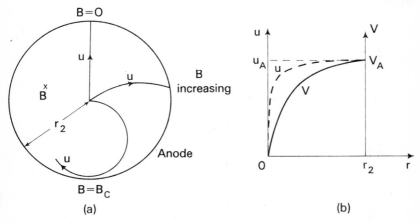

Fig. 6.27 Cylindrical magnetron

(*a*) Electron trajectories   (*b*) Velocity and voltage distributions

269

radius of the cathode and r_2 the radius of the anode it will be seen from Appendix 5, eqn. (A.5.11) that the potential at a distance r from the cathode is

$$V = \frac{V_A}{\log_e(r_2/r_1)} \log_e(r/r_1) \qquad (6.99)$$

V_A being the anode-to-cathode potential. The velocity u of the electrons is given by eqn. (6.12):

$$u = \sqrt{\frac{2eV}{m}}$$

where V is given by eqn. (6.99). The variation of V and u with r is shown in Fig. 6.27(*b*). It will be seen that the electrons almost reach their final velocity u_A very soon after leaving the cathode. Thus it may be assumed that the electrons travel from cathode to anode with approximately uniform velocity u. Their paths will then be circular of radius

$$r = \frac{mu}{Be}$$

and as B is increased r will fall, until when $r = \frac{1}{2}r_2$ the electrons will just graze the anode and the anode current will be zero (Fig. 6.28). For values of B greater than the critical value B_c, the electrons will move in complete circles.

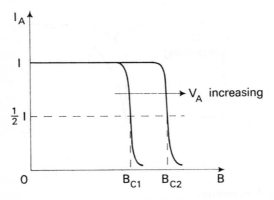

FIG. 6.28 Current/magnetic-field characteristics of a magnetron

At current cut-off,

$$\frac{mu}{B_c e} = \frac{r_2}{2} \tag{6.100}$$

and $u = \sqrt{(2eV_A/m)}$, so that

$$\frac{1}{B_c}\sqrt{\frac{2mV_A}{e}} = \frac{r_2}{2} \tag{6.101}$$

and

$$B_c = \frac{2}{r_2}\sqrt{\frac{2mV_A}{e}} \tag{6.102}$$

A full analysis (Ref. 6.3) shows that eqn. (6.102) is exact when it is multiplied by a factor

$$\frac{1}{1 - \left(\dfrac{r_1}{r_2}\right)^2}$$

so that the expression is true within 1% if $r_1/r^2 < 0\cdot1$. Then, as may be seen from Fig. 6.27(a), the velocity is practically constant near the anode.

In practice the cut-off is not sharp; this may be due to:

(a) Potential drop along the filament, so that the effective anode-to-cathode voltage is not constant for all points on the filament.
(b) Non-uniform electric field.
(c) Initial electron velocities and space charge affect the potential distribution.

(a) may be accounted for by taking B_c as the value of B to halve the anode current (Fig. 6.28). For (b) a uniform electric field can be ensured by mounting guard rings on each side of the anode, thus eliminating edge effects. (c) alters the effective value of V_A, and for determining e/m, V_A may be plotted against $B_c{}^2$, which, from eqn. (6.102) gives a straight line of slope

$$\frac{e}{m}\frac{r_2{}^2}{8}$$

Magnetic Focusing

So far, the magnetic field B has acted at $90°$ to the direction of electron motion. Suppose it acts at angle θ, so that the electron velocity u may be resolved into two components, $u \cos \theta$ parallel to the field and $u \sin \theta$ perpendicular to the field (Fig. 6.29(*a*)). There can be no force on the electron due to the first component, since it has a zero angle with respect to *B*, but the second component will give circular motion of radius *mu* sin θ/Be (Fig. 6.29(*b*)). At the same time the electron is moving parallel to the field with velocity $u \cos \theta$, so that it follows a helical path. Thus the

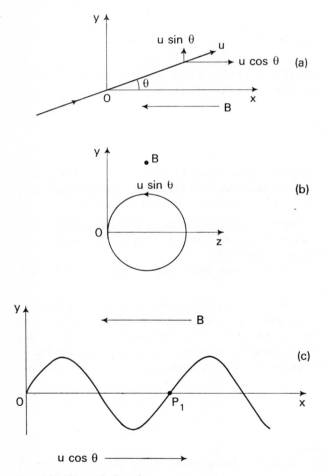

Fig. 6.29 Magnetic focusing

electron will cross the x-axis after a complete revolution at points P_1, P_2, etc. (Fig. 6.29(c)) where

$$OP_1 = Tu \cos \theta \qquad (6.103)$$

or

$$p = \frac{2\pi mu \cos \theta}{Be} \qquad (6.104)$$

T is the period of the circular motion, and p is the pitch of the spiral—the distance between consecutive points such as P_1 and P_2.

When θ is small, $\cos \theta \approx 1$, and for a stream of electrons with the same velocity u but diverging within a small angle, p will be approximately constant. For a circular aperture in the anode a solenoid may be used to focus the beam to a spot on a screen at P, provided the effective source is within the magnetic field of the solenoid (Fig. 6.30).

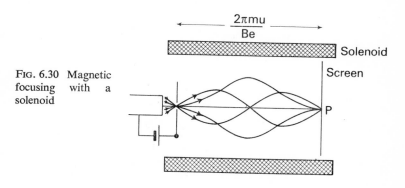

FIG. 6.30 Magnetic focusing with a solenoid

ELECTRON OPTICS

Electron optics is concerned with the production of electron beams which are focused to a narrow diameter by means of magnetic or electric fields. The use of *electron lenses* is analogous to the control of light beams by means of lenses and apertures. The focused electron beam may be used in a cathode-ray tube, in the electron microscope, which has greater resolution than an optical microscope (page 279), or for welding and etching purposes.

Magnetic Lens

Consider a short coil carrying a current as shown in Fig. 6.31(a). The magnetic field due to the current can be resolved into a radial component B_r and an axial component B_x. B_r is directed towards the axis on the left-hand side, becomes zero in the plane of the coil and is directed away from the axis on the right-hand side. An electron moving parallel to the axis with velocity u_x will be unaffected by B_x, but B_r will cause it to move sideways out of the paper (Fig. 6.31(b)). This component of velocity, u_s,

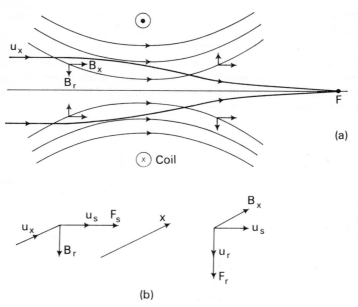

(a)

(b)

FIG. 6.31 Magnetic lens

is normal to B_x, which will give the electron another component u_r towards the axis. As it passes through the centre of the coil B_r reverses and gradually reduces u_s, which becomes zero as the electron leaves the lens. The electron continues to travel in a straight line towards the axis intersecting it at F, the focal point of the lens. A similar argument shows that a diverging beam is first brought parallel to the axis and then converges towards it, intersecting at a point beyond F. The effect of the coil on an electron beam is thus analogous to the effect of a convex lens on a beam of light. Magnetic lenses are used extensively in the electron microscope (see Fig. 6.35).

274

Electrostatic Lens

Consider an electric field with two regions separated by a very small gap s in which the potential changes from a constant value V_1 to a higher value V_2 (Fig. 6.32). The two potential regions are separated by parallel equipotential surfaces between which there is a field E given by

$$E = \frac{V_2 - V_1}{s} \qquad (6.105)$$

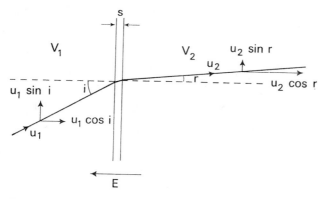

FIG. 6.32 Refraction of an electron beam

Suppose an electron moves in the first region in a direction making an angle i with the normal to the gap and with constant velocity $u_1 = \sqrt{(2eV_1/m)}$. The velocity may be resolved into two components $u_1 \sin i$ and $u_1 \cos i$. At the gap $u_1 \sin i$ will be unaffected, but $u_1 \cos i$ will be increased by the field E so that as the electron enters the second region its velocity will be $u_2 = \sqrt{(2eV_2/m)}$. If the direction of u_2 makes an angle r with the normal, the two components are $u_2 \sin r$, and $u_2 \cos r$, which is greater than $u_1 \cos i$. The components parallel to the gap are unaffected by the field so that

$$u_1 \sin i = u_2 \sin r \qquad (6.106)$$

or

$$\frac{\sin i}{\sin r} = \frac{u_2}{u_1} = \sqrt{\frac{V_2}{V_1}} \qquad (6.107)$$

275

A *refractive index* (n) for electrons may be defined where

$$\frac{\sin i}{\sin r} = \frac{n_2}{n_1} = n_{12} \tag{6.108}$$

or

$$n_{12} = \sqrt{\frac{V_2}{V_1}} \tag{6.109}$$

However, in practice a discontinuous change in V is not possible and V changes more gradually from one region to another. This is analogous to a changing optical refractive index.

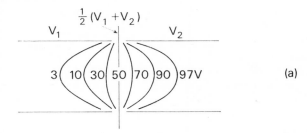

(a)

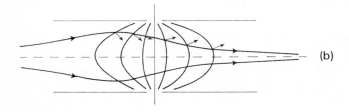

(b)

(c)

Fig. 6.33 Electrostatic lens

The figures at (*a*) refer to equipotentials when $V_1 = 0$, $V_2 = 100\,\text{V}$

A practical electrostatic lens consists of two coaxial cylinders, at different potentials V_1 and V_2, with a gap between them. The electric field is of the type shown in Fig. 6.33(a), where the equipotentials correspond to $V_1 = 0$, $V_2 = 100$ V and may be obtained experimentally using an analogue of the lens (Ref. 6.2). The electron enters the cylinder of lower potential first and the electrostatic force on it acts at right angles to the equipotential line. Thus an electron moving in a path divergent from the axis will experience forces on the low-potential side of the mid-plane which move it towards the axis. Beyond the mid-plane the forces will tend to move it away from the axis, but since they are more nearly parallel to the axis than before the electron continues to move towards it. Again the system acts like a converging lens whose focal length may be reduced by increasing $V_2 - V_1$.

The action of the lens may also be explained by considering the gradual increase in potential through the lens as a corresponding increase in refractive index (Fig. 6.33(c)). The angle through which the electron path is bent at each equipotential is given by eqn. (6.107), where V_1 and V_2 refer to the mean potential between two equipotentials.

An aperture and an anode at a higher potential may also be used as a lens. An example is in the electron gun of a cathode-ray tube, where the aperture is in the *control electrode* held at a negative potential with respect to the cathode (Fig. 6.34(a)). The equipotentials between cathode and anode ensure a focusing action on the beam of electrons leaving the cathode, and the potentials can be adjusted to bring the beam to a focus at the anode. The beam is finally focused at the screen by means of a second anode which with the first one forms a lens (Fig. 6.34(b)).

THE ELECTRON MICROSCOPE

The limit of usefulness of a microscope is set by the distance between two points on a specimen which can be seen as just separated at the eyepiece, which is known as the *resolving power*. Due to the finite wavelength of light the image of a point is spread into a pattern, so that even if the lens system is perfect the patterns of two points very close together will overlap to give a single blurred image. The size of the pattern increases with the wavelength and in practice the resolving power is about one-third of the wavelength. Thus, if yellow light having a wavelength of about 600 nm is used to illuminate a specimen, two points which have a separation less than about 200 nm will appear as a combined image.

277

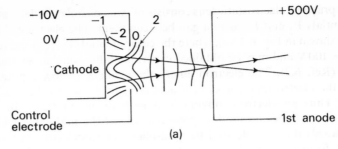

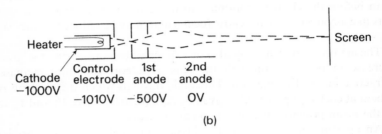

FIG. 6.34 Electron gun

This fundamental limitation can only be overcome by using radiation of a shorter wavelength, provided by a beam of electrons which is associated with a de Broglie wavelength (page 9). The electrons from a gun having an anode voltage of 100 kV have a wavelength of 4 pm, five orders of magnitude less than the wavelength of visible light, so that a great improvement in resolving power is to be expected. The first electron microscope was built in 1932 using magnetic lenses and has been developed considerably since then.

The general layout of a modern electron microscope is shown in Fig. 6.35; it is analogous to an optical microscope. The electron source is a heated tungsten filament within the electron gun and the beam is focused onto the specimen by means of a condenser lens. As in the optical microscope a transparent specimen is used which normally has a thickness between about 10 and 100 nm. The electron beam may be restricted by an aperture of diameter about 25 μm, and after passing through the specimen enters the objective lens to form an image which is further magnified by the projector lens or "eyepiece". Finally the electrons strike a fluorescent screen to give a visible magnified image of the specimen. The screen may be replaced by a photographic plate which is sensitive to electrons as well as to light.

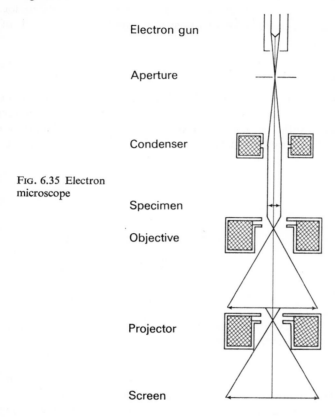

Electron gun

Aperture

Condenser

FIG. 6.35 Electron microscope

Specimen

Objective

Projector

Screen

Each lens consists of a solenoid encased in soft iron and with a soft-iron pole-piece to concentrate the magnetic field. The focal length of the lenses is typically a few millimetres and is adjusted by varying the current through the solenoid. The best resolution is only obtained when both this current and the electron-gun voltage are held constant to within about 0·01%. Then the resolving power is in the range 1–10 nm, which is a great improvement on the optical microscope but far from the theoretical limit due to imperfections in the lenses (Ref. 6.4). Electron microscopes having accelerating voltages up to 1 MV are also available, not only to improve the resolving power but also to allow examination of specimens with thicknesses up to 1 000 nm. These have properties more representative of the bulk material than the very thin specimens. The energy lost by the electron beam in passing through even a thin specimen leads to local heating if it is a poor thermal conductor, so that a replica is often made of a refractory material.

THE EFFECT OF EMISSION VELOCITY

Electrons will be emitted from a practical thermionic cathode with emission velocities ranging from zero to a maximum value, since some of them have more than sufficient energy to surmount the potential barrier ϕ. A typical diode characteristic is shown in Fig. 6.36. When $V_A = 0$ a

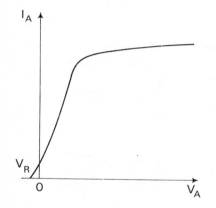

FIG. 6.36 Characteristic of a thermionic diode showing effect of emission velocity

small anode current flows owing to the initial velocities of emission, and a small negative potential, V_R, is required to reduce I_A to zero.

The potential distributions corresponding to various anode voltages are shown in Fig. 6.37(a). When $V_A = -V_R$ the retarding field at the cathode is just sufficient to prevent the fastest electrons reaching the anode. When $V_A = 0$ the faster electrons can reach the anode, but the remainder form a negative space charge between anode and cathode. Hence the potential distribution passes through a minimum where the space charge is densest. As V_A is increased above zero, the anode current is also increased and the space charge density is reduced. Thus the minimum potential is also reduced, since this is a function of space charge density, and it moves back towards the cathode. Finally when the current is temperature limited the minimum has completely disappeared.

The potential distribution under normal space-charge-limited conditions is shown in Fig. 6.37(*b*). The space charge minimum potential is $-V_m$, distant x_m from the cathode, which causes a retarding field between the point x_m and the cathode. Electrons leaving the cathode with velocity $u < \sqrt{(2eV_m/m)}$ will not have sufficient energy to surmount the potential

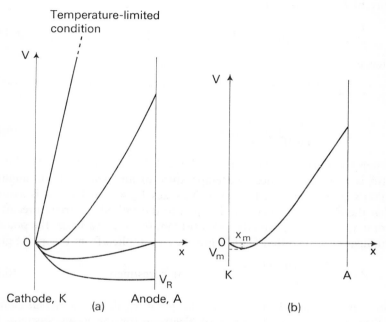

FIG. 6.37 Voltage distributions in a thermionic diode

(*a*) Effect of anode voltage   (*b*) With normal anode voltage

barrier and so will return to the cathode. If $u > \sqrt{(2eV_m/m)}$ they will have energy high enough to overcome the barrier, penetrate the space charge and proceed to the anode. The average velocity of electrons within the space-charge region will be practically zero and electrons will be continually entering it from the cathode and leaving it to maintain the anode current. The magnitude of V_m may be obtained by applying the Boltzmann relationship for the probability of an electron crossing an energy barrier of height W. In this case $W = eV_m$, so that, if N_s is the total

281

number of electrons emitted per second from the cathode and N is the number passing the potential minimum per second,

$$N = N_s \exp\left(-\frac{eV_m}{kT}\right) \tag{6.110}$$

Now, the saturated current $I_s = eN_s$ and the space-charge-limited current $I = eN$ so that

$$\frac{I}{I_s} = \frac{N}{N_S} = \exp\left(-\frac{eV_m}{kT}\right) \tag{6.111}*$$

Hence

$$V_m = -\frac{kT}{e}\log_e\frac{I}{I_s}$$

$$= -\frac{T}{11\,600}\log_e\frac{I}{I_s} \text{ volt} \tag{6.112}$$

When the current is temperature limited $I = I_s$ and $V_m = 0$, while x_m also is zero. In practice the temperature of an oxide cathode might be $1\,000\,\text{K}$ and I/I_s about 10^{-2}, which makes $V_m = -0.4\,\text{V}$ if it is assumed that the electron temperature is equal to the cathode temperature. At the point x_m, which is normally about $0.1\,\text{mm}$ from the cathode, the potential gradient is zero so that Child's law may be applied from this point giving

$$J = \frac{2.33 \times 10^{-6}(V_A - V_m)^{3/2}}{(x_a - x_m)^2} \text{ amperes/metre}^2 \tag{6.113}$$

for a plane parallel diode. Thus the point x_m defines a *virtual cathode*, which is the apparent source of electrons for the anode current. The corrections involved in V_m and x_m are important only for low values of anode voltage, say below $10\,\text{V}$, and both result in an increased value of J, since V_m is negative.

Contact Potential

In general, the work functions of cathode and anode will be different,

---

* A similar reasoning may be used to find the current when the anode voltage is negative. In this case the *retarding-field current* is

$$I = I_0 \exp\left(\frac{eV_A}{kT}\right)$$

where I_0 is the current below which the energy barrier is directly due to the anode voltage and the effect of the space charge is negligible.

giving rise to a contact potential difference V_{CP}. This will demand a second correcting term, which is usually positive and is important at low anode voltages, so that the effective anode voltage is

$$V_A - V_m + V_{CP}$$

THE TRIODE

Since the anode current of a diode is controlled by the space charge, which can be considered as determining the potential minimum V_m, an external control of V_m would provide a convenient means of changing the anode current. This is achieved by interposing a grid of fine wires between cathode and anode, situated just in front of the potential minimum, as illustrated in Fig. 6.38, for a planar structure. Cylindrical structures are

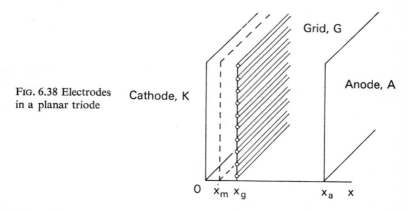

FIG. 6.38 Electrodes in a planar triode

used for triodes as well as diodes. In fact, De Forest found that the current could be controlled by such a grid in 1903, before the presence of the space charge was realized.

In normal operation the grid is held at a negative potential with respect to the cathode, so that its resulting field is added to the retarding field due to the space charge. Hence if the grid is made sufficiently negative the cathode current can be cut off altogether, owing to the combined retarding fields at the cathode. Electrons pass between the grid wires on their way to the anode, but do not strike them except at very small negative grid voltages. Should the grid voltage be made positive, the flow of electrons past the wires will be increased and electrons will also strike the grid, which will act as a second anode.

283

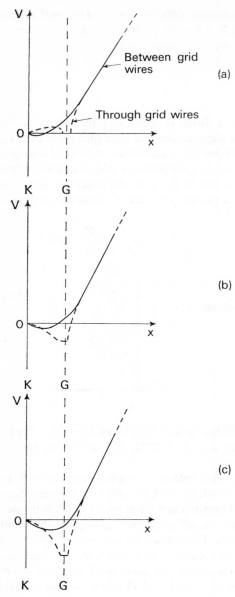

FIG. 6.39 Voltage distributions in a triode

(a) Zero grid voltage
(b) Normal operating grid voltage
(c) Cut-off grid voltage

The potential distributions in a triode are shown in Fig. 6.39 for three different grid voltages. The distributions are due to the combined effects of the anode and grid voltages and the space charge. The slope of a distribution *between* the grid wires, which gives the electric field in that region, will control the flow of anode current. Thus, when $V_G = 0$ (Fig. 6.39(*a*)), V_m is small and close to the cathode, so that the field just outside the cathode is accelerating and a large current will flow. When V_G is a few volts negative, V_m is larger and the potential minimum is closer to the grid; this corresponds to a normal operating condition (Fig. 6.39(*b*)). As V_G is increased more negatively, the field at the cathode becomes more retarding, until when the cut-off voltage is reached no electrons can pass through the grid to the anode (Fig. 6.39(*c*)). The space charge and the grid are most effective in controlling the flow of electrons near the cathode, where their velocity is least, while electrons which have passed through the grid are affected mainly by the accelerating field due to the anode.

The anode current is thus dependent on both the anode and the grid voltages, and their combined effect may be obtained by replacing the triode by an equivalent diode (Fig. 6.40). The anode voltage of this

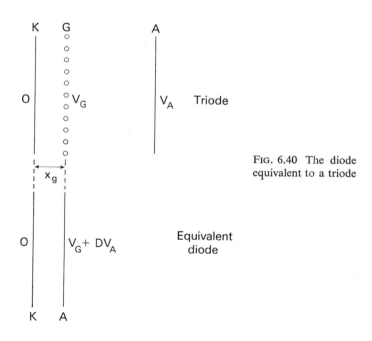

Fig. 6.40 The diode equivalent to a triode

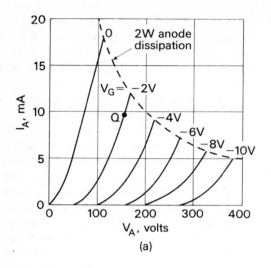

(a)

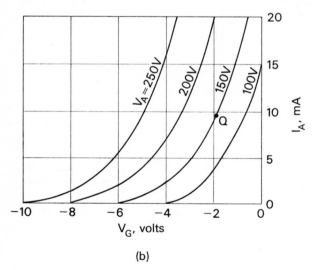

(b)

Fig. 6.41 Electrical characteristics of a triode

(*a*) Output characteristics   (*b*) Transfer characteristics

diode is given by the expression $(V_G + DV_A)$, where D is a constant called the *penetration factor* or *Durchgriff*. The anode of the equivalent diode will be in almost the same position as the grid provided that D is small. D is a factor representing the relative effectiveness of the anode voltage in controlling the anode current and is always less than unity. Thus it is more convenient to use its reciprocal μ, the *amplification factor*, which may have a value up to about 100. The equivalent anode voltage then becomes

$$V_G + \frac{V_A}{\mu} = \frac{1}{\mu}(\mu V_G + V_A) \qquad (6.114)$$

In an ideal triode μ is determined only by the geometry of the valve, i.e. the pitch and diameter of the grid wires and the distance between anode and grid. Thus μ is independent of anode current and voltage, and it may be shown from an electrostatic analysis that

$$\mu = \frac{2\pi(x_a - x_g)}{p \log_e (2 \sin \pi r/p)} \qquad (6.115)$$

where p is the distance between the centres of the grid wires and r is the radius of the grid wires (Ref. 6.3).

Since the anode current of a diode increases as the 3/2 power of the anode voltage, the anode current of a triode is given by

$$I_A = k(V_A + \mu V_G)^{3/2} \qquad (6.116)$$

where k is a constant depending on the grid-to-cathode spacing as well as on μ. The above expression relates anode current and voltage and grid voltage, which define the electrical characteristics of a triode. The value of grid voltage to cut off the anode current, i.e. to make $I_A = 0$, is

$$V_{CO} = -\frac{V_A}{\mu} \qquad (6.117)$$

The characteristics are normally shown in two sets, the output characteristics, I_A versus V_A for various values of V_G, being determined first. From these the transfer characteristics, I_A versus V_G for various values of V_A, can be obtained. These are shown in Fig. 6.41 for a practical valve for which the index in eqn. (6.116) may differ from 3/2. This is due to distortion of the electric fields by the grid supports and variations in electrode spacing along the length of the electrodes.

The small-signal parameters of a triode, which may be obtained from these characteristics, are the *anode slope resistance* r_a, the *mutual conductance* g_m, and the *amplification factor* μ. For the ideal triode the dependence of these parameters on I_A may be obtained from eqn. (6.116):

$$\frac{1}{r_a} = \frac{\partial I_A}{\partial V_A}\bigg|_{V_G \text{const.}} = \tfrac{3}{2}k(V_A + \mu V_G)^{3/2} = \tfrac{3}{2}k^{1/3}I_A^{1/3} \tag{6.118}$$

$1/r_a$ is also the slope of the curves of Fig. 6.41(*a*) at a point such as Q.

$$g_m = \frac{\partial I_A}{\partial V_G}\bigg|_{V_A \text{const.}} = k\mu(V_A + V_G)^{1/2} = \tfrac{3}{2}\mu k^{2/3}I_A^{1/3} \tag{6.119}$$

g_m is the slope of the curves of Fig. 6.41(*b*) at the same point Q.
Combining eqns. (6.118) and (6.119),

$$g_m = \frac{\mu}{r_a} \quad \text{or} \quad \mu = g_m r_a \tag{6.120}$$

which is a very important relationship between the parameters. Finally

$$\mu = -\frac{\partial V_A}{\partial V_G}\bigg|_{I_A \text{const.}} \tag{6.121}$$

μ may be obtained from g_m and r_a through eqn. (6.120). The minus sign indicates that, if V_A is increased positively, V_G must be increased negatively in order to maintain I_A constant. The variation of the parameters with I_A is shown in Fig. 6.42 for a practical valve; it will be seen that μ is independent of I_A except at low currents. Practical values of μ range from 10 to 100 with values of g_m between 5 and 50 mS, while r_a can have values between about 1 and 80 kΩ.

The Operation of a Triode
The triode is used as an amplifier with a load resistance connected between the anode and the positive power supply V_{AA}, in a similar manner to an *n-p-n* transistor (page 141). A load line may then be drawn across the output characteristics given by the equation

$$I_A = -\frac{V_A}{R_L} + \frac{V_{AA}}{R_L} \tag{6.122}$$

i.e. the load line cuts the I_A-axis at the current V_{AA}/R_L and cuts the V_A-axis at the supply voltage V_{AA}. The valve is biased at a suitable negative grid voltage, which fixes a quiescent anode current and voltage, and for large signals the input and output voltages may be found on the characteristic curves.

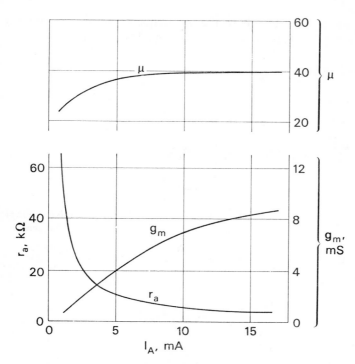

FIG. 6.42 Variation of triode parameters with anode current

For small signals an equivalent circuit is used, again as for the transistor. The equivalent circuit is also useful for large signals, when the characteristics are sufficiently linear for the parameters to be considered constant. From eqn. (6.116) we can write

$$I_A = f(V_A, V_G) \tag{6.123}$$

so that, for small changes,

$$\Delta I_A = \frac{\partial I_A}{\partial V_A} \Delta V_A + \frac{\partial I_A}{\partial V_G} \Delta V_G \tag{6.124}$$

289

Physical Electronics

or with more convenient symbols and using eqns. (6.118) and (6.119),

$$i_a = \frac{v_a}{r_a} + g_m v_g \tag{6.125}$$

This equation leads directly to the low-frequency equivalent circuit of a triode shown in Fig. 6.43(*a*). The anode signal is given by

$$v_a = -g_m v_g \frac{r_a R_L}{r_a + R_L} \tag{6.126}$$

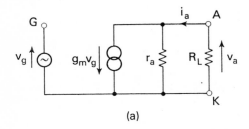

(a)

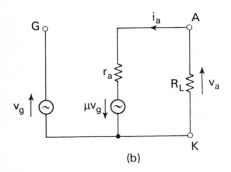

(b)

Fᴵɢ. 6.43 Triode equivalent circuits (low frequency)

(*a*) Current generator
(*b*) Voltage generator

so that the voltage gain of the amplifier is

$$A = \frac{v_a}{v_g} = -g_m \frac{r_a R_L}{r_a + R_L} \tag{6.127}$$

the minus sign indicating phase reversal between the anode and grid signals. Eqn. (6.125) may be rearranged by means of eqn. (6.120) to give

$$i_a = \frac{v_a}{r_a} + \frac{\mu v_g}{r_a} \tag{6.128}$$

which leads to the second form of the low-frequency equivalent circuit shown in Fig. 6.43(*b*). Here

$$v_a = -\mu v_g + i_a r_a$$

$$= -\mu v_g - v_a \frac{r_a}{R_L} \qquad (6.129)$$

so that

$$A = \frac{v_a}{v_g} = -\mu \frac{R_L}{r_a + R_L} \qquad (6.130)$$

which can also be obtained directly from eqn. (6.127). Both forms of the equivalent circuit are used, the circuit of Fig. 6.43(*a*) being preferred when r_a is large as in the pentode, which is considered below.

Inter-electrode Capacitances

The three electrodes of a triode are conductors held at fixed distances from each other by insulating spacers, normally of mica. Hence any two electrodes can be represented by a capacitor with a vacuum dielectric, which gives rise to three interelectrode capacitances, C_{gk}, C_{ga} and C_{ak} (Fig. 6.44(*a*)). Their magnitude is normally a few picofarads, and they set an upper limit to the frequency which the valve can amplify. In particular, since the anode and grid are connected by C_{ga}, energy is fed back from the output to the input circuit. This leads to an apparent increase in the input capacitance between grid and cathode, C_{in}, which is known as the *Miller effect*, and the extra capacitance increases with the gain of the amplifier. An analysis similar to that given for the transistor, involving $C_{b'c}$ (page 172), shows that

$$C_{in} = C_{gk} + C_{ga}(1 + A) \qquad (6.131)$$

THE TETRODE AND PENTODE

If an earthed electrostatic screen is introduced between the grid (here called the *control grid* and labelled g_1) and the anode, C_{ga} is greatly reduced and the input capacitance is reduced almost to C_{gk}. This is achieved by inserting a *screen grid* (g_2) which is held at a positive potential to accelerate electrons through to the anode, but effectively earthed to alternating current through a large-value capacitor. Such a valve, having four electrodes, is known as a *tetrode* (Fig. 6.44(*b*)), and C_{ga} is reduced to about

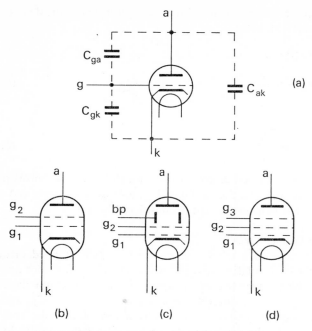

FIG. 6.44 Graphical symbols for thermionic valves
(a) Triode, showing interelectrode capacitances
(b) Simple tetrode
(c) Beam tetrode: the beam-forming plates are connected to each other internally
(d) Pentode

0·01 pF. Owing to the presence of the screen grid, the anode has far less control over the anode current than in the triode. Thus μ is much higher, and since g_m is much the same as in a triode, r_a is about 100 kΩ.

However, owing to the emission of secondary electrons from the anode to the screen grid the anode current falls as the anode voltage rises, so that part of the I_A/V_A characteristic has a negative resistance (Fig. 6.45). This restricts the use of the tetrode as an amplifier to the almost horizontal region of the characteristic. However, the "kink" in the characteristic can be removed by returning the secondary electrons to the anode by means of a fifth electrode, at cathode potential, between the screen and the anode. This can take the form of beam-forming plates or a loosely wound grid known as the *suppressor grid*. The valve is then known as either a *beam*

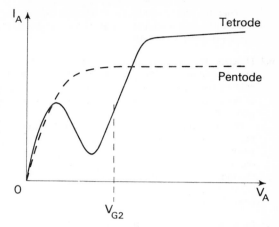

FIG. 6.45 Comparison of the I_A/V_A character-istics of a simple tetrode and a pentode with control-grid and screen-grid voltages constant

tetrode (Fig. 6.44(*c*)) or a *pentode* (Fig. 6.44(*d*)) respectively, and its I_A/V_A characteristic has no negative resistance region. The amplification factor, μ, is even higher than for the simple tetrode owing to the extra shielding of the suppressor, and C_{ga} is further reduced to about 0·001 pF. For a pentode, g_m again depends mainly on the control grid and so has values similar to those for triodes and simple tetrodes. However, r_a is higher than for the simple tetrode owing to the increased value of μ and can go up to a few megohms, so that the pentode is nearly an ideal current generator and is represented by the equivalent circuit of Fig. 6.43(*a*).

REFERENCES

6.1 BECK, A. H. W., and AHMED, H., *An Introduction to Physical Electronics* (Arnold, 1968).

6.2 VITKOVICH, (ed.) *Field Analysis* (Van Nostrand, 1966).

6.3 PARKER, P., *Electronics* (Arnold, 1950).

6.4 WISCHNITZER, S., *Introduction to Electron Microscopy* (Pergamon, 1962).

FURTHER READING

GEWARTOWSKI, J. W., and WATSON, H. A., *Principles of Electron Tubes* (Van Nostrand, 1965).

PROBLEMS

6.1 A photocathode is illuminated with radiation of wavelength 500 nm. The cathode has a work function of 1·2 eV. Calculate the anode voltage required to produce zero anode current.

293

When the anode voltage is $+90$ V find the velocity of the electrons at the anode if the cathode is illuminated with radiation of wavelength 250 nm. (*Ans.* $-1\cdot29$ V, $5\cdot74 \times 10^6$ m/s)

6.2 Describe the principle of operation of (*a*) the gas-filled photocell, bnd (*b*) the secondary-emission photomultiplier. What is the limiting factor in the electron gain of the gas-filled photocell, and what governs the maximum time resolution of both types of cell? What steps can be taken to improve the performance of these photocells?

When a beam of light of 5 000 Å wavelength was shone on to the cathode of a photoemissive cell a photocurrent of 1 mA was measured. The same light beam, when shone into a bolometer, was found to have a power content of 30 mW. Find the quantum efficiency of the photocathode to light of this wavelength. Calculate the long-wave threshold for photoemission if the emitter has a work function of $1\cdot2$ eV. (*IEE*, June 1967)
(*Ans.* $8\cdot3\%$, $1\cdot04\,\mu$m)

6.3 In what respects does the classical theory of electromagnetic radiation fail to account for the photoelectric effect? Show how the quantum theory explains the phenomenon. In an experiment to determine the work function of a surface by means of the photoelectric effect, cut-off voltages of $0\cdot26$, $0\cdot58$ and $1\cdot18$ V were obtained for light with the following wavelengths: 6 615, 5 648 and 4 434 Å. Estimate Planck's constant and the work function of the surface, neglecting contact potential differences. (*L.U.*, *B.Sc.* (*Eng.*), 1966)
(*Ans.* Work function, $1\cdot63$ eV)

6.4 Write a short essay on the subject of thermionic emission from conductors and semiconductors. Describe three of the common forms of thermionic cathode in use and list their advantages and disadvantages.

A thermionic diode gave the following characteristic when tested under temperature-limited conditions:

J_c (A/m^2)	4×10^4	1 385	32	8×10^{-1}
T_c (K)	2×10^3	$1\cdot67 \times 10^3$	$1\cdot43 \times 10^3$	$1\cdot25 \times 10^3$

Calculate for the cathode used the magnitudes of the constant A_0 and the work function E_w, in the Richardson's equation,
$$J_c = A_0 T_c{}^2 \exp\left(E_w/kT_c\right) \qquad\qquad (IEE, May\ 1968)$$
(*Ans.* $2\cdot5 \times 10^6$ A/m^2-K^2, $3\cdot2$ eV)

6.5 Compare tungsten, thoriated-tungsten and oxide-coated cathodes as thermionic emitters.

If the temperature of a tungsten filament changes from 2 400 to 2 420°C, by what percentage does the emission increase? The work function of tungsten is $4\cdot5$ eV. (*G. Inst. P.*, *Part II*, 1967)
(*Ans.* 17%)

6.6 Sketch a curve showing the potential energy of an electron due to the image force as a function of distance from a plane conducting surface. Show that the effective work function at the surface of a material is decreased by an amount

$$\Delta\phi = (eE/4\pi\varepsilon_0)^{1/2}$$

when a uniform field E is applied at the surface.

Calculate the percentage increase in thermionic emission obtained when a field of 10^5 V/m is applied to a cathode operating at $1\,000\,°K$. (You may assume Richardson's equation,

$$J = AT^2 \exp(-e\phi/kT)$$

where ϕ is the work function. (*L.U., B.Sc. (Eng.)*, 1968)
(*Ans.* 15%)

6.7 Describe the mechanism of (*a*) photoelectric emission, (*b*) secondary emission of electrons from a surface.

A photomultiplier has a cathode with a work function of $1\cdot5$ eV and ten dynodes each with a secondary emission coefficient of 6. If radiation is incident on the cathode, calculate (*a*) the maximum wavelength for which collector current will flow, (*b*) the maximum initial electron velocity if the wavelength of the radiation is $0\cdot6\,\mu$m, (*c*) the final collector current if the cathode current is 10^{-10} A. (*L.U., B.Sc. (Eng.)*, 1966)
(*Ans.* (*a*) $0\cdot83\,\mu$m, (*b*) $4\cdot5 \times 10^5$ m/s, (*c*) $6\cdot1$ mA)

6.8 Explain the nature of space-charge-limited current in a thermionic diode and derive the law relating the current density to the applied voltage for a planar diode.

A planar diode has electrode spacing of 1 mm and it operates in space-charge-limited conditions at an anode voltage of 5 V. Determine the electron density at a point midway between the electrodes. (*G. Inst. P., Part II*, 1964)
(*Ans.* $1\cdot95 \times 10^{14}/$m^3)

6.9 Sketch and explain the distribution of potential between cathode and anode of a thermionic diode valve operating under space-charge-limited conditions (*a*) omitting and (*b*) including, the effects of the finite emission velocities of the electrons from the cathode. For case (*b*) indicate the normal position for a control grid and hence explain the action of a triode valve.

The anode current I_a of a triode is related to the grid voltage V_g and the anode voltage V_a by the equation

$$I_a = k \left(V_g + \frac{V_a}{\mu} \right)^{3/2}$$

where k is a constant. If the amplification factor $\mu = 60$ and $I_a = 5$ mA for $V_g = -2$ V, $V_a = 200$ V, calculate the mutual conductance and anode slope resistance under these conditions. (*L.U., B.Sc. (Eng.)*, 1966)
(*Ans.* $4\cdot5$ mS, $13\cdot33$ kΩ)

6.10 Two large parallel metal plates are horizontal with separation $5\cdot0$ mm and the upper plate is held at $+150$ V with respect to the lower plate. An electron with initial velocity 10^6 m/s upwards is released at the centre of the lower plate. Calculate (*a*) the velocity of the electron when it strikes the upper plate, (*b*) the time of flight of the electron, (*c*) the amount of energy conveyed to the upper electrode. (*Ans*. (*a*) 7·32 × 10⁶ m/s, (*b*) 1·2 ns, (*c*) 153 eV)

6.11 Prove from first principles that the velocity of an electron, initially at rest and then accelerated through a potential difference of V, is given by $\sqrt{(2eV/m)}$ where e and m are the electron charge and mass, respectively.

An electron is emitted from the *anode* of a parallel-plate diode with an initial velocity of 2×10^6 m/s at an angle of 60° relative to the normal to the electrode plane. The distance between the electrodes is 1 cm. Ignoring fringing effects, calculate (*a*) the minimum potential difference required between the electrodes to keep the electron from striking the cathode, (*b*) the kinetic energy of the electron just before it strikes the anode on return, with the value of applied voltage calculated in (*a*), (*c*) the distance on the anode surface between the points of departure and return of the electron, (*d*) the time of flight. (*IEE, May* 1968) (*Ans*. (*a*) 2·84 V, (*b*) 2·84 eV, (*c*) 6·9 cm, (*d*) 40 ns)

6.12 Derive an approximate expression for the deflection sensitivity of an electrostatic deflection system such as is used in a cathode-ray tube; state clearly the approximations made.

In a cathode-ray tube the deflecting plates are 2 cm long, 0·5 cm apart and 20 cm distant from the screen. Calculate the value of the final anode voltage if a deflecting p.d. of 50 V is to produce a spot deflection of 2 cm.

With the aid of a sketch describe one way in which a double-beam facility can be provided in a cathode-ray tube. (*G. Inst. P., Part II*, 1966) (*Ans*. 1 kV)

6.13 A solenoid of 1 000 turns is 30 cm long and it is assumed to produce a magnetic field which is uniform over the entire length of the solenoid. An electron travelling in a vacuum is accelerated through a potential difference of 500 V and then enters one end of the solenoid, crossing the axis at an angle of 5°. Calculate (*a*) the minimum and (*b*) the next higher value of solenoid current that will make the electron cross the axis again at the other end of the solenoid. (*L.U. B.Sc. (Eng.)*, 1967) (*Ans*. (*a*) 378 mA, (*b*) 756 mA)

7

Electrons in Gases

ELECTRICAL CONDUCTION IN GASES

The effect of a gas on the passage of electrons between two electrodes depends on its pressure, which determines the mean free path of the electrons within the gas. In a high-vacuum valve, such as a diode or a triode, the mean free path is much larger than the electrode spacing, so that very few collisions occur, and the effect of the residual gas is usually negligible. In a gas-filled valve, however, the mean free path is much closer to the dimensions and many more collisions occur and give rise to electrical characteristics different from those of a high vacuum device.

The expression relating mean free path and pressure for an ideal gas may be obtained by combining eqns. (2.5) and (2.13):

$$l = \frac{kT}{\sqrt{2}\pi d^2 P} \tag{7.1}$$

When considering the motion of an electron through a gas it may be assumed that its effective diameter is much less than that of the gas molecules, d, so that if it is to strike a molecule it must approach within a distance $d/2$ of its centre. Also the speed of the electron due to the applied electric field is much greater than the average thermal speed of the molecules, so that their motion may be neglected in finding the mean free path of the electron. Thus eqn. (7.1) must be modified by substituting $d/2$ for d and removing the $\sqrt{2}$, which leads to

$$l_e = \frac{4kT}{\pi d^2 P} \tag{7.2}$$

for the mean free path of an electron.

THE DISTRIBUTION OF ELECTRON FREE PATHS

It is important to be able to find the fraction of the total number of electrons which experience a collision within a given distance, and this may be expressed in terms of the mean free path. Since l_e is the mean distance between the collisions of a single electron there are on average $1/l_e$ collisions per metre, so that in a small distance dx there will be dx/l_e collisions. Thus dx/l_e also represents the probability of a collision occurring in distance dx. The position of a particular collision may be taken as a reference point within the gas, from which a number of electrons y_0 moves away. If y is the number of electrons which have *not* collided after moving a distance x, then in a further distance dx a fractional decrease dy/y will occur due to collisions. This fraction also represents the probability of a collision occurring in distance dx, so that

$$-\frac{dy}{y} = \frac{dx}{l_e} \tag{7.3}$$

or

$$y = C \exp\left(-\frac{x}{l_e}\right) \tag{7.4}$$

where C is a constant. When $x = 0$, $y = y_0$, so that $C = y_0$, and

$$y = y_0 \exp\left(-\frac{x}{l_e}\right) \tag{7.5}$$

or

$$\frac{y}{y_0} = \exp\left(-\frac{x}{l_e}\right) \tag{7.6}$$

y/y_0 is the fraction of electrons which have *not* collided within distance x, and when $x = l_e$, $y/y_0 = 0.37$. Hence, after moving through a distance equal to the mean free path, 63% of the electrons *have* experienced a collision. The relationship between l_e and y/y_0 is illustrated in Fig. 7.1, which shows that, if the electron can travel a distance $4l_e$, less than 2% of them will have failed to collide.

The Ramsauer Effect

An electron having a kinetic energy of even a few electronvolts will not be in thermal equilibrium with the gas molecules, whose thermal energy is

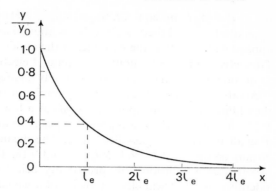

FIG. 7.1 Distribution of electron free paths in gas

only about $0.04\,\text{eV}$ at room temperature (page 35). Thus eqn. (7.2) cannot be applied directly to such electrons; however, its form is retained, $d^2/4$ being replaced by σ, the *collision cross-section* of the gas molecules for an electron. This was found by Ramsauer in 1921 to be a function of the velocity of an incident electron, which would not be so if electrons and atoms were hard spheres, but at high velocities the observed values of σ approach those calculated on this basis. As the velocity is reduced σ rises to a maximum and then falls to a value below that predicted by the kinetic theory, this behaviour being due to the open structure of atoms (Ref. 7). The values of d corresponding to σ for $50\,\text{eV}$ electrons, whose velocity is above that for maximum σ, are given in Table 7.1 for some common gas molecules (Refs. 7.2 and 7.3).

Table 7.1

	Effective collision diameter for $50\,\text{eV}$ electrons	Ionization potential
	nm	V
Helium	0·14	24·6
Neon	0·20	21·6
Xenon	0·37	12·1
Nitrogen	0·36	14·5
Mercury	0·47	10·4

Mean Free Path in High-vacuum and Gas-filled Valves

A typical pressure for a high vacuum valve is $0.13\,\text{mN/m}^2$ ($10^{-6}\,\text{mmHg}$). Assuming that this is mainly nitrogen and taking $T = 293\,\text{K}$, substituting for d, P and T in eqn. (7.2) gives $\bar{l}_e = 300\,\text{m}$. The anode-to-cathode spacing of a small thermionic valve is about $5\,\text{mm}$, so that if eqn. (7.6) is applied with $x = 0$ at the cathode, at the anode $x/\bar{l}_e$ is about $1\cdot7 \times 10^{-5}$. Thus y/y_0 is practically unity with about 1 electron in 60 000 experiencing a collision, so that their motion is virtually unaffected by the gas molecules.

A valve is filled with gas by first pumping it out to a high vacuum and then filling it with the appropriate gas at a low pressure, typically about $1\cdot3\,\text{N/m}^2$ ($0\cdot01\,\text{mmHg}$). Suppose this gas is xenon, which is often used in a small hot-cathode gas-filled valve. The mean free path then becomes $28\,\text{mm}$, so that at the anode $x/\bar{l}_e$ is about $0\cdot18$. Thus about 1 electron in 6 experiences a collision, this probability being sufficient to modify considerably the electrical characteristics of a gas-filled diode compared with those of a high vacuum diode.

These characteristics are illustrated in Fig. 7.2. Initially the current rises

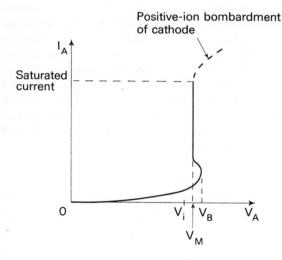

FIG. 7.2 Characteristic of a hot-cathode discharge

slowly with voltage until at a voltage V_B, just above the ionization potential V_i, the current rises very rapidly for a small further increase in voltage. The current is limited to a safe value by means of a resistor in series with the valve. At the same time the gas is seen to glow, the size and intensity of the glow region increasing with the current. These phenomena can be explained by considering in more detail the processes occurring when an electron collides with a molecule.

ELASTIC AND INELASTIC COLLISIONS

When an electron strikes a gas molecule the nature of the collision depends on the kinetic energy of the electron, which is given by eqn. (6.11):

$$\tfrac{1}{2}mu^2 = eV$$

where V is the potential of the point at which the collision occurs. If both particles are considered as hard spheres the collision is elastic, so that this energy cannot be absorbed by the molecule and the electron rebounds without loss of energy but with a change in direction. Thus there is a negligible change in the energy of the molecule, since its mass is much greater than that of the electron, For instance, the mass of a hydrogen atom is 1837 times that of an electron, and an atom of mercury is 201 times heavier than a hydrogen atom.

In a gas the molecules are sufficiently far apart for negligible interaction to occur between them, so that their energy levels are discrete with each atom of the molecule contributing to the total number of levels (Fig. 2.3). Thus energy can only be absorbed by the molecule in discrete amounts corresponding to the difference between two levels. In atoms with many electrons it is normally only an electron from the outermost orbit whose energy is changed by the collision, which is then *inelastic*. Only at very high voltages will the colliding electron have sufficient energy to penetrate to the inner orbits. Hence the zero level in the set of levels illustrated in Fig. 7.3 corresponds to the ground state of the outermost orbit. The highest excited level, W_i, the *ionization energy*, corresponds to removal of an electron from the atom (Fig. 7.3(a)). This will occur when $\tfrac{1}{2}mu^2 \geqslant W_i = eV_i$, where V_i is the *ionization potential* of the gas, some typical

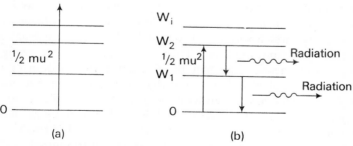

(a)   (b)

Fig. 7.3 Collision processes in a gas
(*a*) Ionization: $\tfrac{1}{2}mu^2 \geqslant W_i$
(*b*) Excitation: $\tfrac{1}{2}mu^2 < W_i$

301

values being given in Table 7.1. When ionization occurs the molecule is split into a positive ion, whose mass is almost equal to that of the original molecule, and the extra electron. It is assumed that all the energy of the colliding electron is given to the molecule when $\frac{1}{2}mu^2 = W_i$.

In the case where $\frac{1}{2}mu^2 < W_i$ ionization cannot occur, except under the special conditions described below. However, an electron may be raised to a higher energy level (Fig. 7.3(b)), a process known as *excitation*. The electron stays in the higher level for about 10^{-8} s, after which it returns to the ground state. It may return in one step or in several steps, staying at intermediate levels for a short time (page 8), a photon of radiation being emitted during each step. The frequency is obtained from eqn. (1.11):

$$hf = W_2 - W_1$$

where W_1 and W_2 are the two energy levels concerned. The radiation may thus occur in the visible part of the spectrum, causing the glow described above.

When $\frac{1}{2}mu^2 < W_i$ elastic collisions normally predominate, with a finite but often small probability that the collision will be inelastic.

METASTABLE STATES

Certain atoms, such as mercury, helium and neon have energy levels which can be occupied for a much greater time than usual, which can be as long as 10^{-2} s. This is because the probability of the spontaneous transition of an electron to any lower level is very low, and little radiation corresponding to these transitions is observable. These energy levels are known as *metastable states*, and an electron in a metastable state can acquire energy which will take it to a higher level of the normal type, or even remove it from the atom. Thus ionization can occur owing to the absorption of energies less than W_i, due either to a photon of suitable energy or to applied voltages less than the ionization potential. For example, mercury has a metastable level at 5·5 eV, while its ionization energy is 10·4 eV. If a 6 V electron collides with a mercury atom this can raise one of its electrons to the metastable state, and if another similar collision occurs within a millisecond the atom will be ionized.

It should be noted that the lifetime of an excited state, including metastable states, in a gas may be determined not by the spontaneous radiation time but by collisions with other particles and the walls of the discharge tube.

THE PLASMA

The current flowing at anode voltages below V_i is due to electrons emitted from the hot cathode and so is space-charge limited, except at very low currents (page 250). Since the collisions between these electrons and the gas molecules are mainly elastic, very little energy is lost and the magnitude of the current is similar to its corresponding magnitude in a high-vacuum valve. When the ionization potential V_i is reached, many ionizing collisions occur and positive ions are created, which distort the potential distribution between cathode and anode.

At the voltage V_B (Fig. 7.2) breakdown occurs and then the gas rapidly reaches a steady state in which the voltage has fallen to a value V_M, just above V_i. Further increase of the current can then occur almost independently of the voltage, and it is found that the potential distribution has become as shown in Fig. 7.4. The potential falls sharply in front of the anode but

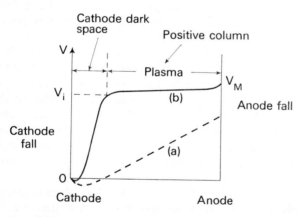

Fig. 7.4 Potential distributions in a gas discharge

    (*a*) Before ionization, V approximately proportional to $x^{4/3}$
    (*b*) After formation of plasma

then is nearly constant up to about 1 mm from the cathode. It then falls rapidly again to a minimum, owing to the electron space charge, and then rises slightly to the cathode. The sharp drops in potential are referred to as the *anode fall* and the *cathode fall* respectively. The region of almost constant potential is called the *plasma* and it consists of almost equal numbers of electrons and positive ions. The field strength within the plasma need only be of the order of 100 V/m in order to maintain it. At the cathode end the potential is approximately V_i, so that positive ions are created between here and the anode. They then move towards the cathode, and on reaching the cathode-fall region they combine with electrons in the space-charge cloud, which is thus partially neutralized. This means that the cathode current, which is limited only by the space charge, is greater than would be the case in the absence of positive ions.

A further effect is that the edge of the plasma acts as a *virtual anode*, much closer to the cathode than the actual anode. Hence a large current can be drawn to it by a much smaller potential difference than for the same valve under high-vacuum conditions. The distance between cathode and virtual anode decreases as the current is allowed to increase, increasing the electric field near the cathode. Thus, since the voltage on the virtual anode remains constant at about V_i, the current is almost independent of the applied anode voltage. This will be true until the discharge current approaches the saturated current available from the cathode. A further increase of current is then only possible by means of secondary emission due to positive ion bombardment of the cathode. The voltage across the diode has to rise in order to achieve this (Fig. 7.2), but operation in this region will much reduce the life of an oxide-coated or thoriated cathode, owing to destruction of the coating by the bombardment. Hot-cathode diodes containing mercury at low pressure can carry currents up to about 100 A with an anode voltage of only about 10 V; they are, however, being replaced by high-current silicon diodes with an effective voltage drop of less than 1 V. The valve contains mercury, which is liquid at room temperature, but it vaporises when the cathode is heated and maintains a vapour pressure of about $1 \cdot 5 \, \text{N/m}^2$ at 50°C.

Although at any instant there will be almost equal numbers of electrons and ions in the plasma, the current through it is carried almost entirely by the electrons. This is because the mass of an electron is very small compared with that of an ion. If V_p is the potential at a point in the plasma, we have for the velocity of electrons

$$u_e = \sqrt{\frac{2eV_p}{m_e}} \tag{7.7}$$

and for positive ions

$$u_i = \sqrt{\frac{2eV_p}{m_i}} \qquad (7.8)$$

where the subscripts e and i refer to electrons and ions respectively. Hence

$$\frac{u_e}{u_i} = \sqrt{\frac{m_i}{m_e}} \qquad (7.9)$$

Also, from eqn. (6.62), the corresponding current densities are

$$J_e = \rho_e u_e \qquad (7.10)$$

$$J_i = \rho_i u_i \qquad (7.11)$$

where ρ_e and ρ_i are the respective charge densities. Thus, assuming $\rho_e = \rho_i$ in the plasma,

$$\frac{J_e}{J_i} = \frac{u_e}{u_i} = \sqrt{\frac{m_i}{m_e}} \qquad (7.12)$$

and since $m_e \ll m_i$, $J_e \gg J_i$. For instance, for mercury $m_i/m_e = 1\,837 \times 201 = 3\cdot69 \times 10^5$, so that $J_e = 606 J_i$.

Both ions and electrons will diffuse out of the plasma to the walls of the discharge tube, where their velocities become zero. More electrons reach the walls initially owing to their much greater speed, so that the walls become negatively charged and an *electron sheath* is formed (Fig. 7.5). This reflects further electrons back into the plasma, which is thus kept out of contact with the walls. Under equilibrium conditions the electron and ion currents to the walls are equal and recombination occurs there rather than within the plasma.

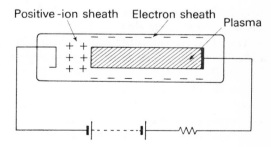

Positive-ion sheath   Electron sheath   Plasma

FIG. 7.5 Containment of a plasma in an electron sheath

Energy within the Plasma

The positive ions which drift out of the plasma are continuously replaced by ionization. The plasma is thus maintained by the transfer of energy from the applied electric field by means of the electrons. After each inelastic collision an electron gains energy from the field and makes elastic collisions with the gas molecules until the next inelastic collision. However, only very little energy is transferred to the molecules at each elastic collision, so that at low pressures the energy of the electrons is much greater than that of the molecules. This energy may be expressed in terms of the *equivalent electron temperature*, T_e, from eqn. (2.8), which gives

$$\tfrac{1}{2}m\bar{u}^2 = \tfrac{3}{2}kT_e \qquad \text{or} \qquad T_e = \frac{m\bar{u}^2}{3k} \tag{7.13}$$

T_e can be determined by means of a probe within the plasma, as described below, and may reach 20 000 K or higher. The electrons are in thermal equilibrium with each other, not with the gas molecules, whose temperature is normally only a few degrees above that of their surroundings. Since the positive ions have a mass practically equal to that of the molecules they can exchange energy efficiently, so that the ion temperature, T_i, is usually close to that of the molecules. However, T_i can be up to 1 000 K above the temperature of the molecules since they may acquire energy due to their motion in the electric field. The high temperature of the electrons cannot affect the surroundings directly, since their mechanism of energy transfer due to collisions is so inefficient owing to their small mass.

The Langmuir Probe

Both the potential distribution in the plasma and the energy and density of the electrons and ions may be obtained by inserting a conducting probe into the plasma. This was one of the earliest methods used to study the plasma, and was introduced by Langmuir in 1923. The probe itself may be in the form of a disc or a wire and its potential is variable with respect to one electrode (Fig. 7.6).

The characteristic obtained is shown in Fig. 7.7, which shows four different regions. The most negative region, AB, has a current almost independent of voltage. This corresponds to randomly moving positive ions reaching the probe, and a space-charge of positive ions forms in front of the probe, which appears as a dark sheath over its surface. The corresponding ion current to the surface of the probe is I_{is} and the potential drop across the sheath increases with the probe potential, so that the

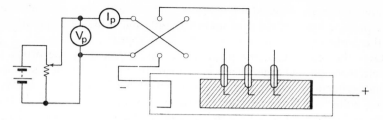

FIG. 7.6 Measurement of plasma potential

The probe is effective only where it is exposed as shown in the centre of the plasma

effect of the probe voltage is just neutralized. The plasma is then unaffected, since ions are neither attracted nor repelled by the probe after the sheath has formed. As the probe voltage is made less negative, some of the higher-energy electrons can penetrate the sheath and reach the probe. The less negative the voltage relative to the plasma, the more electrons will have sufficient energy to overcome the retarding field. Thus over the region BC the current increases steadily, I_{es} being the electron current to the surface of the probe, and where the total current is zero the ion and electron currents are exactly equal. At the point C the probe is

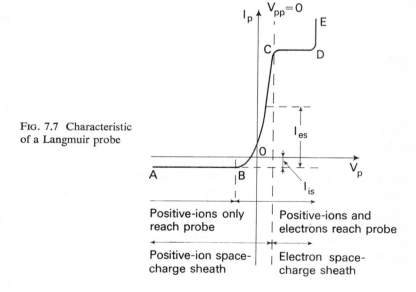

FIG. 7.7 Characteristic of a Langmuir probe

receiving the entire random currents due to both ions and electrons. The probe is then at the potential of the plasma, which extends right up to the probe and the space-charge sheath disappears.

Over the region CD, the positive ions are repelled by the probe and an electron space-charge sheath is formed. In an ideal probe the current is independent of voltage, since all the available ions and electrons are being collected. However, the kinetic energy of the electrons is being increased until at D they have gained sufficient energy to cause ionization by collision. The probe then acts as a secondary anode and the current rises suddenly in region DE.

If the electrons have a Maxwell–Boltzmann distribution of velocities the density of electrons, n_{es}, at the surface of the probe due to random motion from the plasma is related to the electron density in the plasma, n_{ep}, by the Boltzmann relation. Thus in the region BC

$$n_{es} = n_{ep} \exp\left(-\frac{eV_{pp}}{kT_e}\right) \tag{7.14}$$

where V_{pp} is the potential of the probe relative to the plasma, which is also the voltage drop across the sheath. T_e is the absolute temperature of the plasma electrons, and the electron current density to the probe is

$$J_{es} = J_{ep} \exp\left(-\frac{eV_{pp}}{kT_e}\right) \tag{7.15}$$

Taking logarithms of both sides,

$$\log_e J_{es} = \log_e J_{ep} - \frac{eV_{pp}}{kT_e} \tag{7.16}$$

Since V_{pp} is a negative quantity a plot of $\log_e J_{es}$ against V_{pp} will be a straight line of slope e/kT_e, from which T_e may be found. Now $J_{es} = I_{es}/S$, where S is the effective surface area of the probe, so that a graph of $\log_e I_{es}$ against V_p (Fig. 7.8) will have the same form as a graph of $\log_e J_{es}$ against V_{pp}. When the probe has reached the same potential as the plasma, $J_{es} = J_{ep}$ and in the ideal case probe current becomes independent of probe voltage. In practice, true saturation may not occur, but a change in the slope of the graph, as shown in Fig. 7.8, indicates where the probe and plasma currents are equal.

By applying Maxwell–Boltzmann statistics it may be shown that the

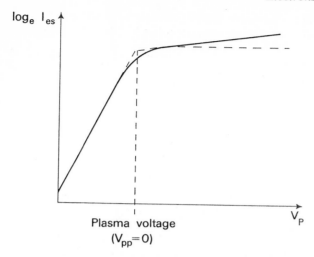

FIG. 7.8 Curve for obtaining plasma potential and electron temperature

density of electrons reaching the probe per second from the plasma is $n_{ep}\sqrt{(kT_e/2\pi m_e)}$ (Ref. 7.3). Hence the probe current density is

$$\frac{I_{es}}{S} = en_{ep}\sqrt{\frac{kT_e}{2\pi m_e}} \tag{7.17}$$

and since T_e has been found, n_{ep} may be obtained from this equation. Similarly for the positive ion current the probe density is

$$\frac{I_{is}}{S} = en_{ip}\sqrt{\frac{kT_i}{2\pi m_i}} \tag{7.18}$$

where the density of positive ions in the plasma, n_{ip}, is equal to n_{ep}. Hence the temperature of the ions, T_i, may be found from eqn. (7.18). Finally a potential distribution of the type shown in Fig. 7.7 may be obtained from a series of probes mounted along the tube (Fig. 7.6).

THE THYRATRON

A practical application of the properties of a plasma occurs in the *thyratron*, a typical form of which is indicated in Fig. 7.9(a). The valve has three electrodes, the grid consisting of a disc with a hole in it mounted

309

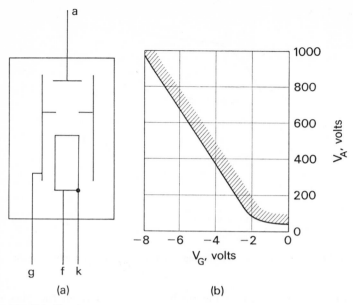

FIG. 7.9 The thyratron

(*a*) Construction: the cathode carries vanes to increase the emitting area
(*b*) Control characteristic

within a cylinder which also encloses the anode and cathode. If the grid
is held at a negative potential of a few volts a significant number of electrons will pass through it only when the anode is raised to a positive
potential of a few hundred volts. The electrons will ionize the gas and
a plasma will be established as in the diode described above. A large
current is then maintained with a much smaller anode voltage, just
above the ionization potential of the gas, and the valve is said to have
"struck". The grid then has no further influence, since if it is negative it is
covered with a positive-ion sheath which just neutralizes the effect of the
voltage. Similarly a positive grid is neutralized by an electron sheath.
In each case, the plasma outside the sheath and hence the anode current
are virtually unaffected by the grid voltage. Thus the valve acts as a
switch, whereby a large current is controlled by a small grid voltage.
The anode current only ceases when the anode voltage returns to zero.

The more negative the grid voltage the higher is the striking voltage
of the anode before conduction commences. The striking voltage and

the grid voltage are related by the control characteristic, as illustrated in Fig. 7.9(*b*). The discharge is initiated by moving into the conduction region of the characteristic. This may be achieved at a fixed grid voltage by increasing the anode voltage, or at a fixed anode voltage by increasing the grid voltage. In either case a finite time is required to establish the plasma, which is usually between 0·1 and 10 μs. Similarly, when the anode voltage is reduced to zero to switch off the current, it must be held at zero or a negative potential until all the ions have been swept out of the gas. The corresponding deionization or *recovery time* is between 100 and 1 000 μs for mercury since ions move much more slowly than electrons. If hydrogen is used as the gas filling this time can be reduced to about 0·01 μs, since hydrogen molecules are much lighter than mercury atoms.

Thyratrons containing mercury vapour are used to control currents up to about 100 A and can withstand a reverse voltage in the kilovolt region, as in the high-voltage direct-current transmission of electrical power. In lower-power applications, they are being replaced by solid-state devices such as the thyristor (page 215), but thyratrons containing an inert gas such as argon or xenon are available to control currents up to about 1 A.

THE TOWNSEND DISCHARGE

Let us consider a planar diode with a photosensitive cathode containing gas at a low pressure. If light falls on the cathode photoelectrons are released, so that as the anode voltage is raised from zero the anode current rises with it. When all the available electrons are flowing to the anode, the current becomes independent of the voltage and so is saturated, at a value I_0 dependent on the light intensity (Fig. 7.10). However, when V_A reaches the value V_i, the ionization potential of the gas, the current begins to rise once more with voltage. This is because when $V_A = V_i$ molecules in the layer of gas next to the anode may be ionized, so that one colliding electron can cause another electron to be produced. When $V_A > V_i$ ionization can occur away from the anode at a point where $V_A = V_i$, and when $V_A > 2V_i$ both electrons can take part in further ionizing collisions (Fig. 7.11). Thus the presence of the gas has caused the increase of current, and gas multiplication is often used to increase the current in a photocell. (Fig. 7.10 for a gas-filled photocell should be compared with Fig. 6.7 for a vacuum photocell).

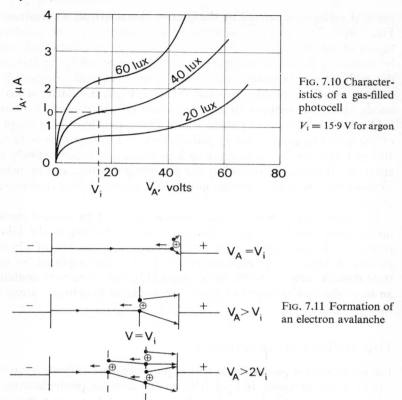

FIG. 7.10 Characteristics of a gas-filled photocell

$V_i = 15·9$ V for argon

FIG. 7.11 Formation of an electron avalanche

The effect is known as a *Townsend electron avalanche*, after J. S. Townsend, who was an early investigator in this field. The first Townsend coefficient, α, is the number of new electrons created per unit length of the path of the original electron. If a plane distant x from the cathode is considered which has n electrons crossing it in one second, in a further short distance δx each electron will create $\alpha \delta x$ new electrons (Fig. 7.12).

FIG. 7.12 Element for electron multiplication

Then, if the total number of new electrons created in this distance is δn, we have

$$\delta n = n\alpha\,\delta x \qquad (7.19)$$

assuming that recombination and diffusion are negligible.

If the number of electrons leaving the cathode in one second is n_0, then as $\delta x \to 0$,

$$\int_{n_0}^{n} \frac{dn}{n} = \alpha \int_{0}^{x} dx \qquad (7.20)$$

or

$$\log_e \frac{n}{n_0} = \alpha x \qquad (7.21)$$

and

$$n = n_0 \exp \alpha x \qquad (7.22)$$

In general $I = en$, and the saturation current $I_0 = en_0$. Hence if the spacing between anode and cathode is d, the anode current is

$$I_A = I_0 \exp \alpha d \qquad (7.23)$$

or, in terms of current densities,

$$J_A = J_0 \exp \alpha d \qquad (7.24)$$

IONIZING POWER

The coefficient α, which is also known as the *ionizing power* of an electron, depends on both the kinetic energy of the electron and the average number of collisions it makes per unit path length. In passing through the gas it makes both elastic and inelastic collisions. At an elastic collision, it loses very little kinetic energy, but after an inelastic collision it loses most of its kinetic energy, which is then regained from the electric field. The average distance it travels between collisions is its mean free path, l_e, so that if E is the electric field within the gas the average energy it can acquire from the field is El_e electronvolts. Also the average number of collisions it makes per unit path length is $1/l_e$. Hence we may write

$$\alpha = f\left(\frac{El_e}{l_e}\right) \qquad (7.25)$$

where f represents a mathematical function. Since $I_e \propto 1/P$ (eqn. (7.2)), this becomes

$$\alpha = Pf_1\left(\frac{E}{P}\right) \qquad (7.26)$$

or

$$\frac{\alpha}{P} = f_1\left(\frac{E}{P}\right) \qquad (7.27)$$

This relationship has been determined experimentally for a number of gases (Ref. 7.3), and some of the results are shown in Fig. 7.13. As expected,

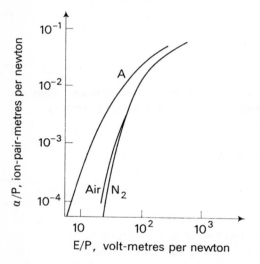

FIG. 7.13 Variation of ionizing power with electric field and pressure

After J. D. Cobine, Ref. 7.3

α increases with E at a given pressure. Assuming negligible space charge we can write E in terms of the anode voltage, and putting $E = V_A/d$, gives

$$\alpha = Pf_1\left(\frac{V_A}{Pd}\right) \qquad (7.28)$$

Substituting this equation into eqn. (7.23),

$$I_A = I_0 \exp\left[Pdf_1\left(\frac{V_A}{Pd}\right)\right] \qquad (7.29)$$

which indicates that I_A rises rapidly with V_A.

THE SELF-SUSTAINED DISCHARGE AND SECONDARY EMISSION

Eqn. (7.29) can explain the rise of anode current up to a certain value of anode voltage. Above this value I_A rises more rapidly, and at $V_A = V_S$, it becomes *self-sustained*, i.e. I_A is maintained even when the source illuminating the cathode is removed. This suggests that electrons are being released from the cathode by a process originating within the gas itself, and it has been found that secondary electrons are being released from the cathode due to the bombardment by positive ions from the gas.

Suppose the total number of electrons leaving the cathode per second is n_c, of which n_0 are the result of photoemission. If n_a electrons per second reach the anode the number of electrons and positive ions created in the gas per second is $n_a - n_c$. Then if each positive ion releases γ electrons from the cathode by secondary emission per second, the number of secondary electrons emitted is $\gamma(n_a - n_c)$. The total number of electrons emitted per second is then

$$n_c = n_0 + \gamma(n_a - n_c) \tag{7.30}$$

But, owing to the avalanche process,

$$n_a = n_c \exp \alpha d$$

so that

$$n_c = n_0 + \gamma n_c (\exp \alpha d - 1) \tag{7.31}$$

or

$$n_c[1 - \gamma(\exp \alpha d - 1)] = n_0 \tag{7.32}$$

and

$$n_a = \frac{n_0 \exp \alpha d}{1 - \gamma(\exp \alpha d - 1)} \tag{7.33}$$

In terms of current,

$$I_A = \frac{I_0 \exp \alpha d}{1 - \gamma(\exp \alpha d - 1)} \tag{7.34}$$

The *secondary emission coefficient*, γ, depends on the gas and the cathode material, and experimental curves of γ as a function of E/P are given in Fig. 7.14 for some gases and metals. γ also depends on the kinetic

315

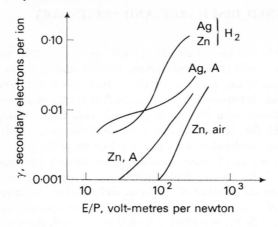

FIG. 7.14 Variation of secondary-emission coefficient with electric field and pressure

After P. Parker, Ref. 6.3

energy of the ions, which collide elastically with gas molecules on their way to the cathode but lose energy since the masses of an ion and a molecule are very nearly equal. Thus the average kinetic energy is the same as that gained in covering the average distance between collisions. This is the mean free path of the ions, l_i, and the gain in kinetic energy is El_i. Since $l_i \propto 1/P$, we can write

$$\gamma = f_2\left(\frac{E}{P}\right) \tag{7.35}$$

where f_2 represents a mathematical function. More electrons are released from the cathode as the energy of the ions is increased so that γ rises with E/P (Fig. 7.14). In terms of the anode voltage,

$$\gamma = f_2\left(\frac{V_A}{Pd}\right) \tag{7.36}$$

Sparking Potential

At low values of V_A the denominator of eqn. (7.34) is approximately unity, since $\exp \alpha d$ is small. But both α and γ increase with V_A and at a certain anode voltage

$$\gamma(\exp \alpha d - 1) = 1 \tag{7.37}$$

and the anode current becomes infinite. In practice, at this point the assumption that the space charge is negligible is no longer valid so that the electric field is no longer uniform and the theory cannot hold exactly. However, a significant change does occur in the I/V characteristic, and the corresponding value of anode voltage is known as the *sparking potential,* V_S (Fig. 7.15). At this potential each electron released from the cathode

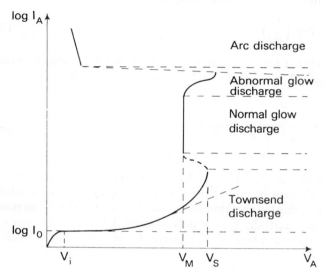

FIG. 7.15 Characteristic of a cold-cathode discharge

causes the release of one more electron by secondary emission, so that the discharge is self-sustained. After breakdown has occurred a plasma is formed, as in the case of the hot-cathode discharge. Consequently, the anode voltage required to maintain the current drops, but to a value V_M well above the ionization potential, and a *normal glow discharge* is set up. The cathode fall in voltage is relatively high, about 100 V, and the current density is up to about 1 A/cm² in the normal glow discharge.

PASCHEN'S LAW

Since $\exp \alpha d \gg 1$ in eqn. (7.37), the breakdown condition becomes

$$\frac{1}{\gamma} = \exp \alpha d \qquad (7.38)$$

From eqn. (7.28), and putting $V_A = V_S$,

$$\alpha = Pf_1\left(\frac{V_S}{Pd}\right) \tag{7.39}$$

Substituting into eqn. (7.38) and taking logarithms,

$$\log_e \frac{1}{\gamma} = Pdf_1\left(\frac{V_S}{Pd}\right) \tag{7.40}$$

Thus the condition for breakdown depends both on V_S and the product Pd. Using eqn. (7.36) with $V_A = V_S$, eqn. (7.40) may be written in the form

$$Pd = \log_e \frac{1}{f_2\left(\dfrac{V_S}{Pd}\right)} \Big/ f_1\left(\frac{V_S}{Pd}\right) \tag{7.41}$$

The left-hand side of this equation does not contain V_S but the right-hand does, so that V_S must be a function of Pd if the equation is to be satisfied. In 1889 Paschen showed experimentally that V_S is a function of the product of pressure and electrode spacing only, a result known as *Paschen's law*. This is illustrated for plane-parallel electrodes in air in Fig. 7.16(a); the

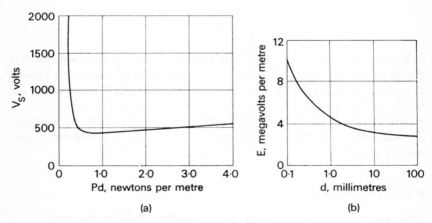

Fig. 7.16 Sparking potential and Paschen's law at 20°C

(a) V_S/P_d curve for air
(b) E/d curve for air at atmospheric pressure

form of the curve is similar for other gases. In general, it passes through a minimum with V_S increasing more rapidly at low values of Pd than at high values. The increase at low pressure or close electrode spacing occurs because the mean free path is long compared with the spacing, so that few collisions can take place. At high pressure or wide spacing the mean free path is short compared with the spacing, so that although many collisions take place only a little energy is gained between collisions and few of them result in ionization. In both cases V_S must be greater than the minimum value. A common consideration in high-voltage applications is the breakdown strength of air at atmospheric pressure, $101 \, kN/m^2$, which involves values of Pd above those given in Fig. 7.16(*a*). The required electric field, E, is given in Fig. 7.16(*b*) as a function of the spacing of plane-parallel electrodes. At a spacing of about 1 cm the corresponding breakdown voltage, Ed, is about 30 kV, while at a spacing of about 0·1 mm only about 1 kV is required for breakdown.

The sparking potential of a gas is increased by raising its pressure, provided the product Pd is above the value corresponding to the minimum sparking potential. This is exploited in high-voltage coaxial cables, for instance, in which the space between the conductors may be filled with nitrogen at about 15 times atmospheric pressure to increase the breakdown voltage. A desirable feature of the gas used is that its molecules should possess less potential energy when an electron is attached to them than when they are in a normal ground state. In this way stable negative ions are formed before the electron can release extra ions and electrons by collision. Such a gas is said to have a high *electron affinity*, a useful example being sulphur hexafluoride.

THE ARC DISCHARGE

A gas discharge in which the potential just outside the cathode is of the order of the ionization potential is called an *arc*. Thus the hot-cathode discharge described on page 303 comes into this category. If the cold-cathode discharge current is allowed to increase until the glow completely covers the cathode, the voltage between the electrodes first rises and then falls to about the ionization potential. The corresponding regions of the characteristic are called the *abnormal glow discharge* and the *arc* (Fig. 7.15). The arc is characterized by a high current density, usually greater than $10^7 \, A/m^2$, and a low voltage of about 20 V, which falls as the current is allowed to rise, giving a negative resistance. The voltage distribution within the tube is similar to that of Fig. 7.4, with a steeper fall in voltage across the plasma.

A process in addition to those described is required to explain the very high current density of the arc. For instance in an arc between carbon electrodes in air the cathode temperature is about 3 700°C, so that electrons are released by thermionic emission. A cold-cathode rectifier may be formed in which an arc is struck between an iron or carbon anode and a cathode of liquid mercury. Here the cathode is much cooler than in the carbon arc and electrons are considered to be released by field emission.

THE CORONA DISCHARGE

When a sufficiently high voltage is applied between plane electrodes, setting up a uniform field, a spark occurs resulting in an arc discharge. However, if the electrodes are curved the electric field is higher near the electrodes than elsewhere. In this case, the gas can break down at a voltage less than the spark breakdown voltage for the given electrode spacing. This is due to partial breakdown near the electrode which has a radius of curvature much smaller than the spacing between the electrodes. At atmospheric pressure a glow discharge occurs which is called a *corona*, and on a high-voltage transmission line it can cause a power loss which is substantial if the line is long. The appearance of the discharge has the form of glowing threads of light between the electrodes.

An application of the corona discharge is in the dust precipitator. If the effluent from a chimney containing particles of solid matter is passed through a corona discharge, electrons adhere to the neutral particles which are then carried to the positive electrode. Here the particles lose their charge and can be collected.

Consider an electrode system consisting of a cylinder and a coaxial wire. The discharge may be maintained with the wire positive or negative, with most of the ionizing collisions occurring near the wire in each case since this is the region of highest electric field strength (eqn. (A.5.6)). For a *negative* wire, positive ions are formed in the region near the wire owing to ionizing collisions between the gas molecules and electrons released from the wire by field- or photo-emission. The positive ions move towards the wire, and since their energy is high they release secondary electrons from the wire and a self-sustained discharge occurs. The current in the main part of the gap is then carried by electrons, which also form negative ions by attaching themselves to neutral gas molecules.

If the wire is *positive* with respect to the cylinder a different mechanism is set up. Here the electrons formed in the space as the result of ionization by external means, such as light and cosmic radiation, move towards the wire, and when they reach the high-field region, electron avalanches are formed. The positive ions from the avalanches move towards the cylinder, but since the field decreases in this direction few ionizing collisions are made and few electrons are released from the cylinder by positive-ion bombardment. However, photons from the discharge may cause ionization of neutral molecules and photoemission from the cylinder. In this case, the current in the main part of the gap is carried by the positive ions.

SPACE-CHARGE-LIMITED CURRENT IN A GAS

When the pressure of the gas in a valve is so high that the mean free path of electrons, l_e, is appreciably less than the anode-to-cathode spacing, electrons leaving the cathode reach a drift velocity dependent on the magnitude of the electric field. The collision processes between the electrons and the gas molecules are similar to those occurring between the current carriers and atoms of a semiconductor (page 48). Thus in general the drift velocity is given by eqn. (2.49): $u_d = \mu E$, where μ is the mobility of electrons in the gas. At low values of E/p the mobility is obtained from eqn. (2.54) and is independent of E, but at high values of E/p the mobility is a function of E as shown in eqn. (2.53).

The current density at any point at a distance x from the cathode is then

$$J = neu_d = ne\mu \frac{dV}{dx} \tag{7.42}$$

where n is the density of electrons and V is the potential at the point x. It is assumed that the field is low enough for μ to be constant and that no ionization is occurring. Using Poisson's equation (6.42) and eqn. (7.42),

$$\frac{d^2V}{dx^2} = \frac{ne}{\epsilon_0} = \frac{J}{\mu\epsilon_0 \frac{dV}{dx}}$$

or

$$\frac{d^2V}{dx^2}\frac{dV}{dx} = \frac{J}{\mu\epsilon_0} \tag{7.43}$$

Integrating,

$$\tfrac{1}{2}\left(\frac{dV}{dx}\right)^2 = \frac{J}{\mu\epsilon_0}x + C_1 \tag{7.44}$$

For space-charge limitation of the current the field at the cathode is zero, or $dV/dx = 0$ when $x = 0$. Hence $C_1 = 0$ and

$$\frac{dV}{dx} = \left(\frac{2Jx}{\mu\epsilon_0}\right)^{1/2} \tag{7.45}$$

Integrating again,

$$V = \left(\frac{2J}{\mu\epsilon_0}\right)^{1/2} \times \tfrac{2}{3}x^{3/2} + C_2 \tag{7.46}$$

When $x = 0$, $V = 0$, so that $C_2 = 0$. The current density is then given by

$$J = \tfrac{9}{8}\mu\epsilon_0 \frac{V^2}{x^3} \tag{7.47}$$

and rises with the square of the voltage instead of the three-halves power as in the high-vacuum case (eqn. (6.53)). Inserting the numerical value of ϵ_0 in eqn. (7.47),

$$J = 9{\cdot}95 \times 10^{-12}\mu \frac{V^2}{x^3} \quad \text{amperes/metre}^2 \tag{7.48}$$

at any point between the electrodes. At the anode, distant d from the cathode, the current density is

$$J_A = 9{\cdot}95 \times 10^{-12}\mu \frac{V^2}{d^3} \tag{7.49}$$

Eqns. (7.48) and (7.49) can be applied to ions as well as electrons, using the appropriate value of mobility, since they do not depend on the charge or mass of the current carriers. Typical values of mobility are about $0{\cdot}2\,\text{m}^2/\text{Vs}$ for electrons in hydrogen, about $6 \times 10^{-4}\,\text{m}^2/\text{Vs}$ for positively charged hydrogen ions and about $8 \times 10^{-4}\,\text{m}^2/\text{Vs}$ for negatively charged hydrogen ions.

REFERENCES

7.1 RAMEY, R. L., *Physical Electronics* (Prentice-Hall, 1962).

7.2 KAYE, G. W. C. and LABY, T. H., *Physical and Chemical Constants*, 13th ed. (Longmans, 1966).

7.3 COBINE, J. D., *Gaseous Conductors* (Dover, 1958).

FURTHER READING

PARKER, P., *Electronics* (Arnold, 1950).

PROBLEMS

7.1 Enumerate the ways in which ionization may be caused in gases, and indicate the part which each of these processes *may* play in affecting the striking voltage of a discharge tube containing a gas at low pressure between large plane electrodes. Discuss the role which the electrodes play and the properties which the electrode material must possess (*a*) to maximize and (*b*) to minimize the striking voltage. Sketch a typical Paschen curve for the gas considered and account for its shape. (*IEE*, June 1964)

7.2 A simple discharge tube consists of a cylindrical glass tube with a large plane electrode at each end and contains neon gas at a pressure of about 1 mmHg [0·133 kN/m²]. Sketch the voltage/current diagram for this type of tube and explain the various processes that control the discharge in the different regions of its voltage/current characteristic.

If the tube has metal electrodes, explain why operation in the abnormal glow region may result in instability and a transition to the arc mode. (*IEE*, June 1966)

7.3 Define Townsend's first, α, and second, γ, discharge coefficients and show that the criterion for the breakdown of a gas is that $\gamma(e^{\alpha d} - 1) \geqslant 1$ if electron-ion recombination is ignored. Hence or otherwise show that for a given gas the sparking potential is proportional to the product Pd, where P is the gas pressure and d is the electrode separation (Paschen's law). Mention and account for any significant departure from this law.

Find the maximum distance that can be tolerated between two electrodes in a low-pressure argon sputtering chamber if electrical breakdown is to be avoided. The Townsend's coefficients are found to be $\alpha = 150$ and $\gamma = 2$ under the relevant conditions. (*IEE*, June 1967)

(*Ans.* 2·7 mm)

7.4 Describe the construction and operation of a modern hot-cathode fluorescent discharge lamp. Explain the difference between the mechanism of the hot-cathode discharge and that of a cold-cathode discharge. Give a diagram showing the potential distribution between the cathode and anode for the hot-cathode case.

In a certain hot-cathode lamp, the space-charge density can be considered zero. If the ion mass is $3 \cdot 6 \times 10^5$ times the electron mass for the gas in the tube, find the ratio of ion current to electron current in the lamp. (*IEE*, Nov. 1966)

(*Ans.* 600)

Physical Electronics

7.5 What is a plasma? Describe its principal characteristics and discuss two engineering systems or devices which contain plasmas.

The electron density in a plasma is $10^{12}/cm^3$ and the drift current density is $100\,mA/cm^2$. What is the electron drift velocity? If the electron temperature is $10\,000°\,K$, what current would flow to a probe of area $0.01\,cm^2$ which is held at the plasma potential? Indicate where approximations are made in calculations, and explain why the probe current density differs from the drift current density. (*LU, B.Sc.* (*Eng.*), 1966)

(*Ans.* $6.25 \times 10^3\,m/s$, $250\,\mu A$)

7.6 A direct high voltage is applied between a cylinder and a coaxial wire, resulting in a corona discharge at the surface of the wire. Discuss briefly how the discharge is maintained when (i) the wire is negative, (ii) the cylinder is negative.

Assuming that the onset of corona occurs when the electric field at the surface of the wire exceeds a value of E_0, calculate the critical applied voltage V_0 for a wire of radius r and a cylinder of radius R. For a cylinder 2 m in diameter and a central wire of $\frac{1}{2}\,mm$ diameter, calculate the critical voltage V_0 if E_0 is $10^5\,V/cm$. (*LU, B.Sc.* (*Eng.*), 1967)

(*Ans.* $20.8\,kV$)

8

Microwave Devices and Electrical Noise

MICROWAVE DEVICES

At microwave frequencies, approximately from 300 MHz to 30 GHz, the simple wire connections between the various parts of an electronic circuit are no longer adequate, and components such as resistors, capacitors and inductors (*lumped* components) no longer behave as predicted at lower frequencies. This is because the dimensions of an amplifier *circuit*, for instance, are comparable to the wavelength of the signals, which ranges from about 1 m down to about 0·1 mm. Thus there are appreciable phase differences between the ends of connections, since a wire 10 cm long represents 30% of a wavelength at 1 GHz but only 0·03% of a wavelength at 1 MHz, so that at the lower frequency phase change along the wire may be ignored. In addition, when lumped components are used the loss of energy by radiation from the circuit becomes excessive and its operation becomes very inefficient.

These problems are overcome by confining the signals within hollow conducting tubes of rectangular or circular cross-section, known as *waveguides*. These prevent energy loss by radiation and can fulfil the circuit functions of resistors, capacitors and inductors by suitable choice of their length and termination (Ref. 8.1). However, up to about 3 GHz coaxial transmission lines have sufficiently low attenuation and provide more flexible interconnections than waveguides, while also preventing radiation loss. Other microwave components, such as the resonant cavity which behaves like a tuned circuit, are introduced below in connection with electronic devices.

The conventional valves and transistors also operate very inefficiently at microwave frequencies, owing to transit time effects and internal reactances. In a small triode the transit time of an electron is about 1 ns from cathode to grid and about 0·1 ns from grid to anode, the second transit time being much shorter than the first since $V_{AG} \gg V_{GK}$. Thus when the operating frequency is a few hundred megahertz the transit time is comparable to the period of the signal. Also the capacitances between electrodes and the inductances of the leads between electrodes and external circuit have appreciable reactance, and it may be shown that both of these effects (*a*) reduce the input impedance and (*b*) reduce the gain of a valve amplifier. Similar considerations apply to a transistor amplifier as discussed in Chapter 4, transit time and interelectrode capacitance becoming effective at frequencies of a few megahertz. Both vacuum and solid-state devices have been developed for use in the microwave region, the vacuum devices being usable at higher powers than the solid-state devices.

MICROWAVE VALVES

The general principle used in microwave valves, such as the klystron and the travelling-wave tube described below is that the kinetic energy of a beam containing a high density of electrons is converted into energy at microwave frequencies. In order to achieve efficient conversion the beam must be prevented from spreading due to the mutual repulsion of the electrons; beam confinement can be achieved by means of a magnetic field. This will be described qualitatively; an analysis of the method (known as Brillouin focusing) is given in Ref. 8.2.

The general arrangement is shown in Fig. 8.1(*a*), the permanent magnet providing a non-uniform magnetic field near its poles and a uniform magnetic field B_0 along its axis. The beam is concentrated from a concave cathode by means of the beam-forming electrode and emerges from the electron gun with a radius equal to that of the anode aperture, *a*. Owing to the shape of the lines of magnetic flux at the ends of the solenoid there will be a radial component of the field B_r (Fig. 8.1(*b*)), so that an electron on the outside of the beam will receive an impulse directing it out of the page as it enters the magnetic field. This will cause the electron to move in a circular path in a plane perpendicular to the page, and this motion combined with its longitudinal motion will cause it to spiral about the axis. The effect is similar to the magnetic focusing described on page 271, but

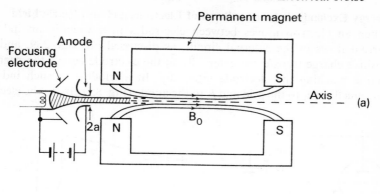

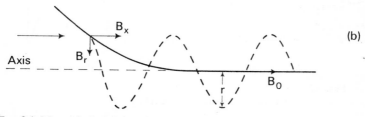

Fig. 8.1 Magnetic focusing

detailed analysis shows that the angular frequency of rotation about the axis is *half* that expected from a magnetic field B_0 in a simple focusing system.

B_0 is given by

$$\frac{6\cdot9 \times 10^{-7}I_0}{V_0^{1/2}r^2} \quad \text{tesla}$$

where I_0 is the beam current, V_0 the potential at the axis of the beam, and $r(\leqslant a)$ the radius of the beam envelope. In practice the optimum field is between $1\cdot5$ and $2\cdot0B_0$ since it is difficult to inject the beam correctly into the magnetic field.

There are now two forces on the electron acting away from the axis, owing to the mutual repulsion of the electrons and their circular motion respectively. The magnetic field B_0 provides an opposing force acting towards the axis, so that the radius of the spiral can equal the radius of the aperture of the electron gun. Such an electron beam, having uniform radius along its length, can be used as a source of energy.

Energy Exchange between a Beam of Electrons and an Electric Field

When an electron moves between electrodes in a vacuum an induced current flows in the external circuit, as discussed on page 239, and the positive charge transferred externally to the electrode equals the negative charge reaching the electrode internally. In a triode two such induced currents flow, i_1 to the grid and i_2 to the anode (Fig. 8.2(a)). As the electron

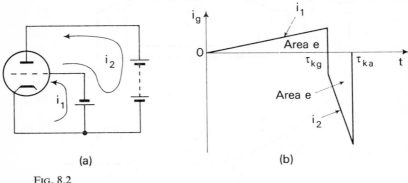

(a)　　　　　　　　　　　　　　(b)

FIG. 8.2

(*a*) Induced currents　　(*b*) Grid current

moves from cathode to grid, i_1 flows, and as it moves from grid to anode, i_2 flows. Provided that the triode has a high amplification factor μ, there is very little charge induced on the anode when the electron is moving in the cathode–grid space and very little induction on the cathode when the electron is moving in the grid–anode space. The variation of i_1 and i_2 with time is shown in Fig. 8.2(*b*), where it may be seen that i_1 reverses as the electron passes through the grid. The effect of space charge is neglected, so that both currents change linearly with time.

The current i_2 flows in the direction which transfers energy from the anode supply to the electron. i_1 flows in the direction in which energy is transferred to the negative grid supply while the electron is approaching the grid, and energy is drawn from the grid supply while the electron is approaching the anode. For a large number of electrons the total energy will be the sum of the individual contributions, so we can say that when a beam of electrons moves towards a negative electrode energy is transferred *to* the supply. When the beam moves away from a negative electrode towards a positive electrode, energy is removed *from* the supply. This principle is also used in the amplification of a.c. power, especially at microwave frequencies.

328

Power Transfer to an External Impedance

Consider a triode with the cathode at a negative potential, the grid earthed directly and the anode earthed through a resistor R (Fig. 8.3(a)).

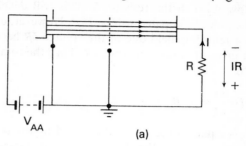

(a)

FIG. 8.3 Energy exchange between an electron beam and electric fields

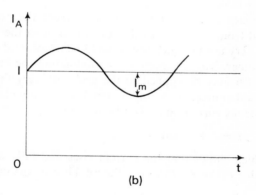

(b)

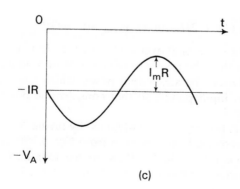

(c)

This arrangement is particularly relevant to the klystron valve, discussed below. Electrons will pass through the grid to strike the anode, and if the rate of flow is dn/dt electrons per second, an induced current $I = e(dn/dt)$ will flow through the resistor R. This will develop a potential drop IR and the power dissipated will be I^2R. However, each electron in the beam meets a retarding field due to the voltage IR between grid and anode, so that it gives up kinetic energy eIR. Thus the total power lost by the electrons is

$$eIR\frac{dn}{dt} = I^2R \tag{8.1}$$

which is equal to the power dissipated by the resistor. The remainder of the kinetic energy of the electrons is dissipated as heat at the anode when they strike it.

Since d.c. power is transferred from the power supply to the beam and from the beam to the resistor through the induced current, it would be very useful if a.c. power could be transferred to the resistor in a similar manner. This is achieved by modulating the current with an a.c. signal, which may be applied to the grid with very little power taken from the signal source. If the input signal has an angular frequency ω the instantaneous current may then be written

$$i = I + I_m \sin \omega t \tag{8.2}$$

which shows that the current rises and falls sinusoidally about a mean value I (Fig. 8.3(b)). Then the total instantaneous power in the resistor R is

$$P = i^2R = (I + I_m \sin \omega t)^2 R \tag{8.3}$$

or

$$P = (I^2 + 2II_m \sin \omega t + I_m^2 \sin^2 \omega t)R \tag{8.4}$$

The mean value of P over a complete cycle is then $I^2R + \frac{1}{2}I_m^2R$, which is greater than the power in the absence of modulation owing to the term $\frac{1}{2}I_m^2R$. This is the a.c. output power which has been obtained for a very small a.c. input power, so that considerable power amplification can be obtained.

The a.c. power has been obtained by further reducing the kinetic energy of the electrons. It may be seen from Fig. 8.3(b) that during the first half-cycle the current is higher than the mean value I, so that more electrons than average are moving towards the anode. During this time the anode voltage is more negative than the mean value $-IR$ (Fig. 8.3(c)), so that a.c. energy is removed from the beam. During the second half-cycle the anode voltage is more positive than $-IR$, so that a.c. energy is supplied to the beam. However, in this half-cycle the current is less than I so that fewer electrons than average are moving towards the anode. Thus over a whole cycle the beam gives up more a.c. energy than it receives and so supplies a.c. power to the resistor. It should be noted that the induced current is independent of the magnitude of R provided that the electron velocity is not significantly changed by the retarding potential $-I_mR$, i.e. provided that the voltage on the electron gun is much larger than $(I + I_m)R$. Also R may be either an actual resistor or the dynamic resistance of a tuned circuit.

VELOCITY MODULATION

In order to obtain the sinusoidal modulation of current shown in Fig. 8.3(b) the velocity of the electrons is modulated in a microwave valve. The single grid of Fig. 8.3(a) is replaced by two closely spaced grids between which an alternating voltage, $V_m \sin \omega t$, is applied. Then the electron velocity is due to the sum of the alternating and direct voltages and is given by

$$u = \sqrt{\frac{2e}{m}} \, (V_{AA} + V_m \sin \omega t)^{1/2} \tag{8.5}$$

Normally $V_m \ll V_{AA}$, so that, using the binomial theorem,

$$u \approx \sqrt{\frac{2eV_{AA}}{m}} \, (1 + \frac{V_m}{2V_{AA}} \sin \omega t) \tag{8.6}$$

or

$$u \approx u_0(1 + \frac{V_m}{2V_{AA}} \sin \omega t) \tag{8.7}$$

where

$$u_0 = \sqrt{\frac{2eV_{AA}}{m}} \tag{8.8}$$

Thus, from eqn. (8.7), the velocity is sinusoidally modulated and the velocity modulation has again been obtained with a small a.c. power input to the grids.

Physical Electronics

The effect on the beam may be illustrated by means of an Applegate diagram (Fig. 8.4). Since the distance from the modulating grids is plotted against time, the velocities of electrons leaving the grids at different times are shown by straight lines of different slopes. An electron leaving when the input signal is zero will have the normal velocity u_0, but if it leaves during a positive half-cycle it will travel faster, and if it leaves during a negative half-cycle it will travel slower, than u_0. The faster and normal-velocity electrons will catch up the slower ones and they will all arrive at a particular plane in the valve, the *bunching plane*, at the same time. Thus the electron density in the beam can be a maximum here and these maxima arrive at the bunching plane at time intervals equal to the period of the input signal. At times between the electron bunches the electron density is a minimum, and the actual position of the bunching plane will move slightly up and down with the amplitude of the input signal. This effect will be very small when $V_m \ll V_{HT}$ since the change in velocity is then also negligible. The two grids are known as a *buncher*, and a geometrical construction will show that there are a series of equally spaced bunching planes along the beam.

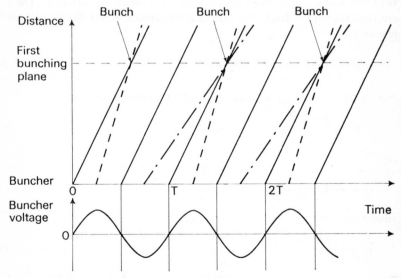

FIG. 8.4 Applegate diagram for a klystron amplifier

Buncher voltage $= V_m \sin 2\pi \dfrac{t}{T}$

Resonant Cavities

In practice the buncher is part of a resonant cavity, which is the microwave equivalent of an LC parallel resonant circuit. As the resonant frequency f_0 is raised L and C must be reduced, since $f_0 = 1/2\pi\sqrt{(LC)}$. Ultimately the inductance becomes a single turn and the capacitance may be in the form of two discs (Fig. 8.5). A further reduction in inductance is obtained by

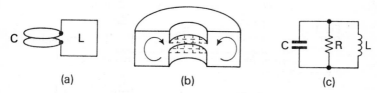

(a)   (b)   (c)

FIG. 8.5 Formation of a resonant cavity

    (*a*) Circuit
    (*b*) Section through resonant cavity
    (*c*) Equivalent circuit

considering more turns in parallel until a solid of revolution is formed. The cavity is completed by replacing the discs with wire grids so that an electron beam can pass through it, while in a practical cavity the solid of revolution is made from continuous copper sheets of low resistivity.

Inductive and capacitive parts of the cavity can thus be identified which store energy in a magnetic field and an electric field respectively. When the cavity is excited into oscillation at its resonant frequency the electromagnetic energy is exchanged between the electric and magnetic fields, just as in an LC circuit at resonance (Fig. 8.5). Energy is also dissipated as heat at the surface of the cavity walls, since at microwave frequencies currents flow only in the surface layer, and this energy is represented as being dissipated in a parallel resistance R. The Q-factor of the cavity is defined as

$$Q = 2\pi \times \frac{\text{Maximum stored energy}}{\text{Energy dissipated per cycle}} \tag{8.9}$$

and is typically about 10 000 when the cavity is unloaded. Electromagnetic power is supplied to the cavity or removed from it by means of a small wire loop which is threaded by the magnetic field in the inductive part. When the cavity is coupled to a resistive load as in the klystron amplifier (Fig. 8.6(*a*)) the dissipated energy is increased by the power supplied to the load and Q falls to below 1 000. Another definition of Q is the ratio of resonant frequency to bandwidth, so that for a cavity resonant at 900 MHz with $Q = 300$ the bandwidth is 3 MHz. The cavity may be tuned mechanically over a small range of resonant frequencies by means of a screwed

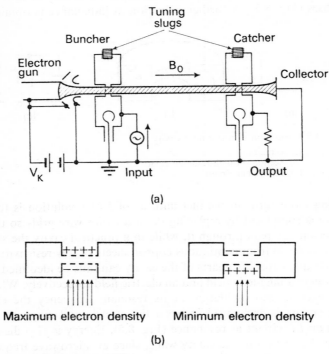

FIG. 8.6 Klystron amplifier

(*a*) Two-cavity klystron amplifier
(*b*) Oscillation conditions in a resonant cavity

plunger in the side. Moving the plunger into the cavity effectively increases the length of the inductive part and reduces the resonant frequency, the opposite effects occurring when the plunger is moved outwards.

THE KLYSTRON AMPLIFIER

Two cavities are used having the same resonant frequency, the one nearer to the cathode being the buncher; the other, known as the *catcher*

(Fig. 8.6(*a*)), is mounted at a bunching plane. The electron beam is confined by a magnetic field as described above and is velocity modulated by a signal fed into the buncher at its resonant frequency, while the load, represented by a resistance, is coupled into the catcher. The beam will induce currents in the walls of the catcher cavity and excite it into oscillations, which will be maintained if the a.c. field across the catcher grids can extract energy from the beam. This occurs when the bunches of maximum electron density arrive at the catcher to meet a retarding field, and the conditions of minimum electron density coincide with an accelerating field (Fig. 8.6(*b*)). The waveform of the current at the catcher is then of the form shown in Fig. 8.7. The amplitude of the oscillations in the

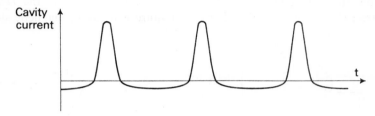

FIG. 8.7 Cavity current waveform

catcher cavity builds up at the expense of the kinetic energy of the electrons. The amplitude becomes constant when the energy lost by the electrons is equal to the sum of the energy supplied to the load and the energy dissipated in the cavity and load.

The input power at the buncher cavity is much less than the output power obtained at the catcher cavity, and a power gain of about 10 000, or 40 dB, is obtainable. Since the input voltage is much less than the anode supply voltage, V_{AA}, of the electron gun, the amplifier is insensitive to changes in V_{AA}, but it can operate over only a narrow band of frequencies owing to the high Q of the cavities. The effect of transit time through either cavity is small, since electrons are accelerated to a high velocity *before* passing through the cavity. This contrasts with conventional valves where most of the acceleration takes place *after* they have passed through the control grid, so that the electron transit time through the modulating region is much longer and becomes a limiting factor at much lower frequencies.

Klystron amplifiers are used in transmitters for television broadcasting and are available to cover a frequency range from 0·4 to 1·2 GHz with a power gain of 40 dB and an output power of 25 kW, the electron gun voltage being about 18 kV in this case. For high-power amplifiers of this type two intermediate cavities are mounted at bunching planes between the input and output cavities, making a four-cavity amplifier. The alternating voltages due to oscillations in these extra cavities increase the density of the electron bunches at the output cavity and so increase the output power.

THE REFLEX KLYSTRON

The microwave signals used in the amplifier described in the preceding section are generated by another type of valve, known as the *reflex* klystron (Fig. 8.8(*a*)), which does not require a focusing magnetic field

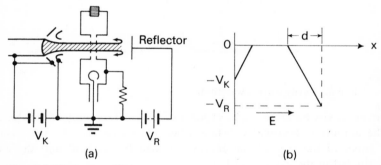

Fig. 8.8 Klystron oscillator

    (*a*) Reflex klystron oscillator
    (*b*) Voltage distribution in the reflex klystron

since it is much shorter than the klystron amplifier. The reflex klystron has only one cavity to which the load is coupled and electrons passing through it are returned to the cavity when they meet the retarding electric field provided by the *reflector* electrode (Fig. 8.8(*b*)). When the voltage on the electron gun is switched on and the first electrons pass through the cavity they induce currents in its walls and excite it into oscillation. These oscillations will be maintained if the electrons meet a retarding a.c. field when they return to the cavity, as shown in the Applegate diagram of Fig. 8.9. Here the alternating voltage has caused velocity modulation of the electrons on their first passage, and their trajectories have been adjusted by means of the reflector voltage so that bunching

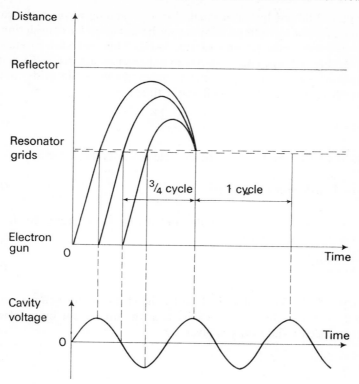

Fig. 8.9 Applegate diagram for the reflex klystron

occurs three-quarters of a cycle after the normal-velocity electrons have passed through the cavity. At this time there is a retarding alternating potential acting on the electrons which have returned to the cavity, so that they give up their energy to the a.c. field and maintain oscillations in the cavity. The cavity then acts as both a buncher and a catcher. Ideally the returning electrons give up all their energy to the cavity, but some may pass through it and rejoin the oncoming electrons. It is apparent that oscillations will also be maintained if electrons are bunched $1\frac{3}{4}$ cycles after passing through the cavity, by reducing the retarding potential on the collector, and in general bunching must occur after $(n + \frac{3}{4})$ cycles, where $n = 0, 1, 2, 3, \ldots$. Each integral value of n corresponds to a *mode* of oscillation, called the $\frac{3}{4}$ mode, the $1\frac{3}{4}$ mode and so on.

The path followed by the normal electrons passing through the cavity when the alternating voltage is zero may be obtained by considering the voltage between reflector and cavity, $-V_R$. This provides a retarding field $-E$ (Fig. 8.8(*b*)), and an electron experiences a force $-Ee$ and an acceleration $-Ee/m$. If the electron passes through the grids at time $t = 0$ with velocity u_0, the velocity after time t is

$$u = u_0 - \frac{Ee}{m} t \tag{8.10}$$

and the distance travelled beyond the grid towards the reflector is

$$s = \int_0^t u\, dt = u_0 t - \tfrac{1}{2}\frac{Ee}{m} t^2 \tag{8.11}$$

This is the equation of a parabola, and the electron returns to the grid when $s = 0$, given by a time

$$t = \frac{2mu_0}{eE} \tag{8.12}$$

from eqn. (8.11). Thus, if u_0 is fixed by the electron gun voltage and the distance between reflector and cavity is d, the time spent by an electron between reflector and cavity is given by

$$t_f = \frac{2mu_0 d}{eV_R} = \frac{n + \frac{3}{4}}{f} \tag{8.13}$$

where f is the oscillation frequency to be maintained. This need not necessarily be at the resonant frequency of the cavity, although it must be fairly close to it, since t_f, and hence the oscillating frequency, *can* be adjusted electronically by changing the reflector voltage V_R without changing the mode of oscillation.

The maximum output power is obtained at the resonant frequency, but where this is 100 mW at 6 GHz, for example, it is possible to obtain an output greater than 40 mW over a tuning range of 60 MHz, with a sensitivity of about 1·5 MHz per volt of change in reflector voltage. The reflex klystron is the most widely used microwave valve, although it is being superseded in many applications by solid-state devices, and typical versions cover the frequency range 6–12 GHz with output powers between 0·1 and 1 W. The anode and cavity voltage is commonly about 300 V, and the reflector voltage, which may be varied for electronic tuning, is between about −100 and −200 V. These voltages are quoted with respect to the cathode, but in practice the cavity is normally earthed.

THE CAVITY MAGNETRON

A device which uses a number of resonant cavities for the generation of power at microwave frequencies is the *cavity magnetron*, illustrated in Fig. 8.10. The anode is a circular block of copper, which commonly

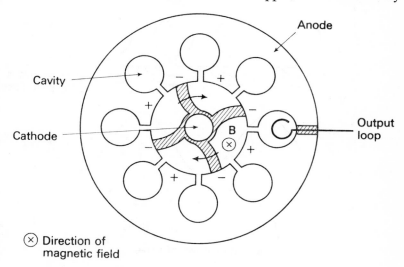

Fig. 8.10 Cavity magnetron

contains eight identical cavities separated by eight poles and opening onto an annular space with the cathode at the centre. A strong magnetic field, provided by a permanent magnet, acts at right angles to the d.c. electric field between anode and cathode. Thus electrons leaving the cathode would move in approximately circular paths near the anode as in the cylindrical magnetron (page 269) and would be prevented from reaching the anode if the magnetic field were strong enough. However, the electrons will induce currents in the cavities and set up oscillations which create a.c. electric fields between the poles. Then at any instant alternate poles are at potentials which are positive and negative with respect to the direct anode voltage. There is a phase difference of 180° between each pair of poles and so this method of oscillation is called the *π-mode*. As each cavity goes through a complete cycle of oscillation it appears that a wave travels round the anode, such that the wavelength equals the distance between alternate poles round the internal circumference. If the oscillations are to be maintained an integral number of wavelengths must occur round the anode, so that an even number of cavities is required.

The a.c. electric force acting on the electrons and the form of the travelling wave at a particular instant are shown in Fig. 8.11, in which

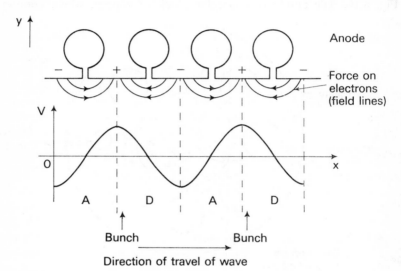

FIG. 8.11 Travelling wave in cavity magnetron

the cavities are drawn in a linear arrangement for convenience. In the actual magnetron the *x*- and *y*-directions correspond to tangential and radial directions respectively. The force acting on the electrons in the *y*-direction due to the d.c. fields is given by

$$F_y = Ee - Beu_x \qquad (8.14)$$

where u_x is the component of velocity in the *x*-direction parallel to the anode. The electric field E is the resultant of fields due to the anode voltage and the space charge and acts to increase the velocity of electrons *towards* the anode. The a.c. field at the anode causes the space between the electrodes to be divided into accelerating regions, A, and decelerating regions, D, which move round the anode with the travelling wave. In the A-regions the velocity parallel to the anode, u_x, is increased and in the D-regions u_x is decreased.

Electrons leave the cathode continuously to follow curved paths to the anode. Those entering an A-region experience an increase in u_x, so that F_y becomes negative (eqn. (8.14)) and they return to the cathode. Here they cause secondary emission as they give up their energy, which comprises about 90% of the total cathode emission in a practical magnetron. Electrons entering a D-region have u_x reduced which makes F_y positive, so that they continue towards the anode. Thus there is a tendency for electrons to collect in the decelerating regions, where bunches are formed with the wave. Since the bunches are moving in a region of retarding field they give up the energy they have gained from the d.c. field to the a.c. field, so that oscillations are maintained in the cavities. The bunches are in a state of dynamic equilibrium, since they lose electrons to the anode and gain electrons from the cathode. The space charge in the bunches then appears like the spokes of a wheel rotating round the cathode (Fig. 8.10).

The density of the space charge in a magnetron is high so that it is normally operated with the anode voltage applied in short pulses, although continuous wave (c.w.) magnetrons are also available. The output then consists of bursts of microwave power at the oscillation frequency, which can be varied over a relatively small range by mechanical tuning of the cavities. The anode is normally cooled either by air forced past it or by means of a water jacket. Applications include the transmitters of radar and navigation systems and microwave heaters. A typical communication magnetron oscillates between 9·3 and 9·4 GHz, with a peak anode voltage of 6 kV applied in 1 μs pulses at a rate of 1 000 times a second, having a microwave output of 8 kW for each pulse.

THE TRAVELLING-WAVE TUBE

In a klystron amplifier an electron beam interacts with resonant cavities which operate over a narrow frequency range. On the other hand, in a travelling-wave tube an electron beam interacts with a form of coaxial transmission line, which operates over a wide frequency range, so that a high gain is obtained with a much greater bandwidth than before. The inner conductor of the line is usually in the form of a wire helix and the electron beam moves along its axis (Fig. 8.12).

An electromagnetic wave travels down a loss-free transmission line at

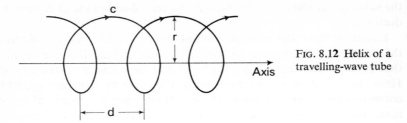

Fig. 8.12 Helix of a travelling-wave tube

about the velocity of light. The axial velocity of propagation is thus effectively reduced by the helix, so that the wave travels slightly more slowly than the electrons in the beam and in the same direction. This causes bunches to be formed which move in the decelerating region of the wave, as in the magnetron. Thus energy is continuously extracted from the beam by the wave, which therefore grows in amplitude as it travels down the tube. The wave is guided by the helical conductor, which has radius r and pitch d. Thus in one turn the wave travels distance $2\pi r$ along the helix and d along the axis of the tube. Its velocity round the helix may be taken as the velocity of light, c, so that the time for one turn is

$$t = \frac{2\pi r}{c} \tag{8.15}$$

The velocity along the axis is

$$u = \frac{d}{t} = \frac{cd}{2\pi r} \tag{8.16}$$

Considering the whole tube,

$$u = c \times \frac{\text{Axial length}}{\text{Wire length}} \tag{8.17}$$

and typically u might be $0\cdot 1c$, which corresponds to the velocity of a $2\cdot 6\,\text{kV}$ electron beam. The maximum operating frequency is determined by the condition that r is much less than the wavelength of the maximum operating frequency, which ensures that all parts of the wave may be considered in phase round a particular turn. The minimum operating frequency is determined by the length of the tube, which becomes too large to be practical for frequencies below about $500\,\text{MHz}$.

Bunching Conditions

The a.c. electric force acting on the electrons at a particular instant is shown in Fig. 8.13 together with the form of the travelling wave; V

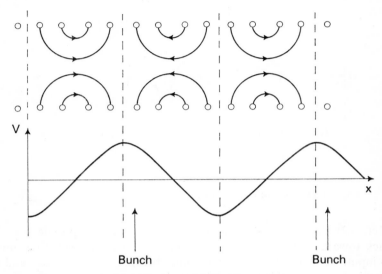

FIG. 8.13 Bunching conditions in a travelling-wave tube

represents the voltage distribution along the helix. Accelerating and decelerating regions occur, as in the magnetron, and electrons tend to bunch where the a.c. field is decelerating ahead and accelerating behind. If the bunches travelled with the peak of the wave, where $\partial V/\partial x$ is zero, there would be no exchange of energy between the beam and the wave. Thus their velocity is adjusted until they travel slightly faster than the wave, in the decelerating region where $\partial V/\partial x$ is negative. Then, since the a.c. field is retarding, it extracts energy from the electrons, which are therefore slowed down as each bunch occurs along the tube. Eventually the velocity of the electrons relative to the wave becomes zero, so that they move with the peak of the wave and energy exchange ceases.

The amplitude of the wave on a normal transmission line decreases exponentially due to the losses in the line. In the travelling-wave tube there is power gain along the line instead of power loss, so that the wave amplitude *increases* exponentially, as shown in Fig. 8.14. A full analysis

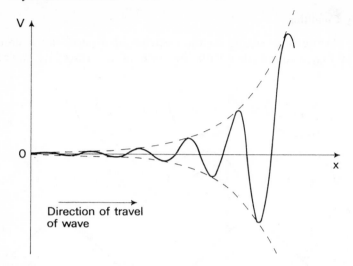

FIG. 8.14 Growth of wave along a travelling-wave tube

(Ref. 8.2) shows that there is a wave of constant amplitude travelling back towards the input. This constitutes positive feedback from output to input, which can lead to the amplifier becoming an oscillator, and in a practical amplifier this backward wave is attenuated.

An amplifier using a travelling-wave tube is illustrated in Fig. 8.15. The electron beam is confined by a magnetic field, as described on page 327, which is usually provided by a permanent magnet. The beam leaves the electron gun with a velocity determined by the anode voltage V_A and is extracted from the tube by the collector, which is held at a lower voltage V_C. The helix is at a higher voltage V_H, which is adjusted for maximum gain, all voltages being measured with respect to the cathode. The signal is fed into and out of the amplifier by means of waveguides, and the input and output couplers are carefully designed to ensure that the maximum bandwidth is obtained. The inside of the walls of the glass tube containing the helix is coated with colloidal graphite (Aquadag). This has the effect of attenuating the backward wave sufficiently to prevent oscillation.

Practical travelling-wave tubes are used in the gigahertz region, and typically have a bandwidth of 500 MHz centred on 6 GHz, although octave bandwidths are possible, which makes them very useful in microwave communication amplifiers. Typical values of operating voltage are

$V_A = 2\,\text{kV}$, $V_H = 2{\cdot}25\,\text{kV}$ and $V_C = 1{\cdot}5\,\text{kV}$, and a gain of 36 dB with an output power of 10 W is readily obtainable.

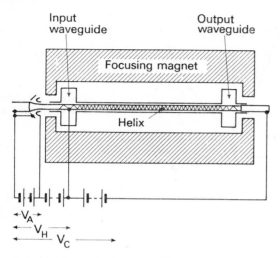

FIG. 8.15 Travelling-wave amplifier

THE TUNNEL DIODE

If the doping levels in a *p-n* junction are increased above those in a Zener diode the breakdown voltage is reduced until it reaches zero. However, this is not the limit, which is set by the solubility of impurities in the material and is reached when about 1 atom in 10^4 has been replaced by an impurity atom. This gives a doping density of about $10^{25}/\text{m}^3$, and under these conditions the semiconductors are highly degenerate (page 46). The diode is then still in a breakdown condition at a small forward bias, with the current dropping to the normal value for a *p-n* junction at a slightly larger forward bias. This leads to the current/voltage characteristic shown in Fig. 8.16, which has a negative-resistance region between V_P and V_V in which the diode is normally operated.

The high density of impurities reduces the width of the depletion layer to about 10 nm so that the potential barrier is sufficiently narrow for an electron to tunnel through it, since it also has a finite height (Appendix 2). This effect is responsible for the shape of the characteristic below the voltage V_V, above which forward current flows by the normal process.

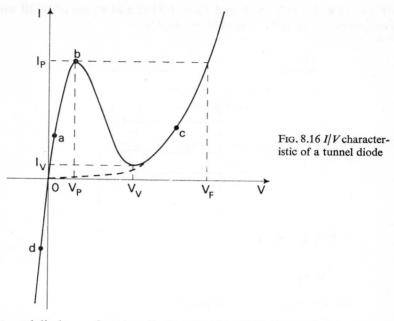

FIG. 8.16 I/V characteristic of a tunnel diode

The tunnel diode was first described by Esaki in 1958 (Ref. 8.3). The energy diagram at zero bias is shown in Fig. 8.17. The Fermi level is continuous through the junction and lies within the conduction band in the n-region and within the valence band in the p-region since the materials are degenerate. Thus even at room temperature the lower levels in the conduction

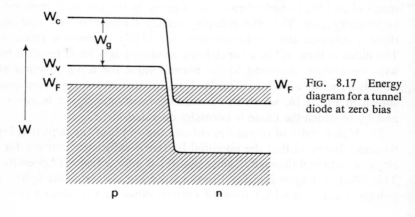

FIG. 8.17 Energy diagram for a tunnel diode at zero bias

band are filled and the upper levels in the valence band are empty, to about the Fermi level in each case. There are no empty levels available opposite filled levels in either the *p*- or *n*-regions, so that no tunnelling occurs and the current is zero.

When forward bias is applied the height of the barrier is reduced and the energy levels in the *n*-region move upwards on the energy diagram (Fig. 8.18(*a*) and point *a* on Fig. 8.16). Electrons in the conduction band of the *n*-region face empty levels in the *p*-region, and since energy is conserved during tunnelling, electrons move horizontally on the energy diagram. Thus an electron tunnel current flows from the *n*- to the *p*-region *through* the potential barrier, and at the same time electrons diffuse from the *n*-region and holes diffuse from the *p*-region *over* the potential barrier. Hence the total current is the sum of the tunnel and diffusion currents. As the forward bias is increased a point is reached at which all the conduction-band electrons face empty levels in the *p*-region and the tunnel current reaches a maximum (Fig. 18(*b*) and point *b* on Fig. 8.16), while for a slightly larger bias the filled and empty levels move away from each other and the current falls. The diffusion current increases slowly with bias until when the filled and empty levels are completely separated the tunnel current has become zero and current flows only by diffusion to give the normal diode characteristic (Fig. 18(*c*) and point *c* on Fig. 8.16). At the forward voltage V_F the diode current due to diffusion has become equal to the peak tunnel current I_P.

Under reverse bias conditions, filled levels in the valence band of the *p*-region move opposite empty levels in the conduction band of the *n*-region, so that an electron tunnel current flows from the *p*- to the *n*-region. This leads to a high current at a low reverse voltage (Fig. 18(*d*) and point *d* on Fig. 8.16), and it may be said that breakdown due to tunnel current extends from reverse bias to the forward bias V_V, which occurs at the valley point where the tunnel current ceases.

The current scale on Fig. 8.16 depends on the junction area and the doping level, diodes being obtainable with peak currents, I_P, between $50\,\mu A$ and $5\,A$. Typical values of peak-to-valley current ratio range from 5 to 15. Materials used for making tunnel diodes include germanium, silicon and gallium arsenide, whose electrical properties are compared with those of germanium and silicon in Table 5.1 on page 228. The voltages V_P, V_V and V_F depend on the material used for the diode, and mean values for germanium, silicon and gallium arsenide are compared in Table 8.1. It may be noted that these voltages increase with the energy

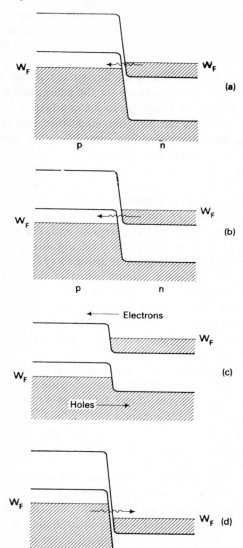

FIG. 8.18 Energy diagrams for a tunnel diode

(*a*), (*b*), (*c*) and (*d*) refer to the corresponding points on Fig. 8.16

Table 8.1 Tunnel Diode Characteristics

	Ge	Si	GaAs
Peak voltage V_P, mV	55	65	150
Valley voltage V_V, mV	320	420	500
Forward voltage V_F, mV	480	720	980
Maximum operating temperature, °C	100	200	175
Energy gap, eV	0·72	1·10	1·35

gap of the material and that the dynamic ranges, $V_V - V_P$, are about equal for silicon and gallium arsenide diodes and twice the value for germanium diodes.

Since the propagation of electrons by tunnelling proceeds at a velocity near that of light, the transit time is extremely short, given approximately by $10^{-8}/3 \times 10^8 \approx 10^{-17}$ s, so that the frequency response is limited only by stray inductance and capacitance and can extend into the gigahertz region. Only majority carriers are involved in the operation of a tunnel diode, so that it is relatively insensitive to changes in temperature. It can operate at temperatures as low as 4 K (page 47) and up to several hundred degrees Celsius.

An equivalent circuit for a tunnel diode is given in Fig. 8.19. R represents

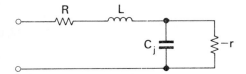

FIG. 8.19 Equivalent circuit of a tunnel diode

the resistance of the contacts and leads to the *p-n* junction and has a typical value of about 1 Ω, while L represents the self-inductance of the leads and is typically about 10 nH. The junction itself is represented by the capacitance C_j, which is a function of the bias voltage and is typically about 10 pF, and by the resistance $-r$, typically about $-100\,\Omega$. This is the slope resistance of the region between the voltages V_P and V_V where the current falls for an increase in voltage, so that $\partial I/\partial V$ is a negative quantity, $-1/r$. The self-resonant frequency of the equivalent circuit is given approximately by $1/2\pi\sqrt{(LC)}$ (see Problem 8.7), and for the typical values given above is about 500 MHz. For frequencies below this the diode may be represented by R in series with the parallel combination of

$-r$ and the effective capacitance. The diode can then be used in parallel with a tuned circuit or mounted in a resonant cavity to form either an amplifier or an oscillator.

Tunnel Diode Amplifier and Oscillator

The equivalent circuit of the amplifier is given in Fig. 8.20, where R_s

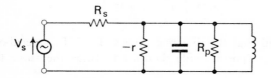

FIG. 8.20 Equivalent circuit of a tunnel diode amplifier

includes the source resistance and the diode lead resistance. At resonance the impedance of the tuned circuit or cavity is equal to its parallel resistance R_p, which includes any externally connected resistive load. The output voltage becomes

$$V_o = -V_s \frac{rR_p}{R_sR_p - r(R_s + R_p)} \tag{8.18}$$

and the voltage gain is

$$\frac{V_o}{V_s} = -\frac{\dfrac{rR_p}{R_s + R_p}}{\dfrac{R_sR_p}{R_s + R_p} - r} \tag{8.19}$$

dividing through by $R_s + R_p$. For stable amplification there is no phase change introduced by the diode, so that the gain is positive; this occurs when

$$r > \frac{R_sR_p}{R_s + R_p} \tag{8.20}$$

from eqn. (8.19). Since it is required that $V_o > V_s$, from eqn. (8.18),

$$rR_p > r(R_s + R_p) - R_sR_p \tag{8.21}$$

or

$$R_p > r$$

so that, for stable amplification,

$$R_p > r > \frac{R_s R_p}{R_s + R_p} \tag{8.22}$$

For given values of r and R_p the gain increases with R_s since the denominator of eqn. (8.19) becomes smaller as $R_s R_p/(R_s + R_p)$ tends to r.
When

$$\frac{R_s R_p}{R_s + R_p} = r \tag{8.23}$$

the gain becomes infinite, so that an output voltage is obtained for no input voltage. The amplifier has then become an oscillator, and since $V_s = 0$, R_s now represents the resistance of the power supply providing the bias current and voltage to the tunnel diode. The total loss resistance of the circuit is R_s in parallel with R_p, which is exactly compensated by the negative resistance of the diode. The tuned circuit has then become effectively loss free, so that when an oscillatory current has been started in the circuit (by the transient occurring when the bias supply is switched on) it will continue indefinitely.
   If we put

$$\frac{R_s R_p}{R_s + R_p} = R_B \tag{8.24}$$

where R_B is the effective bias resistance, the d.c. circuit becomes as shown in Fig. 8.21(a), and R_B may be represented by a load line on the diode characteristics (Fig. 8.21(b)). R_B includes both the source and load resistances since the tunnel diode amplifier has a direct connection between the input and output circuits. It is clear that the conditions $R_B < r$ and $R_B = r$, representing stable amplification and oscillation respectively, are shown by a load line cutting the characteristics at one point only, Q. If $R_B > r$ the load line cuts the characteristics in the three points Q, Q_1 and Q_2, and the current can adjust itself either to Q_1 or to Q_2, which define the limiting values of current and voltage. This is an unstable biasing condition and is used in switching applications. For instance, in a logic circuit the operating point Q_1 could represent a "1" and Q_2 a "0". The diode can also be used to generate pulses by switching between Q_1 and Q_2; these applications are discussed in more detail in Ref. 8.4.

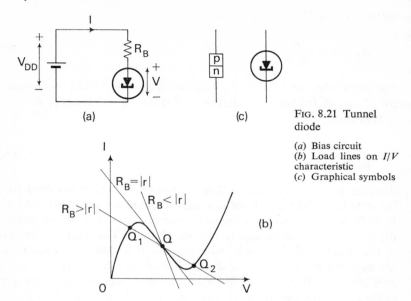

(a)

(c)

(b)

Fig. 8.21 Tunnel diode

(a) Bias circuit
(b) Load lines on I/V characteristic
(c) Graphical symbols

THE GUNN DIODE

In 1963 Gunn reported that a homogeneous sample of *n*-type gallium arsenide generated microwave oscillations when a steady electric field was set up in it. If the field exceeded a threshold value of about 300 kV/m the current through the sample was found to flow in pulses with a period proportional to its length *l*, which can lie between about 0·01 and 2·5 mm. This is the *Gunn effect*, which takes place in the bulk of the semiconductor instead of at a junction. Since the frequency of the current pulses lies in the gigahertz region it has led to the production of a simple microwave oscillator, requiring a supply of only a few volts for operation, with an output power of a few milliwatts.

The production of pulsed currents from a steady electric field depends on the availability of a material with two conduction bands. These are normally shown on an energy/momentum-vector diagram, which is described in Appendix 2 for one dimension. In a three-dimensional crystal the *k*-axis is divided into sections each of which corresponds to a particular direction in the crystal lattice and is called a *Brillouin zone*. The two conduction bands shown in Fig. 8.22 for gallium arsenide correspond to two different directions in the lattice and are parabolic in form, with their minima coinciding with a zone boundary. This is the same form as for an electron in free space, but in a solid the electron has an effective mass m^* related to the curvature of the W/k graph (eqn. (A.46)). In the lower energy band, $m^* = 0.067m$, while in the upper energy band $m^* \approx 0.35m$ (Ref. 8.5), with a separation $\Delta W = 0.36\,\text{eV}$ between

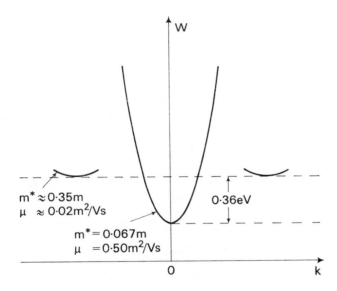

FIG. 8.22 Energy/momentum-vector diagram for gallium arsenide

the minimum energies of the bands. Since the effective mobility may be considered to fall as mass rises (eqn. (2.54)), the corresponding mobilities are found to be about $0.50\,\text{m}^2/\text{Vs}$ in the lower band and about $0.02\,\text{m}^2/\text{Vs}$ in the upper band. The density of allowed energy levels is greater in the upper band since it rises with effective mass (eqn. (A.59)).

When the field E in the device is low most of the electrons are in the lower conduction band since their thermal energy $(3/2)kT$ is less than ΔW at room temperature. As the field is increased the electrons gain energy El, where l is their mean free path, and more electrons are able to transfer to the upper conduction band, until at a sufficiently high field all the electrons will be in the upper band. Thus at low fields the electron mobility is high and at high fields the mobility falls as all the electrons are transferred to the upper conduction band. This is illustrated in Fig. 8.23, which relates electron velocity and electric field. The slope du/dE at a particular point gives the differential mobility. The regions of high and low positive mobility described above are separated by a third region of *negative* mobility, which is the cause of the Gunn effect.

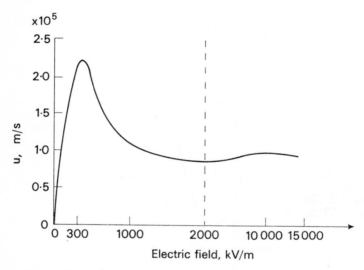

FIG. 8.23 Drift-velocity/field characteristic of a Gunn diode

In fact, Fig. 8.23 also represents a current/voltage characteristic, since for a device of given cross-sectional area and length the current is proportional to the electron velocity (eqn. (2.60)) and the voltage is proportional to the electric field. Thus the characteristic has a region of negative resistance, similar to that in the tunnel diode characteristic (Fig. 8.16). However, unlike the tunnel diode it is not possible to bias the device to a stable operating point in this region, since the current flows in pulses (Fig. 8.24(a)), and so the curve is a calculated one (Ref. 8.5). Gunn showed by means of a capacitively coupled electric probe that the electric field within the semiconductor was not uniform when the threshold field E_T was exceeded but was distributed as in Fig. 8.24(b). A narrow region of high electric field, known as a *domain*, was found to move along the specimen in the direction of electron flow with a constant drift velocity

u_D. While the domain was travelling from cathode to anode the current in the external circuit remained constant at the valley level I_V, but as the domain moved into the anode contact the current rose to the threshold level I_T. As soon as the whole domain had moved into the anode the current fell to I_V again, while a new domain formed at the cathode to repeat the cycle. The velocity u_D was found to remain constant at about 10^5 m/s even when the applied voltage was increased above the threshold value.

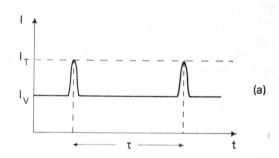

(a)

Fig. 8.24 Gunn diode

(a) Current pulses
(b) High-field domain

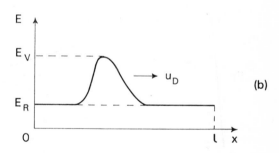

(b)

At present, it is considered that a domain begins to form at a *nucleating centre*, which may be found where there is a change in crystal structure such as occurs near the cathode contact. Then, if the diode is biased at a constant voltage just above the threshold voltage $V_T = E_T l$, the electric field is equal to E_T everywhere except inside a nucleating centre of length x, where it is just above E_T. Electrons within the centre are then transferred to the upper conduction band, so that they have a lower mobility than electrons outside the centre, which are still in the lower conduction band. Thus the transferred electrons are retarded relative to the electrons drifting both in front of them and behind them. This leads to the accumulation of space charge in the centre due to both the low-mobility electrons and the high-mobility electrons catching them up from behind. The space charge causes a further increase of the field inside the centre until it reaches a value E_V, while the field in the remainder of the diode falls to a value E_R (Fig. 8.24(*b*)), which correspond to the currents I_V and I_R respectively. The domain has now formed and moves towards the anode with velocity u_D.

Theory of Operation

A simple analysis of a Gunn diode, based on that suggested by Hilsum (Ref. 8.6), brings out some of the important operating features; more comprehensive theories are given in Ref. 8.7. The characteristic is idealized into three parts (Fig. 8.25), consisting of two straight lines, having slopes

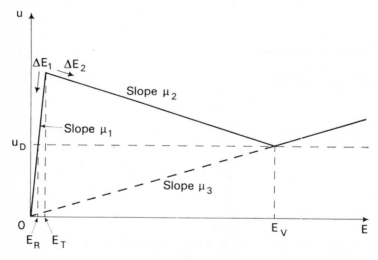

FIG. 8.25 Idealized drift-velocity/field characteristic of a Gunn diode

μ_1 at low fields and μ_3 at high fields, which are joined by a third straight line of slope $-\mu_2$. The slope μ_1 corresponds to the mobility of electrons in the lower conduction band, and μ_3 to the mobility in the upper conduction bands. The following approximate values are suggested by Hilsum for gallium arsenide: $\mu_1 = 0.50\,\text{m}^2/\text{Vs}$, $\mu_2 = 0.02\,\text{m}^2/\text{Vs}$, $\mu_3 = 0.02\,\text{m}^2/\text{Vs}$.

The field E_V occurs at the valley point on Fig. 8.25 where the mobility changes from a negative to a positive value, and the point of intersection of the domain velocity and low-field lines defines E_R, approximate values being $E_R = 200\,\text{kV/m}$, $E_T = 300\,\text{kV/m}$, and $E_V = 5\,000\,\text{kV/m}$.

If the increase in field within the centre is ΔE_2 due to the accumulation of space charge, the corresponding increase of potential across it is $\Delta E_2 x$. Since the bias voltage is constant this must be accompanied by a decrease in potential in the remainder of the diode, $\Delta E_1(l - x)$, where ΔE_1 is a reduction in the field E_T (Fig. 8.25). Then

$$\Delta E_2 x = \Delta E_1(l - x)$$

and

$$\frac{\Delta E_2}{\Delta E_1} = \frac{l - x}{x} \tag{8.25}$$

Since x is considerably less than l, $\Delta E_2 > \Delta E_1$ so that the field within the centre continues to rise rapidly until it has reached E_V and the field within the remainder of the diode falls to E_R. At this stage the domain has formed and drifts towards the anode with velocity u_D, the carriers both inside and outside it moving with the same velocity, so that, from Fig. 8.25,

$$u_D = \mu_1 E_R = \mu_2 E_V \tag{8.26}$$

The changes in drift velocity from the threshold value are $\mu_1 \Delta E_1$ outside the domain and $\mu_2 \Delta E_2$ inside it, and since both drift velocities start and finish at the same values,

$$\frac{\mu_1 \Delta E_1}{\mu_2 \Delta E_2} = \frac{\mu_1 x}{\mu_2(l - x)} = 1 \tag{8.27}$$

using eqn. (8.25), which leads to

$$\frac{x}{l} = \frac{\mu_2}{\mu_1 + \mu_2} \tag{8.28}$$

for a diode biased at the threshold voltage V_T. Inserting typical values for μ_1 and μ_2 gives $x/l = 1/26$. Thus x is much less than l, but it does increase with the bias voltage, since for an applied voltage V greater than V_T the equation

$$V = E_V x + E_R(l - x) \tag{8.29}$$

must be satisfied. Differentiation of this equation yields

$$\frac{dx}{dV} = \frac{l}{E_V - E_R} \approx \frac{l}{E_V} \tag{8.30}$$

which indicates that an increase in voltage does not affect the drift velocity u_D given by eqn. (8.26), but instead it causes the width of the domain to increase. Inserting typical values in eqn. (8.30) suggests a rate of increase of $0.2\,\mu\text{m/V}$ for voltages above V_T.

Only one domain is formed at a time, since when the field starts to grow in a nucleating centre the field external to it is reduced below E_T and nucleation can occur nowhere else until the domain has reached the anode. Then the field rises to E_T everywhere in the diode and a new domain can form. It is assumed that the domain field builds up to its steady-state amplitude in a time much shorter than the transit time, l/u_D. If this were not so it would grow as it drifted along the diode and this could prevent its complete formation. Such a situation occurs in short diodes of high-resistivity material, which have a low value of free electron density n_0. A minimum value of the product of doping density and length, $n_0 l$, is thus obtained for an oscillator, and has been found to be about $10^{16}/\text{m}^2$.

When $n_0 l \approx 10^{16}/\text{m}^2$ the domain length is about the same as the sample length, so that the above analysis applies only to samples where $n_0 l \gg 10^{16}/\text{m}^2$. However, for samples in which $n_0 l$ is much less than $10^{16}/\text{m}^2$, domain formation is prevented and a stable *microwave amplifier* has been produced (Ref. 8.7).

Modes of Operation

The behaviour of a Gunn oscillator depends on the circuit conditions in which it operates. When the circuit is resistive, or the voltage across the diode is constant, the period of oscillation is fixed at the transit time of a domain, l/u_D. However, when it is mounted in a resonant cavity the frequency of oscillation can be varied over about an octave by changing the cavity dimensions (Ref. 8.7). In both cases the diode is said to be operating in a *transit-time* mode, with the oscillation frequency controlled mainly by the sample length. Another mode of operation in which the frequency is controlled by the circuit rather than the sample length is the LSA (*limited space-charge accumulation*) mode. Here the field acting on the diode consists of the sum of the d.c. and the a.c. fields, and

$$E = E_0 + E_1 \sin 2\pi ft \tag{8.31}$$

The diode is biased at a field E_0 well above threshold (Fig. 8.26), and it is found that space charge can only accumulate into a domain in about the time it spends between E_T and E_0. This is clearly much shorter than the time spent below threshold, where space charge decays so that a domain never forms. In these circumstances the current carriers are considered as independent charges and the diode behaves as a negative resistance, in a similar manner to the tunnel diode (page 345). However, there is a minimum frequency below which the time spent between E_T and E_0 is long enough for space charge to accumulate, and there is a minimum

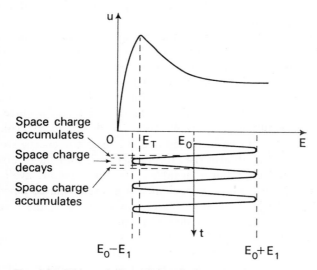

FIG. 8.26 LSA mode for a Gunn diode

doping level below which the negative resistance is too high. Thus operation in the LSA mode is found to be restricted to values of n_0/f between 2×10^{10} and $2 \times 10^{11} \mathrm{m^{-3} Hz^{-1}}$, giving a frequency range of about a decade for a device with a given doping density. Other modes of operation are also possible, as discussed more fully in Ref. 8.7.

An upper limit to the operating frequency is set by the length of the diode. Since gallium arsenide can be deposited epitaxially (page 186), the highest operating frequencies are obtained with a diode consisting of an epitaxial layer only a few micrometres thick deposited on a gallium-arsenide substrate, with the Gunn effect taking place through the layer.

PARAMETRIC AMPLIFICATION

In the most usual type of amplifier, such as that described on page 154, the energy for operation is obtained from a direct voltage supply and an active device, such as a transistor or a valve, provides the essential amplifying element. In a *parametric amplifier*, however, the energy to operate the amplifier is obtained from an alternating voltage supply whose frequency is different from the signal frequency. In addition, the essential amplifying element is a circuit parameter, such as capacitance or inductance, which is varied at the supply or *pump* frequency.

Consider a capacitor C which has a charge q on its plates so that the resulting voltage across it is $v = q/C$. If the plates are pulled apart very quickly, C will fall and q will remain unchanged, so that v must rise proportionately. Thus the energy stored by the capacitor, $\frac{1}{2}Cv^2$, will rise; this energy is supplied by the agency causing movement of the plates. If v is the instantaneous value of an alternating voltage, the plates may be pushed towards each other when $v = 0$ without any change in voltage.

In principle an inductor L may also be used, which has a current i flowing through it resulting in a flux Φ. Then $i = \Phi/L$, and if L can be reduced very quickly, i will be increased and the magnetic energy stored by the inductance, $\frac{1}{2}Li^2$, will also rise. In practice, it is difficult to obtain a suitable variable inductor, but a variable capacitor is readily available in the form of the depletion-layer capacitance of a reverse-biased junction diode (page 110). The junction capacitance is a function of the reverse voltage, which may be varied at the desired pump frequency.

Suppose such a variable capacitance $C(t)$ forms part of a series circuit with a fixed inductance (Fig. 8.27) and is supplied with a signal V_s at the resonant frequency of the circuit. The output from the circuit is taken as the voltage V across $C(t)$. If the capacitance is constant initially at a value

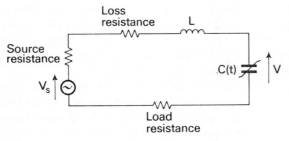

FIG. 8.27 Simple parametric amplifier circuit

C and the instantaneous value of V is $v = V_m \sin \omega t$ then the mean energy W stored in the capacitance is $\frac{1}{2}CV_m{}^2$ (Fig. 8.28(a)). However, if $C(t)$ is reduced to $C - \delta C$, where $\delta C \ll C$, when v reaches its peak value, both v and the stored energy will be increased (Figs. 8.28(b) and (c)). The capacitance then returns to C when $v = 0$ without affecting the stored energy, and $C(t)$ is reduced again when v reaches the next negative peak. Thus if

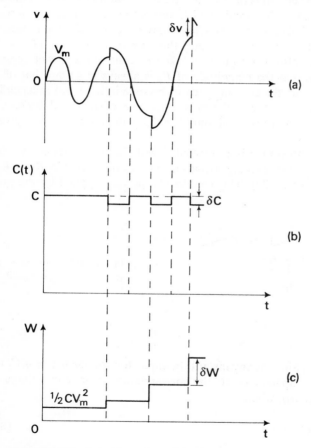

FIG. 8.28 Waveforms for a degenerate parametric amplifier

    (*a*) Capacitor voltage
    (*b*) Capacitance variation
    (*c*) Mean stored energy

361

$C(t)$ is changed at *twice* the signal frequency, each time $C(t)$ is reduced both the voltage and the stored energy are increased, such an arrangement being called a *degenerate* parametric amplifier.

A practical circuit contains series resistance R_1, which is the sum of the resistances due to the signal source, the losses in the inductor and capacitor and the load resistance. At resonance, with the pump source disconnected, the current in the circuit is then V_s/R_1. When the pump source is connected power is supplied *to* the circuit, which means that a *negative* series resistance, $-R_2$, is introduced.* The total circuit resistance then becomes $R_1 - R_2$, which is still positive, and the current rises to $V_s/(R_1 - R_2)$, causing v to rise also. Care must be taken to control the power supplied by the pump so that the magnitude of R_2 is always less than that of R_1, because if $R_2 = R_1$ the current tends to become infinite and independent of V_s. Thus the circuit would generate its own current and become an oscillator, so that in a practical amplifier the gain is normally not greater than about 30 dB.

When the pump source is connected, v will rise until the energy added per cycle of capacitance change equals the energy dissipated in $R_1 - R_2$ in the same time (Fig. 8.29). This enables the power supplied by the pump

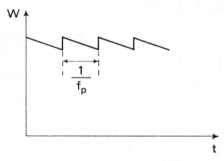

FIG. 8.29 Energy stored in a varactor diode at equilibrium

to be calculated. Since at any instant the capacitor voltage is $v = q/C$, then for a change in capacitance δC the corresponding change in voltage will be δV (Fig. 8.28(a)), where

$$\delta v = \frac{dv}{dC}\delta C \tag{8.32}$$

$$= -\frac{q}{C^2}\delta C = -\frac{v}{C}\delta C \tag{8.33}$$

* Power is absorbed by a positive resistance and so power is supplied by a negative resistance.

and v is the voltage at the instant at which C is changed. At any instant the energy stored in the capacitor is $\frac{1}{2}Cv^2$, so that when both C and v are changed together the corresponding change in energy will be δW, where

$$\delta W = \frac{\partial W}{\partial C}\delta C + \frac{\partial W}{\partial v}\delta v$$

$$= \tfrac{1}{2}v^2\,\delta C + vC\,\delta v$$

$$= \tfrac{1}{2}v^2\,\delta C - v^2\,\delta C \qquad\qquad (8.34)$$

or

$$\delta W = -\tfrac{1}{2}v^2\,\delta C \qquad\qquad (8.35)$$

It should be noted that δW is positive when δC is negative. Then at equilibrium the energy supplied is dissipated before the next capacitance change in time $1/f_p$, where f_p is the pump frequency. Hence the power supplied, p_p, is given by

$$p_p = -\frac{v^2}{2}f_p\,\delta C \qquad\qquad (8.36)$$

and for a pump waveform which is rectangular, as in Fig. 8.28(b), $v = V_m$. Thus power is supplied by the pump when δC is negative and this power is proportional to the pump frequency. When δC is positive, power would be absorbed by the pump if v were not zero at the same time.

When the frequency of the pump source is exactly twice that of the signal, as in the above circuit, the amplifier is said to be *degenerate*. It is difficult to maintain the correct phase relationship between the pump and signal waveforms in practice, owing to lack of control over the signal frequency. However, greater flexibility in operation and improved power gain are obtained if a second tuned circuit is connected across the variable capacitance (Fig. 8.30).* This is known as an *idler* circuit, since it contains no source of energy, but it does allow the pump to work at a frequency other than twice the signal frequency by modifying the voltage v across $C(t)$. If the signal source is made zero and the pump source applied on its own the idler and signal circuits are excited into oscillation at their respective resonant frequencies, so that v is the sum of the voltages due to the signal and idler circuits. Suppose these voltages are equal, with $v_i =$

---

* A pump circuit resonant at f_p can only be used when the pump waveform is sinusoidal, as in the practical form of the amplifier.

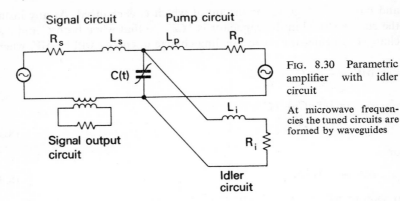

Signal circuit   Pump circuit

Fig. 8.30 Parametric amplifier with idler circuit

At microwave frequencies the tuned circuits are formed by waveguides

$V_m \sin \omega_i t$ due to the idler circuit and $v_s = V_m \sin \omega_s t$ due to the signal circuit. Then

$$v = v_i + v_s \tag{8.37}$$

$$= V_m (\sin \omega_i t + \sin \omega_s t) \tag{8.38}$$

$$= 2 V_m \sin \frac{(\omega_i + \omega_s)t}{2} \cos \frac{(\omega_i - \omega_s)t}{2} \tag{8.39}$$

The waveform corresponding to eqn. (8.39) is illustrated in Fig. 8.31(*a*) and it may be seen that the voltage is zero when either

$$\sin \frac{(\omega_i + \omega_s)t}{2} = 0 \quad \text{or} \quad \cos \frac{(\omega_i - \omega_s)t}{2} = 0$$

Thus if $\omega_p = \omega_i + \omega_s$, $C(t)$ can be increased without loss of energy as before, and similarly for $\omega_p = \omega_i - \omega_s$. In either case $C(t)$ may not be decreased exactly at the peak of the waveform, and the amplitudes of consecutive peaks will not be the same. The power supplied by the pump source is then divided between the idler and signal circuits, so that for any given cycle of the pump waveform when $\omega_p = \omega_i + \omega_s$,

$$p_p = p_i + p_s \tag{8.40}$$

$$= -\frac{v^2}{2} f_p \, \delta C = -\frac{v^2}{2} (f_i + f_s) \, \delta C \tag{8.41}$$

364

The power supplied from the pump will depend on the value of v at which $C(t)$ is reduced, but on every occasion

$$\frac{p_i}{p_s} = \frac{f_i}{f_s} \tag{8.42}$$

and pump power is divided between idler and signal circuits in the same ratio as their respective frequencies. Suppose now that the signal source

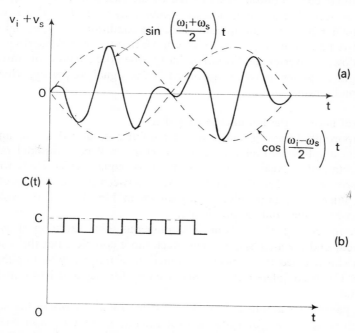

Fig. 8.31 Waveforms for a non-degenerate parametric amplifier

is applied and supplies power P_s at frequency f_s. Then, in order to maintain the condition implied in eqn. (8.42), the pump must supply extra power P_i at frequency f_i, so that

$$\frac{P_i + p_i}{P_s + p_s} = \frac{f_i}{f_s} \tag{8.43}$$

and this equation can be satisfied only if $P_i/P_s = f_i/f_s$ also. Thus power gain from the signal to the idler circuit will occur when $f_i > f_s$, together with an increase in frequency, and in this condition the amplifier is known as an *up-converter*. In some applications it may be necessary to reduce the signal frequency, and this will be achieved when $f_i < f_s$, together with a power loss from signal to idler circuit, the amplifier then becoming a *down-converter*. Finally power *at* the signal frequency can be extracted from the signal circuit, so that the power gain is $(P_s + p_s)/P_s$, in which case it is operating as a *straight amplifier*.

If similar considerations are applied to the case when $\omega_p = \omega_i - \omega_s$, it will be seen that amplification can occur only when $f_i > f_s$, i.e. when the amplifier is used as an up-converter. The condition $f_i < f_s$ would give a negative pump output power, which implies that the pump absorbs power in the down-converter condition. Similarly, when the circuit is operated as a straight amplifier, power is absorbed at the signal frequency, which is again undesirable.

Practical Form of the Parametric Amplifier

An important difference between the amplifier described above and a practical amplifier is that the pump waveform will be sinusoidal rather than rectangular. Also v_s and v_i need not be equal, which leads to the waveform of Fig. 8.32(*a*). The relationships between the frequencies of the signal, idler and pump voltages are shown in Fig. 8.32(*b*), the voltage relationships being only approximate.

Since the change of $C(t)$ is not instantaneous some variation of charge will occur and the analysis becomes much more complex, but the conclusion is still reached that power is a function of frequency as implied by eqn. (8.42). A complete analysis was given by Manley and Rowe in 1956 (Ref. 8.8).

The parametric amplifier is very useful in situations where the signal strength is very low, for instance in a satellite communication receiver where the signal power received from the satellite may be only about 10^{-13} W. Thus the noise power introduced by the amplifier must be even lower if the signal is not to be lost in the noise. Since its essential amplifying element is a capacitance, which would introduce no noise at all if it were perfect (page 374), a parametric amplifier operated at a low temperature is used as the first stage in the earth receiver. The earth's atmosphere allows electromagnetic waves within certain ranges of frequency to pass through it more easily than others, and so satellite communication systems are operated between 4 and 8 GHz, since this frequency range corresponds to a "window" of easy transmission (page 376). The bandwidth of a typical

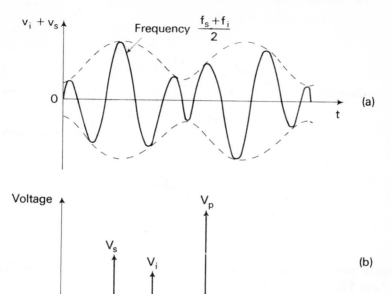

Fig. 8.32 Conditions for a practical parametric amplifier with
$$f_p = f_s + f_i$$

amplifier is 500 MHz, which is similar to that of a travelling-wave-tube amplifier. This corresponds to a straight amplifier operating with $f_s = 4\,\text{GHz}$, $f_p = 28\,\text{GHz}$ and $f_i = 24\,\text{GHz}$, providing a power gain of 400 times, or 26 dB.

The capacitance $C(t)$ is obtained from a reverse-biased junction diode mounted in a pill-shaped container (Fig. 8.33(a)). It is called a *varactor* diode and may be constructed from silicon, or from gallium arsenide which is preferred for applications at frequencies above about 20 GHz. An equivalent circuit is shown in Fig. 8.33(b), where L represents the inductance of the internal leads to the p-n junction, and C_p the effective capacitance across the terminals due to the encapsulation. R_{sc} is due to the resistance of the semiconductor regions external to the junction, and to the resistance of the contacts, and may be about $0.5\,\Omega$. There is also a resistance in parallel with $C(t)$ which depends on the bias applied to the diode and is very high for reverse bias (page 78). These resistances introduce a

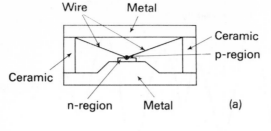

FIG. 8.33 Varactor
diode

(*a*) Practical form
(*b*) Equivalent circuit

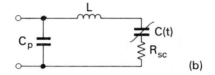

small amount of noise, which is reduced by cooling the amplifier to about 5 K in liquid helium.

$C(t)$ can be a function of $V_R^{-1/2}$ or $V_R^{-1/3}$, depending on whether the junction is abrupt or graded (page 115). V_R consists of the externally applied reverse bias voltage with a pump voltage alternating at frequency f_p, and owing to the non-linear relationship between $C(t)$ and V_R for either type of junction, the capacitance may be written in the form

$$C(t) = C_0 + C_1 \cos \omega_p t + C_2 \cos 2\omega_p t + C_3 \cos 3\omega_p t + \ldots \quad (8.44)$$

where C_0 is the capacitance for zero pump voltage. This equation includes capacitance changes at harmonics of the pump frequency, so that the varactor diode can also be used as a harmonic generator. In the parametric amplifier the varactor diode is effectively operated in a tuned circuit resonant at f_p, so that eqn. (8.44) becomes

$$C(t) = C_0(1 + \gamma \cos \omega_p t) \quad (8.45)$$

where $\gamma = C_1/C_0$. In practice γ has values up to about 0·3 and C_0 is typically 1 pF, both γ and C_0 being functions of V_R. It may be seen from the equivalent circuit (Fig. 8.33(*b*)) that there are two resonant frequencies. The series resonant frequency is

$$f_0 = \frac{1}{2\pi\sqrt{(LC_0)}}$$

which is in the gigahertz region since L is about 10^{-9} H, while the parallel resonant frequency is about $3f_0$. The diode may be operated at f_0 for maximum current, or below f_0 where the overall combination is capacitive. The pump may be a reflex klystron or a solid-state source of microwave power.

FREQUENCY MULTIPLICATION

At the present time, a common technique for generating low-power microwave signals using solid-state devices is to start with a transistor oscillator working at about 100 MHz. The frequency of the output from the oscillator is then multiplied in stages until the desired frequency is achieved. One method uses the oscillator output as the pump waveform of a varactor diode. This is mounted in a cavity resonant at 400 GHz, say, from which is extracted the power at $4f_p$ generated by the varactor diode (eqn. (8.44)). Two similar frequency multipliers then raise the frequency to 1·6 GHz and 6·4 GHz respectively.

A second method uses a *step recovery diode* as a frequency multiplier. When a p^+-n diode is used as a rectifier the excess charge established within the n-region during the conduction half-cycle needs a short time to disperse when the diode is cut off. This depends on the lifetime of holes in the n-region, and at microwave frequencies conduction continues after the alternating voltage has reversed polarity (Fig. 8.34). Such a waveform has a high harmonic content, and the step recovery diode can give useful frequency multiplication up to ten times the signal frequency. The storage property is undesirable in a normal diode rectifier, so that a Schottky diode (page 81) is used for rectification at microwave frequencies. Here the current is carried by majority carriers so that no time is required for recombination and the current can be switched off within about 0·1 ns.

Finally, microwave power can be generated directly by means of a reverse-biased p-n junction. This is known as an IMPATT (IMPact Avalanche and Transit Time) diode, or Read diode, since it was suggested by W. T. Read in 1958 (Ref. 8.9). It is also called a p-n-i-n or a p-i-n diode, depending on its construction.

THE READ DIODE (IMPATT DIODE)

A Read diode consists of a lightly doped n-region together with an intrinsic region between more heavily doped p- and n-regions. This gives

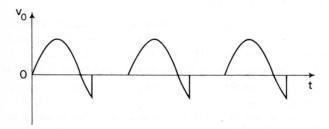

FIG. 8.34 Step recovery diode

(*a*) Circuit   (*b*) Current waveform

a *p-n-i-n* structure, where *i* stands for "intrinsic" (Fig. 8.35(*a*)). When such a diode is reverse biased the field is high at the *p-n* junction and falls towards the end *n*-region (Fig. 8.35(*b*)). The depletion layer extends between the end *p*- and *n*-regions and so is of fixed width. During operation the maximum field exceeds 10^7 V/m and avalanche multiplication of the current occurs (page 103). Electron-hole pairs are generated at the *p-n* junction and the electrons travel across the depletion layer, where the field is always above about 10^6 V/m, in which range they will move with a constant velocity of about 10^5 m/s in silicon (page 51).

The diode is biased so that the d.c. field is just below E_B, the critical field for avalanche breakdown. When an alternating voltage is superimposed on the bias the total field at the *p-n* junction exceeds E_B and causes an electron avalanche. However, the ionization of the atoms by the impact of electrons is relatively slow and the current will reach a maximum value only *after* the total field has reached its maximum. At the correct signal frequency, which is of the order of gigahertz, the avalanche current at the *p-n* junction lags $90°$ behind the field. The time taken by the electrons to cross the depletion layer to the end *n*-region at constant velocity can then be adjusted by choosing the correct spacing between the *p*-region and the far *n*-region. This is normally about $10\,\mu$m, which ensures that the current leaving the end *n*-region lags 90° behind the current at the *p-n* junction.

If the space charge due to the current is small the signal voltage is in phase with the field, so that the signal current through the diode then lags

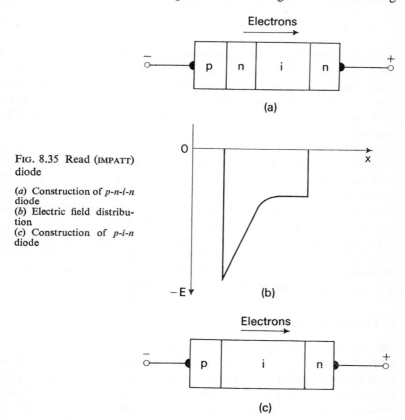

(a)

FIG. 8.35 Read (IMPATT) diode

(a) Construction of *p-n-i-n* diode
(b) Electric field distribution
(c) Construction of *p-i-n* diode

(b)

(c)

180° behind the voltage and the diode appears to have a negative resistance. The Read diode is mounted in an inductive cavity which is tuned to resonance by the junction capacitance. The negative resistance of the diode then compensates the loss resistance of the cavity, so that oscillations can occur (see also the tunnel diode oscillator, page 350).

The optimum operating frequency is obtained from the transit time, τ, of holes across the depletion layer. Where the width of the layer is w

and the velocity of the holes is u, $\tau = w/u = 10^{-10}$ s, inserting the typical values for silicon. The transit time τ should correspond to one half-period of oscillation, so that the frequency is given by $1/2\tau$, typically 5 GHz. The capacitance per unit area of the diode is given by

$$C = \frac{\epsilon_r \epsilon_0}{w} \tag{8.46}$$

and for a silicon diode with a depletion layer width of 10 μm, C is 10·6 pF/mm^2. If the inductance of the cavity is about 1 nH the diode area is chosen to give a capacitance which will resonate with 1 nH at 5 GHz, the required area being about 0·1 mm^2. A continuous power output of 0·5 W at 10 GHz and a pulsed output of 0·4 W at 50 GHz have been obtained from *p-n-i-n* diodes.

A simpler device is the *p-i-n* diode (Fig. 8.35(*c*)), in which avalanche breakdown occurs at the *p-i* junction, but which otherwise operates in a similar manner to the Read diode. A single *p-i-n* diode has a power output measured in milliwatts, but they can be connected in series or parallel to increase the total power. Thus five *p-i-n* diodes connected in parallel can give a continuous output of 2·8 W at 12 GHz.

The electrical noise produced by an IMPATT diode is large compared with that from a Gunn diode, for instance, owing to the random nature of the avalanche process, so that at the present time its widest application is in switching circuits at microwave frequencies.

ELECTRICAL NOISE

The microwave devices described above are often used in situations where the signal power is very small, for instance in the receiver of a satellite communication system or a radio telescope. The minimum signal that can be observed is set by the level of *electrical noise* in the system, which is a random fluctuation of voltage and current or electromagnetic fields. It is present in all electronic devices and components and also in the atmosphere. An indication whether satisfactory amplification can be obtained is given by the ratio of signal power to noise power, because if this ratio is only just above unity it will be difficult to decide whether the signal is present at all. Thus when weak signals are to be amplified it is desirable that the noise power introduced by the devices and components should be as small as possible.

Two of the main sources of noise in electronic devices are *thermal noise* and *shot noise*. Thermal noise is due to the random motion of the current carriers in a metal or semiconductor (page 49), which increases with temperature. If a resistor is short-circuited, the mean current is zero but the instantaneous current varies about zero in a random fashion and with very small amplitude. Measurements show that all frequencies within the bandwidth B of the measuring instrument are present in the random current. The phenomenon was studied in detail by Johnson and Nyquist in 1928. They showed that the thermal noise power available from a resistor is given by the expression

$$P_n = kTB \qquad (8.47)$$

where k is Boltzmann's constant and T is the absolute temperature.

The resistor can be represented by a noise-free resistor R in series with a generator producing $\overline{v^2}$, the mean-square noise voltage (Fig. 8.36(a)). If

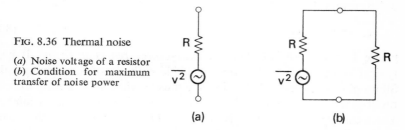

FIG. 8.36 Thermal noise

(a) Noise voltage of a resistor
(b) Condition for maximum transfer of noise power

(a)   (b)

the resistor is short-circuited the mean-square noise current flowing is $\overline{v^2}/R$, which represents the sum of the individual noise currents at all the frequencies present within the bandwidth considered. Maximum noise power will be extracted if a second noise-free resistor R is connected across the first (Fig. 8.36(b)). The total power is then $\overline{v^2}/2R$ and the power dissipated in the second resistor is half the total power, given by

$$P_n = \frac{\overline{v^2}}{4R} = kTB \qquad (8.48)$$

so that

$$\overline{v^2} = 4kTRB \qquad (8.49)$$

Thus for a $1\,000\,\Omega$ resistor operating at room temperature over a bandwidth of 5 MHz, the mean-square noise voltage is $8{\cdot}1 \times 10^{-11}\,\text{V}^2$ and the r.m.s. noise voltage is $9\,\mu\text{V}$. Eqn. (8.49) applies to the effective resistance of any circuit however complicated and represents its noise contribution in terms of voltage fluctuation. Pure reactances, having zero resistance, contribute no electrical noise.

Shot noise is due to the random flow of electrons in an electric current and is due to the particle nature of electric charge. Thus, for example, in a saturated diode at a given cathode temperature, the current has a fixed average value with very small random fluctuations about this value. These constitute a mean-square noise current, which was shown by Schottky to be

$$\overline{i_n^2} = 2eI_S B \tag{8.50}$$

where I_S is the average saturated current. Under space-charge-limited conditions $\overline{i_n^2}$ is reduced, since the electrons no longer move independently, the current being controlled by the electrons in the space charge. Then the noise current is

$$\overline{i_n^2} = \Gamma^2 2eI_S B \tag{8.51}$$

where $\Gamma \leqslant 1\cdot0$ and is called the *space-charge reduction factor*. It has a typical value of about $0\cdot3$ for the oxide cathode of a receiving valve. Shot noise is also extended to include the electrical noise arising from the random nature of the diffusion of current carriers across a *p-n* junction. The mean-square current is given by eqn. (8.50), I_S now being the current flowing through the junction.

Another source of noise in semiconductor devices is due to the random nature of the generation and recombination of charge carriers, which in effect causes a small fluctuation of resistance and is known as *recombination noise*. A further source appears as the operating frequency is reduced, since the total noise rises above the level of shot noise with a power proportional to $1/f$. This is known as *flicker noise* and is strongly dependent on surface conditions. It becomes apparent below about 1 kHz for bipolar transistors and JUGFETs, and below 100 kHz for MOSTs, which rely upon surface phenomena for their operation. *Partition noise* is due to the random manner in which a stream of electrons divides amongst a number of electrodes, for instance between the screen grid and anode of a pentode.

Noise Factor

The noise introduced by an amplifier will be due to a combination of some or all of the sources described above. They will give rise to an output noise voltage $\overline{v_o}^2$, which may be conveniently represented by an input noise voltage given by

$$\overline{v_r}^2 = \frac{\overline{v_o}^2}{A^2} \tag{8.52}$$

where A is the voltage gain of the amplifier. If the noise voltage introduced with the signal is $\overline{V_g}^2$, and the mean-square signal voltage is $\overline{v_s}^2$, the signal/noise ratio at the input is

$$\frac{S}{N} = \frac{\overline{v_s}^2}{\overline{v_r}^2 + \overline{v_g}^2} \tag{8.53}$$

This ratio depends on the signal level, which is not necessarily constant, and the bandwidth of the amplifier. A useful parameter which expresses the degradation of the signal/noise ratio when the amplifier is introduced is the *noise factor* (or *noise figure*), which is defined by

$$F = \frac{\overline{v_r}^2 + \overline{v_g}^2}{\overline{v_g}^2} = 1 + \frac{\overline{v_r}^2}{\overline{v_g}^2} \tag{8.54}$$

Then in a noise-free amplifier $\overline{v_r}^2 = 0$ and $F = 1$. The noise factor in decibels is $10 \log_{10} F$, which becomes $0\,\text{dB}$ for a noise-free amplifier.

Noise Input Temperature, T_i

When the noise factor is close to unity, as in the cooled parametric amplifier and the maser (page 386), it is useful to express F in terms of absolute temperature. $\overline{v_g}^2$ is standardized as the noise from a source at a temperature of 290 K and the amplifier noise voltage $\overline{v_r}^2$ can be obtained in terms of the noise input temperature of the amplifier, T_i. Then, from eqn. (8.54),

$$\frac{\overline{v_r}^2}{\overline{v_g}^2} = F - 1 = \frac{P_r}{P_g} \tag{8.55}$$

But $P_g = 290kB$ and $P_r = T_i kB$, so that

$$F - 1 = \frac{T_i}{290} \tag{8.56}$$

and

$$T_i = 290(F - 1) \quad \text{kelvins} \tag{8.57}$$

A noise-free amplifier has $T_i = 0\,\text{K}$. Representative values of noise factor in decibels and noise input temperature in kelvins are given in Table 8.2 for low-noise microwave amplifiers. The klystron amplifier is not included

Table 8.2 Characteristics of Low-noise Microwave Amplifiers

	Gain	Bandwidth	Noise factor	Noise input temperature
	dB	MHz	dB	K
Travelling-wave amplifier	25–28	500	5–8	630–1 540
Tunnel-diode amplifier	15	500	5	630
Cooled parametric amplifier	26	500	0·3	20
Maser	30	50	0·1	6·5

since it has a high noise factor and so is not used as an amplifier of very small signals. The electrical noise present in the atmosphere may also be described in terms of noise temperature, which reaches a minimum value of a few kelvins for signals of frequency between about 4 and 8 GHz. Thus this frequency range is very suitable for satellite communications, since the noise received with the signal is smaller than at frequencies outside the range.

REFERENCES

8.1 HARVEY, A. F., *Microwave Engineering* (Academic Press, 1963).

8.2 GEWARTOWSKI, J. W. and WATSON, H. A., *Principles of Electron Tubes* (Van Nostrand, 1965).

8.3 ESAKI, L., "New phenomenon in narrow germanium *p-n* junctions", *Physical Review,* **109**, p. 603 (1958).

8.4 GENTILE, S. P., *Basic Theory and Applications of Tunnel Diodes* (Van Nostrand, 1962).

8.5 BUTCHER, P. N., "The Gunn effect", *Reports on Progress in Physics*, **30**, p. 97 (1967).

8.6 HILSUM, C., "A simple analysis of transferred electron oscillators", *British Journal of Applied Physics*, **16**, p. 1401 (1965).

8.7 *Transactions IEEE*, **E.D. 14** (Sept. 1967): special issue on semiconductor bulk effect and transit time devices.

8.8 BLACKWELL, L. A. and KOTZEBUE, K. L., *Semiconductor Diode Parametric Amplifiers* (Prentice-Hall, 1961).

8.9 READ, W. T., "A proposed high-frequency negative-resistance diode", *Bell System Technical Journal*, **37**, p. 401 (1958).

FURTHER READING

KING, R., *Electrical Noise* (Chapman & Hall, 1966).
BENNETT, W. R., *Electrical Noise* (McGraw-Hill, 1960).

PROBLEMS

8.1 An electron beam is formed by accelerating electrons to a uniform velocity of 10^7 m/s. The beam then passes through two grids spaced 0·5 mm apart, between which is applied a 6 V r.m.s. sinusoidal voltage of frequency 1 GHz. (*a*) Stating any assumptions made, show that the beam leaving the output grid is velocity modulated, and calculate the depth of modulation.

(*b*) What would be the effect of increasing the signal frequency to 20 GHz?

(*Ans.* (*a*) 1·5 %, (*b*) Transit time = 1 cycle, so there is no velocity modulation)

8.2 Explain the terms velocity modulation and beam bunching and draw a time/distance (Applegate) diagram to show how a velocity-modulated electron beam would bunch (*a*) in a field-free region, (*b*) in a retarding-field region. In both cases indicate the relative phase between the modulating voltage and the first peak of the bunched beam current and indicate the importance of this phase relationship in the operation of a reflex klystron oscillator.

The reflector region of a reflex klystron oscillator is 0·3 mm long and the beam (resonator) voltage of the valve is 1 kV. What is the largest voltage that can be applied to the reflector if the valve is to oscillate at 10 000 Mc/s? (*IEE*, June 1966) (*Ans.* 855 V)

8.3 A reflex klystron oscillator has a reflector mounted 2 mm away from the resonator grids and is mechanically tuned to a frequency of 5 GHz. The resonator is maintained at 300 V and the reflector voltage can be varied between 0 and −500 V with respect to the cathode. Determine the number of possible modes of oscillation that can be obtained and the corresponding values of the reflector voltage.

(*Ans.* Three modes, with $V_R = 367$, 125 and 11 V)

8.4 Explain the principle and describe the constructional form of a travelling-wave tube for amplification at microwave frequencies.

The helix of a travelling-wave tube has 20 turns/in, a diameter of 0·15 in and an axial length of 9 in. *Estimate* what final-anode voltage must be applied to obtain useful gain. Also calculate the average transit time of an electron through the helix. The ratio of charge to mass for the electron is $1·76 \times 10^{11}$ C/kg. (*L.U.*, *B.Sc.* (*Eng.*), 1963)

(*Ans.* 2·87 kV, 7·2 ns)

8.5 Draw a diagram to show the essential details of a travelling-wave-tube amplifier, and describe the function of each part. Explain briefly how the energy is transferred from the electron beam to the r.f. circuit.

The helix in such an amplifier has a pitch of 0·75 mm, an angle of 6° 7′ and has 192 turns. Estimate the minimum electron-beam accelerating voltage, and determine the operating frequency when the number of r.f. wavelengths on the helix is 30.

What happens when the beam voltage is less than the value estimated?

(*Ans.* 2·91 kV, 6·67 GHz)

8.6 With the aid of electron energy-band diagrams give a basic explanation of the shape of the forward characteristic of the tunnel diode. For the circuit shown in Fig. 8.37 plot on graph paper the output voltage for one cycle of the low-frequency a.c. signal.

The tunnel diode characteristic is given by

V (mV)	0	20	40	60	100	160	200	300	400	460	500
I (mA)	0	1·4	2·6	3·0	2·4	1·1	0·6	0·4	0·6	1·6	3·0

Explain your result. (*G.Inst. P., Part II*, 1968)

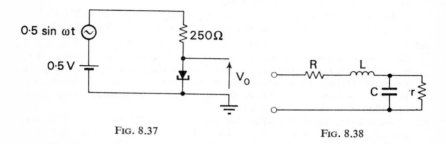

FIG. 8.37   FIG. 8.38

8.7 Sketch the characteristics of an Esaki or tunnel diode indicating approximate scales of voltage and current. Explain briefly its importance as a circuit element.

Fig. 8.38 indicates a possible equivalent circuit for the tunnel diode. Derive an expression for the input impedance and determine the frequencies for which the real and quadrature components become zero if $R = 1·5\,\Omega$, $L = 0·01\,\mu H$, $C = 50\,pF$, $r = -25\,\Omega$. What is the significance of the results? (*IEE*, June 1963)

Answer

$$R - \frac{r}{1 + (\omega Cr)^2} + j\left[L - \frac{\omega Cr^2}{1 + (\omega Cr)^2}\right]$$

503 MHz: self-oscillation occurs
186 MHz: diode is a pure negative resistance

8.8 A coil and a variable-capacitance diode are connected in series with a sinusoidal source of e.m.f. The capacitance of the diode is modulated by a "pump" source having an e.m.f. of square waveform.

If the signal frequency is 50 MHz, the diode sensitivity is 2 pF/V, the peak-to-peak amplitude of the square wave is 0·5 V and the power dissipated in the load and losses at equilibrium is 0·1 mW, estimate (*a*) the pump frequency, and (*b*) the peak voltage across the variable capacitance.

(*Ans.* (*a*) 100 MHz, (*b*) 1·41 V)

8.9 Define the terms *noise factor* and *signal-to-noise ratio* and explain a relationship between them in respect of a two-port network.

Fig. 8.39 shows a two-port network with its noise properties represented by two equivalent input generators, one of mean-square voltage $\overline{v^2}$ and the other of mean-square current $\overline{i^2}$. The network is driven from a source having an internal resistance R_g. Assuming the generators to be uncorrelated, deduce an expression for the noise factor of the network with this source and show it is a minimum when $R_g{}^2 = \overline{v^2}/\overline{i^2}$.

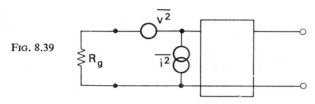

FIG. 8.39

Give expressions for $\overline{v^2}$ and $\overline{i^2}$ if the network consists of (a) a single shunt-connected resistor R, (b) a single series-connected resistor R. (*L.U., B.Sc. (Eng.)*, 1967)
(*Ans.* $F = 1 + (\overline{v^2} + \overline{i^2}R_g{}^2)/4kTR_g$; (a) $\overline{v^2} = 0$, $\overline{i^2} = 4kT/R$; (b) $\overline{v^2} = 4kTR$, $\overline{i^2} = 0$)

8.10 An amplifier with a gain of 20 dB has a bandwidth of 2 MHz and at the output the mean-square noise voltage is found to be 10^{-10} V^2. It is supplied from a source of resistance 50 Ω at a temperature of 20°C. Calculate the noise temperature of the amplifier and explain briefly how it might be achieved at an operating frequency near 1 GHz.
(*Ans.* 180 K; parametric amplifier (or maser) required)

9

Masers and Lasers

When electromagnetic radiation interacts with matter, transitions occur between the atomic energy levels, as first suggested by Bohr in 1913 (page 3). However, it was not until 1917 that Einstein considered the *probabilities* of absorption and emission of radiation by an atom, and he showed that a second emission process, *stimulated emission*, must also occur. This process is more significant at microwave frequencies than at optical frequencies, and in 1954 a device was introduced called the *maser* (Microwave Amplification by Stimulated Emission of Radiation). The principle of the maser was extended to optical frequencies with the advent of the *laser* in 1960, the *l* standing for "light".

In 1901 Planck introduced the quantum theory to account for the observed spectrum of the energy radiated from a black body. He showed that the energy could not change continuously but only in units of one quantum, *hf* joules. For a black body radiating energy within a frequency range f to $f + df$ this leads to an expression for the energy radiated per unit volume, $E_f \, df$. E_f is the energy density and is given by

$$E_f = \frac{8\pi f^2}{c^3} \frac{hf}{\exp\left(\dfrac{hf}{kT}\right) - 1} \tag{9.1}$$

This equation is illustrated in Fig. 9.1 for a black body at 1 500 K and is in agreement with the observed experimental values of E_f.

Now suppose that the radiation from a black body at temperature T is in thermal equilibrium with atoms having energy levels W_1, W_2, and so on. The energy of an atom can be increased if an electron is excited

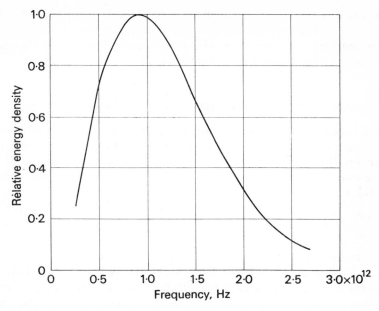

FIG. 9.1 Energy density and radiation frequency for a black body at a
temperature of 1 500 K

from a level W_1 to a higher level W_2 when a photon of radiation of
frequency f_{21} is absorbed, where

$$hf_{21} = W_2 - W_1$$

(from eqn. (1.11)). The photon is re-radiated after a short time when the
atom returns to level W_1, a process known as *spontaneous emission*.
Hence at any instant the atoms will be distributed between the available
energy levels, and if N_1 atoms have energy W_1 the number of atoms with
any other energy level W_n is determined by Boltzmann statistics, so that

$$N_n = N_1 \exp\left(-\frac{W_n - W_1}{kT}\right) \tag{9.2}$$

For atoms in the levels W_1 and W_2, then,

$$N_2 = N_1 \exp\left(-\frac{hf_{21}}{kT}\right) \tag{9.3}$$

381

which means that N_2 is less than N_1 when the exponential term is less than unity. A dynamic equilibrium exists between the two levels such that in the same time interval the number of transitions from W_1 to W_2 equals the number of transitions from W_2 to W_1.

EINSTEIN'S THREE RADIATION COEFFICIENTS

The number of atoms raised from W_1 to W_2 in a time interval dt by absorption of photons will be proportional both to the number of atoms of energy W_1 and to the energy density of the radiation at frequency f_{21}. If the constant of proportionality is B_{12}, the rate of change of the number of atoms in level W_2 is

$$\frac{dN_2}{dt} = B_{12}N_1E_{f21} \tag{9.4}$$

B_{12} is called the *coefficient for absorption of radiation*.

The number of atoms falling from W_2 to W_1 by radiation of a photon is similarly proportional to the number of atoms in level W_2. Then if the constant of proportionality is A_{21} the rate of change of the number of atoms in level W_1 is

$$\frac{dN_1}{dt} = A_{21}N_2 \tag{9.5}$$

A_{21} is called the *coefficient of spontaneous emission*. Then, if only these two processes exist, at equilibrium

$$\frac{dN_1}{dt} = \frac{dN_2}{dt} \tag{9.6}$$

Hence

$$E_{f21} = \frac{A_{21}}{B_{12}}\frac{N_2}{N_1} \tag{9.7}$$

$$= \frac{A_{21}}{B_{12}}\exp\left(-\frac{h_{f21}}{kT}\right) \tag{9.8}$$

using eqn. (9.3). However, it is clear that the form of eqn. (9.8) does not correspond to eqn. (9.1), which has been derived theoretically and confirmed experimentally. Einstein therefore suggested a second emission process which is *stimulated* by the presence of radiation of frequency f_{21}. The number of transitions from W_2 to W_1 by this process is then proportional both to the number of atoms in W_2 and the energy density of radiation at frequency f_{21}. Thus an additional term is required in eqn. (9.6) which gives, for transitions from W_2 to W_1,

$$\frac{dN_1}{dt} = A_{21} N_2 + B_{21} N_2 E_{f21} \tag{9.9}$$

where B_{21} is the *coefficient of stimulated emission*. Again at equilibrium $dN_1/dt = dN_2/dt$, so that, from eqns. (9.4) and (9.9),

$$E_{f21}(B_{12}N_1 - B_{21}N_2) = A_{21}N_2 \tag{9.10}$$

Substituting for N_1 from eqn. (9.3),

$$E_{f21}\left[B_{12} \exp\left(\frac{hf_{21}}{kT}\right) - B_{21} \right] = A_{21}$$

and

$$E_{f21} = \frac{A_{21}}{B_{12} \exp\left(\dfrac{hf_{21}}{kT}\right) - B_{21}} \tag{9.11}$$

This equation then agrees with Planck's equation (9.1) if

$$B_{12} = B_{21} \tag{9.12}$$

and

$$\frac{A_{21}}{B_{21}} = \frac{8\pi h f^3_{21}}{c^3} = E_{f21}\left[\exp\left(\frac{hf_{21}}{kT}\right) - 1 \right] \tag{9.13}$$

Eqn. (9.12) shows that the probabilities of absorption and stimulated emission of a photon are equal; Einstein's three coefficients are illustrated in Fig. 9.2.

THE APPLICATION OF STIMULATED EMISSION

A photon of energy hf_{21} which interacts with an atom of energy W_2 will reduce its energy to W_1 and cause the stimulated emission of a second photon, also of energy hf_{21}. Thus the number of photons has doubled and amplification of the original signal has occurred. However, photons of

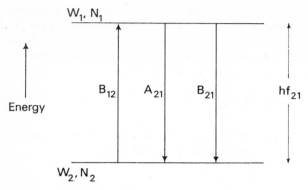

FIG. 9.2 Einstein's radiation coefficients

energy hf_{21} will also be emitted at random time intervals by spontaneous emission, and these constitute a source of electrical noise which interferes with the signal. The relative powers of the signal and noise depend on the relative number of photons emitted per second. Thus from eqn. (9.10) the ratio of stimulated to spontaneous emission under conditions of thermal equilibrium is

$$\frac{B_{21}E_{f21}}{A_{21}} = \frac{1}{\exp\left(\dfrac{hf_{21}}{kT}\right) - 1} \tag{9.14}$$

In the microwave region $hf_{21} \ll kT$ (see Table 1.1 and page 35 for relative values in electronvolts). Hence the higher terms in the exponential series may be ignored, and

$$\exp\left(\frac{hf_{21}}{kT}\right) \to 1 + \frac{hf_{21}}{kT}$$

From eqn. (9.13) the ratio of stimulated to spontaneous emission then becomes kT/hf_{21}, which at room temperature and a frequency of 5 GHz has a value of about 1 200. Even if the temperature is reduced as described below a satisfactory signal/noise ratio at microwave frequencies can thus be achieved.

Now consider a stream of photons passing through a material which has a difference in energy levels corresponding to the energy of the photons. If the photons strike unexcited atoms they may be absorbed and removed from the stream, which thus loses energy. However, if the photons strike excited atoms, more photons can be produced which are added to the stream and increase its energy. Since the probabilities of absorption and stimulated emission are the same (eqn. (9.11)), both attenuation and amplification of the stream occur simultaneously and amplification can only predominate if there are more atoms in the higher level than in the lower level.

Population Inversion

Under equilibrium conditions the lower level has more atoms than the higher level (eqn. (9.2)). Thus for amplification to be possible the population of atoms, N_2, in the upper level, must exceed the population, N_1, in the lower level, the process leading to $N_2 > N_1$ being known as *population inversion*. This can be achieved in a material where the lifetime of the upper level exceeds that of the lower level, for example when the upper level is metastable (page 302).

If there is a third normal level, W_3, above the metastable level, W_2 (Fig. 9.3), a strong microwave signal of frequency f_{31}, where

$$hf_{31} = W_3 - W_1 \tag{9.15}$$

can sometimes be used to raise the energy of electrons in some of the atoms from W_1 to W_3 (Fig. 9.4). This will continue until the populations

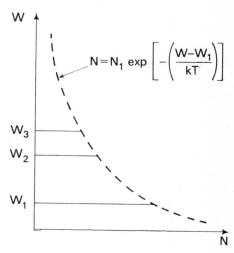

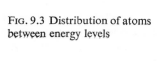

FIG. 9.3 Distribution of atoms between energy levels

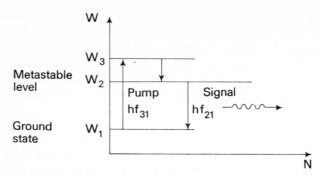

FIG. 9.4 Population inversion

of W_3 and W_1 are equal, since then the absorption of radiation is just balanced by the stimulated emission. Since W_3 is a normal level, atoms will remain in it only for a very short time, and will then return to W_1, either directly or by way of W_2, which is chosen to be a metastable level with a much longer lifetime. Thus the number of atoms in W_2 can be arranged to exceed the number in W_1 because a "head" of atoms is maintained in W_3. f_{31} is known as the *pump* frequency, and the amplified signal has a frequency f_{21}, typical values being 10 and 4 GHz respectively. In fact, avalanche multiplication of the photons is occurring since one photon colliding with an atom removes a second photon and both of these remove two more. Thus the number of photons multiplies in a similar manner to the electron avalanche (page 103).

The Ruby Maser

Suitable energy levels are found in chromium ions present as an impurity in an aluminium oxide crystal, about 1 in 5000 of the aluminium ions having been replaced by chromium. This combination is a version of the precious stone, ruby, being pink instead of deep red in colour. The levels are associated with states of the electron having a magnetic moment, which means that the actual energies can be adjusted by means of an external magnetic field. This provides a means of tuning the amplifier by controlling the strength of the magnet. Since the levels are in fact narrow bands, owing to interaction with surrounding atoms, a useful bandwidth is provided of about 50 MHz (see Table 8.2, page 376).

The amplification is increased by cooling the ruby crystal down to about 4 K by means of a liquid-helium bath. This has the effect of increasing the population of W_1 at the expense of W_3 before the pump signal is applied (from eqn. (9.2)), so that many more atoms can be raised to W_3 when pumping commences. This results in the population of W_2 being greatly increased over that of W_1, and makes possible a power amplification up to about 10 000 times, or 40 dB.

In practice, the ruby crystal is supplied with power at the pump frequency from one waveguide and is coupled by slots to a second waveguide terminated in a short-circuit (Fig. 9.5). Power at the signal frequency is

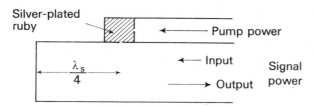

Fig. 9.5 Maser mounted on a waveguide

supplied to the second waveguide and the power reflected back from the short-circuit is amplified by the crystal. The waveguide is resonant at the signal frequency, so that power at frequency f_{32} (Fig. 9.4) is rejected. The maser may be regarded as a straight parametric amplifier operating in a similar way to the varactor diode version (page 366) and so may be represented by a negative resistance in the microwave circuit.

COHERENCE OF EMISSION

Since a photon incident on an atom causes the emission of another photon, their associated electromagnetic waves are in phase (Fig. 9.6(*a*)). Thus the waves reinforce each other and the emission is said to be *coherent*. However, the spontaneous emission of photons is a random process, so that the phase relationships between their waves and the incident wave is also random (Fig. 9.6(*b*)). Thus the waves tend to cancel one another out and the emission is said to be *incoherent*.

Coherent sources with frequencies up to the microwave region are comparatively easy to obtain, an example being a laboratory oscillator which can provide an output at any selected frequency within its working range. It is far less easy to obtain a coherent source for frequencies extending into the infra-red and visible regions, since conventional light sources such as a tungsten filament lamp emit a continuous range of frequencies, which appear as white light.

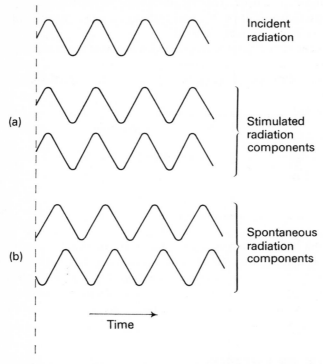

FIG. 9.6 Phase relationships for (*a*) coherent and (*b*) incoherent emission

In the visible region $hf_{21} \gg kT$ at room temperature, as may be seen by substituting a value of frequency between 4 and 7×10^{14} Hz (Table 1.1). Thus in thermal equilibrium the spontaneous emission is much greater than the stimulated emission (eqn. (9.14)), so that in an optical amplifier the signal would be completely lost in the noise. Similarly in a light source such as a gas discharge tube, which emits discrete frequencies, the spontaneous emission predominates and the light output is incoherent. However, in a laser the stimulated emission is amplified and separated from the spontaneous emission to give an optical oscillator of coherent radiation at high intensity.

The Ruby Laser

The chromium ion in a ruby crystal provides another suitable set of three energy levels in the ruby laser, and in this case no magnetic field is required. The energy difference $W_3 - W_1$ corresponds to a band of wavelengths between 520 and 600 nm. This represents *green* light, which is used for pumping, and since $hf_{31} \gg kT$, $N_1 \gg N_2$ even at room temperature, so no cooling is required for efficient population inversion. The metastable level W_2, which has a lifetime of 3 ms at room temperature, gives rise to transitions to W_1 corresponding to a wavelength of 694·3 nm, which represents *red* light. If the crystal is irradiated with green light, atoms will be excited to W_3 with a lifetime of less than 10^{-9} s, so they will fall very quickly to W_2. The transition from W_2 to W_1 will occur spontaneously, so that under these conditions the radiation will be incoherent.

To obtain laser action the transition from W_2 must be stimulated by light of wavelength 694·3 nm ($4·32 \times 10^{14}$ Hz), which will result in coherent radiation. The ruby is formed into a cylindrical rod a few centimetres long and the ends are ground flat and parallel to a high degree of accuracy. The ends are then silver coated so that one is fully reflecting and the other partially reflecting, allowing up to about 5% of the light to be transmitted (Fig. 9.7(*a*)). When the atoms have been excited, photons are emitted in all directions (Fig. 9.7(*b*)) with a few of them travelling along the axis of the cylinder. These photons are reflected by the end mirrors and travel to and fro between them, causing the stimulated emission of more photons and hence a photon avalanche (Fig. 9.7(*c*)). The laser beam then emerges through the partially reflecting mirror (Fig. 9.7(*d*). In order that the light waves may reinforce each other at a particular wavelength the distance between the mirrors must be an integral multiple of the *half-wavelength* of the radiation, as shown below. Furthermore the number of photons gained in the avalanche must exceed the number lost by processes such as emission out of the cylinder and absorption at the mirrors. Since only those photons moving exactly perpendicular to the mirrors can take part in light amplification, the angular divergence of the beam is very small, within an angle of 0·5°.

The initial excitation of green light can be obtained from a xenon flash tube (Fig. 9.8) in which an arc is set up by discharging a bank of capacitors. The ruby crystal and the flash tube are mounted at the foci of an elliptical internally reflecting cavity to ensure an efficient transfer of energy to the ruby crystal. Only a small proportion of the output from the tube is of the correct wavelength; the remainder causes heating of the

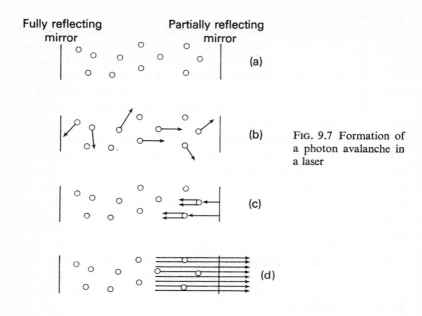

FIG. 9.7 Formation of a photon avalanche in a laser

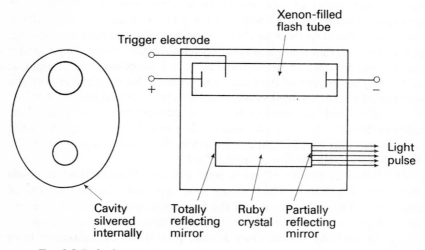

FIG. 9.8 Ruby laser

crystal. Thus the laser is normally operated with a pulsed output to prevent overheating and destruction of the flash tube. The output energy from the flash tube must be above a threshold value which is required to maintain the population of W_2 above that of W_1, and typical values are between 500 and 1 000 J per pulse. The corresponding light output from the laser is between 0·1 and 1·5 J for a 4 cm × 1 cm crystal, with the output occurring in a burst of irregular pulses (Fig. 9.9). These occur because the

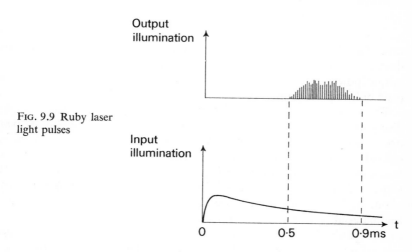

FIG. 9.9 Ruby laser light pulses

photon density builds up so rapidly in the crystal that stimulated transitions occur faster than the rate at which population inversion is maintained. Thus at the end of each small pulse the population of W_2 has fallen below the threshold value for sustained emission and the next pulse only appears after population inversion has been restored (Ref. 9.1).

The energy of all the small pulses may be concentrated into a single high-power pulse by the technique known as Q-switching. As discussed below, the mirror system can be considered as a cavity resonator whose Q-factor depends on the dissipation within it (eqn. (8.9)). The mirrors are detached from the ruby and a fast-acting shutter is mounted between the crystal and one mirror (Fig. 9.10), so that when the shutter is closed losses are introduced and the Q-factor is degraded. Stimulated emission cannot occur and the excitation greatly increases the population of W_2. When the shutter is opened the photon avalanche builds up rapidly and an intense burst of radiation, known as a *giant pulse*, occurs in a very short

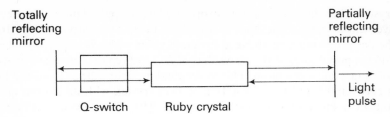

FIG. 9.10 Arrangement for *Q*-switching

time. The shutter is opened about 0·5 ms after the initiation of the exciting flash and a single output pulse is obtained lasting about 0·1 μs. The peak output power can be increased by up to 100 times by *Q*-switching, and peak output powers up to 100 MW have been reported using large crystals.

The shutter can take several forms: a mechanical one is an opaque disc with an aperture which is rotated at high speed, the shutter being open each time the aperture coincides with the crystal axis. A simple optical shutter uses a material which absorbs low-intensity radiation but becomes transparent when the intensity has reached a high value. The transmission coefficient of such a bleachable absorber can increase from about 10^{-6} to 0·3 in a few nanoseconds, and the effect is reversible so that the material recovers in time for the next pulse. For ruby lasers the dye kryptocyanine has been used in an optical cell (Ref. 9.2). A less convenient optical shutter is the Kerr cell, which rotates the plane of polarization of light when a voltage of about 10 kV is applied to it (Ref. 9.1).

THE GAS LASER

A continuous light output can be obtained from a discharge in a gas or a mixture of gases, laser action being achieved by mounting the discharge tube between two mirrors. A commonly used mixture is that of the inert gases helium and neon, approximately in the proportion of 7 to 1 by volume. The discharge can be maintained by a direct voltage of a few hundred volts between an anode and a cathode, as shown in Fig. 9.11. An earlier arrangement uses an alternating voltage of frequency about 30 MHz applied through electrodes on the outside of the tube (electrodeless discharge).

Helium atoms are excited in the discharge to the two $2S$ levels, which are metastable (Fig. 9.12). Neon also has two metastable levels, $2s$ and

392

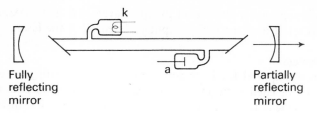

Fully
reflecting
mirror

Partially
reflecting
mirror

FIG. 9.11 Gas laser

3*s*, whose energies are close to the two helium levels. Thus, when a helium atom collides with a neon atom, energy can be transferred to the neon atom by a process known as a *resonant transfer*. Spontaneous emission then occurs due to neon atoms dropping from the metastable levels to the lower levels. The strongest spectral lines occur for the transitions shown in Fig. 9.12, at wavelengths of 632·8 nm in the visible red region and at 1·153 μm and 3·39 μm in the infra-red. The neon levels are in fact split into multiple levels, due to the various possible combinations of spin and angular momentum, and so far 136 lines have been observed with wavelengths between 585·2 nm and 132·8 μm (Ref. 9.1).

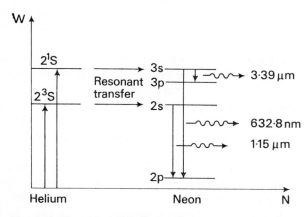

FIG. 9.12 Transitions in helium and neon

The discharge tube is sealed at each end by optically flat fused-quartz windows mounted at an angle of 55·5° to the tube axis (Fig. 9.11). This is the *Brewster angle* at which reflection of the 632·8 nm ($4·74 \times 10^{14}$ Hz) light is a minimum and transmission of this wavelength is a maximum, which helps to select it as the main output wavelength of the laser.

The Mirror System

The two mirrors required for laser action are spherical, with the centre of curvature, C, of each lying in the surface of the other, so that their focal points coincide in the centre of the tube at F (Fig. 9.13). The distance

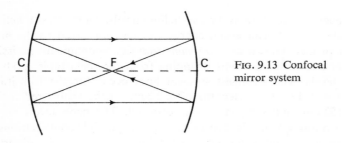

FIG. 9.13 Confocal mirror system

between the mirrors has to be an exact multiple of a half-wavelength for reinforcement of reflected waves to occur. The system is termed *confocal*, and even if the mirrors are slightly tilted a regenerative path similar to the one shown is still possible. Thus the adjustment of spherical mirrors when the laser is set up is much less critical than for a plane mirror system, or *Fabry–Perot interferometer*. Again one of the mirrors allows up to about 5% of the light to be transmitted, so that a light output is obtained at one end. The mirrors are coated with alternate layers of high- and low-refractive-index materials, which have a thickness of $\lambda/4$ at 632·8 nm. These reflect 99% of the light at this wavelength but much less at other wavelengths.

In effect the mirrors form a resonant cavity, in both the gas and solid-state lasers. Standing waves are set up in the cavity when the optical spacing between the mirrors is an integral number of half-wavelengths, the situation being analogous to the standing waves for an electron in a potential well, discussed in Appendix 2. If the optical spacing between the mirrors is L then resonance occurs when

$$L = \frac{n\lambda}{2} = n\frac{c}{2f} \tag{9.16}$$

or the resonant frequency is

$$f = n\frac{c}{2L} \tag{9.17}$$

Each integral value of n corresponds to a frequency at which oscillations may occur and constitutes a resonance or *mode*. A number of values of L are possible, depending on the path followed by the light between the mirrors. Where L is the optical spacing along the longitudinal axis of the discharge tube the *longitudinal* or *axial modes* are given by eqn. (9.17). Then, for a helium-neon laser with $L = 37\,$cm, n is $1\cdot2 \times 10^6$, while for a ruby laser with $L = 4\,$cm, n is $1\cdot3 \times 10^5$.

The laser will only generate those modes which correspond to optical paths very close to the axis, but even then a very large number are possible and the cavity has to be designed with only a few resonances having a high Q-factor.

These modes are separated by a frequency $c/2L$ hertz, which is 406 MHz for a laser of length 37 cm. The energy levels of a gas have a finite width, which leads to a broadening of the spectral lines, and for the 632·8 nm line this broadening corresponds to a frequency spread of about 1·5 GHz. Thus output frequencies are possible centred on $4\cdot74 \times 10^{14}$ Hz, but within a band 1·5 GHz wide which will include about three modes.

This is illustrated in Fig. 9.14, which shows a typical variation of the

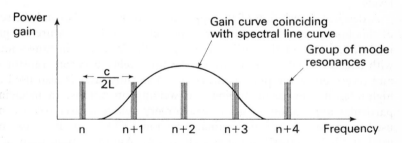

FIG. 9.14 Modes of a typical cavity compared with laser
power-gain curve

power gain of the laser as a function of frequency. The gain curve and the spectral line curve coincide, and the bandwidth is taken where the gain has dropped to half its maximum value, which includes three modes in the diagram. Each mode contains a number of resonances corresponding to optical paths between the mirrors, whose lengths satisfy eqn. (9.17). The mode separation is increased by reducing L, and using the example given above, if the spacing L is reduced to about 10 cm the mode separation rises to 1·5 GHz, so that only one mode is possible within the spectral linewidth.

The interaction of the various modes causes the output beam to form a characteristic pattern instead of a single spot. However, careful cavity design and the use of a short tube have resulted in lasers being commercially available with only one mode amplified, which will provide a spot source.

LASER APPLICATIONS

Among the many gases in which an inverted population of atoms can be established, argon and carbon dioxide have emerged as among the most important for laser applications. Argon provides coherent radiation in the visible spectrum at 488 nm and 514·5 nm, and an output of 1 W is obtainable at an efficiency of about 0·1%. Where high continuous power is required the carbon dioxide laser is used. This can provide output powers of many kilowatts at 10·6 μm in the far infra-red. The efficiency is increased by the addition of helium and nitrogen, with which a value of up to 15% is obtainable, and it is thought that resonant transfer occurs between metastable levels in the nitrogen and the upper CO_2 levels.

A detailed description of the applications of lasers is beyond the scope of this book, but a large number of references to the literature are given in Ref. 9.3. One of the earliest uses of a ruby laser was in range-finding with a Q-switched pulse, in which the time delay between transmission and reception of the pulse is measured. The ruby laser is also used as a high-intensity source in drilling and welding operations. In medicine a particular application occurs in eye surgery, where a laser beam can be used to reattach a detached retina. Argon and CO_2 lasers have also been used as "light knives" in surgery (Ref. 9.4). Another application of the gas laser is holography, in which a three-dimensional photograph of an object can be obtained and for which a coherent source is essential (Ref. 9.5).

The Semiconductor Laser (Injection Laser)
The production of a light output from a heavily doped *p-n* junction was described on page 227 in connection with the gallium-arsenide lamp.

The gallium-arsenide laser, or *injection* laser, is a development of the lamp in which a reflecting cavity is formed merely by polishing or cleaning the ends of the crystal perpendicular to the plane of the *p-n* junction (Fig. 9.15). These end surfaces reflect about 30% of the light falling on

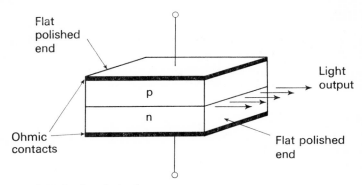

FIG. 9.15 Semiconductor laser

them, so that as the forward bias on the junction is increased some of the light emitted by the direct recombination of electrons and holes is fed back into the junction. This causes stimulated emission, and a photon avalanche builds up in the region of the junction. Light is then emitted in a thin sheet parallel to the plane of the junction, with a wavelength less sharply defined than that of a ruby or gas laser and with less coherence. Before the feedback of light from the mirrors is sufficient for laser action the critical current density for population inversion must be exceeded. The critical current falls with temperature, and gallium-arsenide lasers are therefore operated near the temperature of liquid nitrogen, 77 K. Here the critical current density is in the region of $10^7 \, \text{A/m}^2$, and it is applied in microsecond pulses at repetition rates up to a few kilohertz. The light output is much higher from the laser than from the lamp, being measured in kilowatts per square millimetre, and the efficiency can be as high as 35%. Laser action has also been observed for the other III–V compounds mentioned on page 228.

REFERENCES

9.1 LENGYEL, B. A., *Introduction to Laser Physics* (Wiley, 1966).
9.2 SOFFER, B. H., "Giant pulse operation by a passive reversibly bleachable absorber", *Journal of Applied Physics*, **35**, p. 2551 (1964).

9.3   PATEK, K., *Lasers* (Iliffe, 1967).
9.4   McGUFF, P., *The Surgical Applications of Lasers* (Witney, 1965).
9.5   DEVELIS, J. B. and REYNOLDS, G. O. *Theory and Applications of Holography* (Addison-Wesley, 1967).

FURTHER READING

WALLING, J. C. and SMITH, F. W., *Philips Technical Review*, **25**, p. 289 (1965).
ROSS, M., *Laser Receivers* (Wiley, 1966).

APPENDIX 1

Travelling Waves

The expression for a wave motion travelling with velocity u in the positive x-direction is given in general by

$$\psi = f(x - ut) \tag{A.1.1}$$

This equation is a function both of distance and time, ut being the distance travelled by the wavefront in time t. For a sinusoidal waveform the wave repeats itself after a distance $x = \lambda$, the wavelength, so that eqn. (A.1.1) becomes

$$\psi = \psi_m \sin\left(\frac{2\pi}{\lambda}x - ut\right) \tag{A.1.2}$$

where ψ_m is the peak value (Fig. A.1.1). The travelling wave can be represented by a *complexor** of length ψ_m rotating clockwise. The locus of the tip of the complexor projected onto the vertical axis gives the value of the displacement ψ at a particular time, and the waveform may be considered

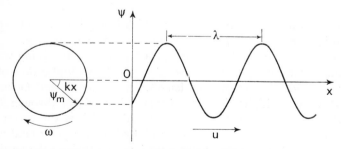

Fig. A.1.1 Vector representation of a travelling wave

* Previously called a *phasor* and before that a *rotating vector*.

to move to the right with velocity u as the complexor rotates with angular frequency ω in radians per second. If f is the frequency of the wave in cycles per second, or hertz (Hz),

$$\omega = 2\pi f \qquad\qquad (A.1.3)$$

a complete cycle corresponding to rotation of the complexor through 2π radians, or 360°. As the complexor rotates and the distance x travelled by a point on the wave increases, the angle kx increases *negatively* and when $x = \lambda$, $kx = 2\pi$, or

$$k = \frac{2\pi}{\lambda} \qquad\qquad (A.1.4)$$

where k is the *phase-change coefficient*. Similarly, movement of the wave to the *left* occurs for anticlockwise rotation, with the angle kx increasing *positively*. In either case the time for a complete cycle is $1/f$, so that

$$\lambda = \frac{u}{f} \qquad\qquad (A.1.5)$$

We need to express eqn. (A.1.2) in a standard form in terms of distance only, and differentiating twice leads to

$$\frac{\partial^2 \psi}{\partial x^2} + \left(\frac{2\pi}{\lambda}\right)^2 \psi = 0 \qquad\qquad (A.1.6)$$

or

$$\frac{\partial^2 \psi}{\partial x^2} + k^2 \psi = 0 \qquad\qquad (A.1.7)$$

These are two forms of the *wave equation*, which is applicable to all types of wave motion with suitable interpretation of the displacement ψ.

A more convenient expression than eqn. (A.1.2) for a travelling wave may be obtained by using the operator $j = \sqrt{-1}$. Multiplication of a complexor by j does not change its length, but rotates it *anticlockwise* through 90°, while multiplication by $-j$ rotates it *clockwise* through 90°. In Fig. A.1.2 the complexor OP has unit length and it is shown on an *Argand* diagram, where real numbers are plotted on the horizontal axis and imaginary numbers, which incorporate j, are plotted on the vertical axis. OP then is fully described in amplitude and phase angle θ by the expression

$$a + jb = 1 \tag{A.1.8}$$

where $a + jb$ is a *complex number*, with a real part, a, and an imaginary part, b. Then, by Pythagoras, the length OP, known as the *modulus* of the complexor, is given by

$$1 = \sqrt{(a^2 + b^2)} \tag{A.1.9}$$

and the phase angle is

$$\theta = \tan^{-1}(b/a) \tag{A.1.10}$$

Using trigonometry and the exponential series so that the complex number may be expressed in exponential form,

$$a + jb = \cos\theta + j\sin\theta = \exp(j\theta) \tag{A.1.11}$$

Then at a given time the expression for a wave travelling to the left is $\psi_m \exp(jkx)$, and for a wave travelling to the right it is $\psi_m \exp(-jkx)$. The complete solution of the wave equation (A.1.7) is given by

$$\psi = A \exp(-jkx) + B \exp(jkx) \tag{A.1.12}$$

which may be confirmed by differentiating this equation twice and substituting back into eqn. (A.1.7). A and B are constants depending on the boundary conditions of a particular problem and are determined in Appendix 2 in selected cases; either A or B may be zero.

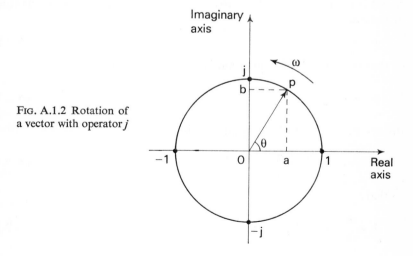

Fig. A.1.2 Rotation of a vector with operator j

APPENDIX 2

Wave Mechanics: an Introduction

De Broglie's concept of a guiding wave for a moving particle was developed by Schrödinger in 1926 into a comprehensive theory called *wave mechanics*, which includes the equation named after him. Before deriving Schrödinger's equation by a less rigorous method, we must relate the properties of the particle to its wavelength. Einstein's theory of relativity led to the result that energy and mass were related by

$$W = mc^2 \qquad \text{(A.2.1)}$$

where c is the velocity of electromagnetic waves. Hence for a quantum of radiation,

$$hf = mc^2 \qquad \text{(A.2.2)}$$

so that the "mass" of a quantum is

$$m = \frac{hf}{c^2} \qquad \text{(A.2.3)}$$

But the momentum of a quantum, p, is

$$p = mc = \frac{hf}{c} = \frac{h}{\lambda} \qquad \text{(A.2.4)}$$

so that

$$\lambda = \frac{h}{p} \qquad \text{A.2.5)}$$

—the expression applied by de Broglie (page 9). The kinetic energy of a particle moving with velocity u is

$$K = \tfrac{1}{2}mu^2 = \frac{p^2}{2m} = \frac{h^2}{2m\lambda^2} \tag{A.2.6}$$

Since the total energy of a system is the sum of its kinetic and potential energies

$$W = K + V \tag{A.2.7}$$

and from eqn. (A.2.6),

$$\frac{1}{\lambda^2} = \frac{2mK}{h^2} = \frac{2m}{h^2}(W - V) \tag{A.2.8}$$

Substituting into the wave equation (A.1.6),

$$\frac{\partial^2 \psi}{\partial x^2} + \frac{8\pi^2 m}{h^2}(W - V)\psi = 0 \tag{A.2.9}$$

This is Schrödinger's wave equation in one dimension, and the question arises of the interpretation of the displacement of the wave, ψ.

With other types of wave motion the energy density of the wave is proportional to the square of the displacement, so that, if eqn. (A.2.9) represents the guiding wave of a beam of electrons, say, then the intensity of the beam in a volume element $d\tau$ must be proportional to $\psi^2 d\tau$, where $d\tau = dx\,dy\,dz$. The more intense the beam the greater is the probability of finding an electron in the element $d\tau$, so that ψ^2 may be interpreted as being proportional to the probability of finding an electron at the point represented by allowing $d\tau$ to become very small. Since multiplication of ψ by a constant does not alter eqn. (A.2.9), ψ^2 may be chosen so that it is *equal* to the probability of finding the electron, and ψ is then the displacement of a wave whose intensity gives the probability of finding the electron at a particular point.

ELECTRON IN FREE SPACE

The solution of Schrödinger's equation is very difficult except in a few cases. The simplest of these is that of an electron in free space, which has zero potential energy whatever its position, or $V = 0$. Schrödinger's equation then reduces to

$$\frac{\partial^2 \psi}{\partial x^2} + \frac{8\pi^2 m}{h^2} W\psi = 0 \tag{A.2.10}$$

which becomes the same as eqn. (A.1.7) when

$$k = \frac{(8\pi^2 mW)^{1/2}}{h} \tag{A.2.11}$$

and a solution of eqn. (A.2.10) for a wave travelling to the right is

$$\psi = A \exp(-jkx) \tag{A.2.12}$$

where A is a constant. Since k is a real number, W can only have positive values, all of which are allowed. Thus in free space an electron may have any value of kinetic energy and is free of quantum restrictions, which agrees with experiment.

Another way of describing this continuous variation of energy is to express W in terms of k, using eqn. (A.2.11), so that

$$W = \frac{h^2 k^2}{8\pi^2 m} \tag{A.2.13}$$

k is a function of momentum p, since

$$k = \frac{2\pi}{\lambda} = \frac{2\pi}{h} p \tag{A.2.14}$$

using eqn. (A.2.5), so that for matter waves k is termed the *momentum vector* or *k-vector*. Since in one dimension p can be positive or negative, k considered as a vector can also be positive or negative, so that W varies parabolically with k, as shown in Fig. A.2.1.

ELECTRON IN A POTENTIAL WELL

On the other hand, an electron in an atom will be affected by a variation of potential energy such as that shown in Fig. 1.3 for a hydrogen atom. This Coulomb potential well may be very crudely approximated by the rectangular potential well of Fig. A.2.2. Let us suppose for simplicity that the electron moves between the wall of the well in straight lines

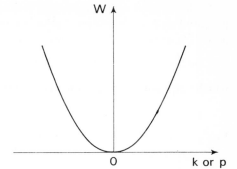

FIG. A.2.1 Energy/momentum-vector diagram for an electron in free space

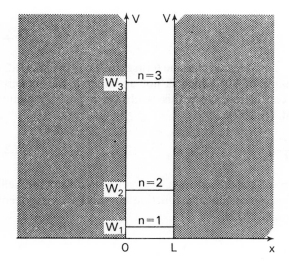

FIG. A.2.2 Energy levels of an electron in a rectangular potential well

parallel to the *x*-axis and that the walls are perfectly reflecting; also that the potential energy is given by

$V = 0$ for $0 < x < L$

$V \to \infty$ for $x \leqslant 0$ or $x \geqslant L$

This implies that the electron is to be found only within the well, or that

$\psi = 0$ for $x \leqslant 0$ or $x \geqslant L$

Again Schrödinger's equation reduces to eqn. (A.2.10) within the potential well, with a general solution given by eqn. (A.1.12), from which at $x = 0$, $\psi = A + B$, and since $\psi = 0$, $A = -B$. Then at $x = L$,

$$\psi = B\left[\exp{(jkL)} - \exp{(-jkL)}\right] \tag{A.2.15}$$

which, expressed in trigonometric terms, gives

$$\psi = j2B \sin kL \tag{A.2.16}$$

For $\psi = 0$ at $x = L$ either $B = 0$, which implies no displacement, or

$$kL = \frac{2\pi L}{\lambda} = n\pi$$

This makes

$$L = n\frac{\lambda}{2} \tag{A.2.17}$$

where $n = 1, 2, 3 \ldots$, and makes the sine term zero. The j in eqn. (A.2.16) indicates a phase angle of 90° which is not apparent if the magnitude (modulus) of ψ is considered, since this is always a positive quantity, $|\psi|$. Then, using eqn. (A.2.17),

$$|\psi| = 2B \sin\left(\frac{n\pi x}{L}\right) \tag{A.2.18}$$

which is the equation of a *standing wave*, the sum of two waves travelling in opposite directions, since the sine function contains two exponential terms. But, from eqn. (A.2.17),

$$\frac{n\pi}{L} = \frac{(8\pi^2 m W_n)^{1/2}}{h} = k \tag{A.2.19}$$

or

$$W_n = \frac{n^2 h^2}{8mL^2} \tag{A.2.20}$$

This means that the energy of the electron is restricted to discrete levels such as W_1, $W_2 \ldots$ shown in Fig. A.2.2, which are similar to the discrete energy levels of the Bohr hydrogen atom (Fig. 1.3). The value of B depends on the probability that a particular level will be occupied, which in turn depends on the statistics used such as Fermi–Dirac or Maxwell–Boltzmann. For a level where it is certain that an electron will be found $\int |\psi|^2 dx = 1$, since $d\tau \to dx$ in the one-dimensional case.

ELECTRON TUNNELLING

Let us now suppose that an electron is incident upon an energy barrier of finite height V_0 and finite width d (Fig. A.2.3(a)). If the energy of the

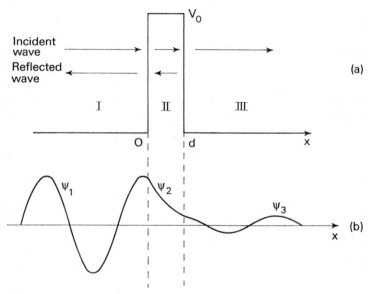

FIG. A.2.3 Electron tunnelling

    (a) Potential barrier and de Broglie waves
    (b) Wave functions

electron is less than V_0 and d is large it cannot penetrate the barrier, any more than it can escape from a potential well. However, if d is made sufficiently small then there is a corresponding small probability that the electron will penetrate the barrier, a phenomenon known as *tunnelling*. In regions I and III, outside the barrier $V = 0$ and Schrödinger's equation then becomes

$$\frac{\partial^2 \psi}{\partial x^2} + \frac{8\pi^2 m}{h} W\psi = 0 \qquad (A.2.21)$$

while in region II, inside the barrier,

$$\frac{\partial^2 \psi}{\partial x^2} + \frac{8\pi^2 m}{h}(W - V_0)\psi = 0 \tag{A.2.22}$$

where $W - V_0$ is negative since $V_0 > W$.

The solution of eqn. (A.2.21) for region I is then

$$\psi_1 = A_1 \exp(-jk_1 x) + B_1 \exp(jk_1 x) \tag{A.2.23}$$

where

$$k_1 = \left(\frac{8\pi^2 m W}{h^2}\right)^{1/2} \tag{A.2.24}$$

The first term represents a wave of amplitude A_1 travelling towards the barrier, while the second term represents a wave of amplitude B_1 reflected from the barrier (Fig. A.2.3(a)). For region II the solution of eqn. (A.2.22) is

$$\psi_2 = A_2 \exp(-k_2 x) + B_2 \exp(k_2 x) \tag{A.2.25}$$

where

$$k_2 = \left[\frac{8\pi^2 m(V_0 - W)}{h^2}\right]^{1/2} \tag{A.2.26}$$

with the j's absent in eqn. (A.2.25), and again represents two waves travelling in opposite directions. Finally for region III we have

$$\psi_3 = A_3 \exp(-jk_1 x) \tag{A.2.27}$$

since there is no reflected wave.

Considering only the waves travelling to the right represented by the A terms, it may be seen that the amplitudes of ψ_1 and ψ_3 remain constant with distance. However, the amplitude of ψ_2 falls exponentially with distance and will tend to zero when the barrier width d is sufficiently large. Since both the wave functions and their derivatives must be continuous they will appear as in Fig. A.2.3(b), so that, for a narrow barrier, ψ_3 will possess a finite amplitude representing a probability that an electron will penetrate the barrier and appear on the other side. In practice electron tunnelling is only significant for barriers less than a few electronvolts in height and less than a nanometre in width.

ELECTRON DIFFRACTION

The atoms in a crystal form a regular 3-dimensional array, known as a *space lattice*. All the atoms may then be included in a set of parallel and equally spaced planes (Fig. A.2.4(*a*)), which can be chosen in a large

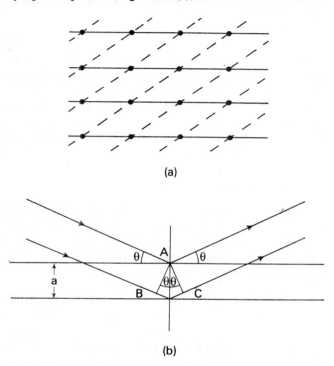

(a)

(b)

FIG. A.2.4 Electron diffraction

(*a*) Crystal planes   (*b*) Reflection of electron beam

number of ways. The sets containing a large number of atoms will then be more widely spaced than those containing fewer atoms (shown dotted in Fig. A.2.4(*a*)) and will tend to correspond with the external faces of the crystal.

Suppose that a parallel beam of electrons is incident in a direction making an angle θ with a particular set of planes with spacing a (Fig. A.2.4(*b*)). Each plane may then be considered as reflecting a small fraction of the incident beam, but for a reflected beam to exist the waves reflected from different planes must be exactly in phase. If there is a phase difference then complete destructive interference will occur owing to the large number of planes and their regular spacing. The path difference between the incident wavefront AB and the reflected wavefront AC is $2a \sin \theta$, where a is the spacing between the planes, so that for reinforcement to occur

$$2a \sin \theta = n\lambda \qquad (A.2.28)$$

where $n = 1, 2, 3 \ldots$. This equation was first derived by Bragg in 1912 to explain the diffraction of X-rays by a crystal. The incident beam is said to be *diffracted* at a number of different angles, each of which corresponds to one value of n. Bragg's law was used to interpret the results of experiments in which a beam of electrons was diffracted by a crystal, either from its surface or by a film about 10 nm thick, and in this way the wave nature of electrons was confirmed.

ELECTRON IN A CRYSTAL

An electron moving through a crystal will be influenced by the periodic potential due to the atomic cores, which are regularly spaced at distance r_0 (Fig. 2.4). The electron may then be considered as a wave moving at right angles to a set of planes with spacing a (not necessarily equal to r_0), so that Bragg's law becomes

$$n\lambda = 2a \qquad (A.2.29)$$

At values of λ where this equation is satisfied, reinforcement of the waves will occur, so that a standing wave will be set up between the planes. Thus the electron will be unable to penetrate the lattice and its particle velocity u will be zero. At all other values of λ, standing waves will not be set up and the wave will travel through the lattice with period λ. Since V in Schrödinger's equation (A.2.9) will be periodic with a period equal to the plane spacing (compare Fig. 2.4(b) where $a = r_0$), it was shown by Bloch that the amplitude of the wave is modulated also at period a.

Eqn. (A.2.13) for the energy of an electron in free space may then be applied to an electron in a crystal, with two modifications. Firstly m becomes m^*, the effective mass of the electron, in order to account for the effect of the periodic lattice on the travelling wave. Secondly, when

$$\lambda = \frac{2a}{n} \qquad \text{or} \qquad k = \frac{n\pi}{a} \qquad (A.2.30)$$

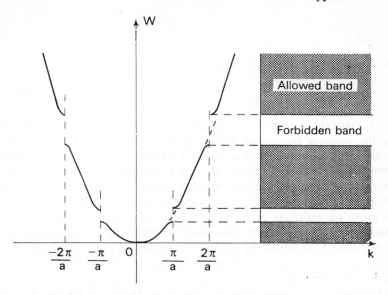

FIG. A.2.5 Energy/momentum-vector diagram for an electron in a crystal

a discontinuity occurs in the $W - k$ parabola, which corresponds to the centre of a forbidden band of energies (Fig. A.2.5). The particle velocity of the electron is related to the slope of the curve, since from eqn. (A.2.13),

$$\frac{dW}{dk} = \frac{h^2 k}{4\pi^2 m^*} \tag{A.2.31}$$

and

$$u = \frac{p}{m^*} = \frac{h}{\lambda m^*} = \frac{hk}{2\pi m^*} \tag{A.2.32}$$

so that

$$\frac{dW}{dk} = u \frac{h}{2\pi} \tag{A.2.33}$$

But in the forbidden band $u = 0$, so that $dW/dk = 0$ also, as shown at the band edges in Fig. A.2.5. The effective mass of the electron is related to the rate of change of slope with k, since from eqn. (A.2.31),

$$\frac{d^2W}{dk^2} = \frac{h^2}{4\pi^2 m^*}$$

or

$$m^* = \frac{h^2}{4\pi^2} \bigg/ \frac{d^2W}{dk^2} \qquad\qquad (A.2.34)$$

m^* is constant wherever energy increases with k^2 (eqn. (A.2.13)), which applies towards the top of the energy/k-vector graph. This corresponds to the conduction band of a material where the electron will behave as though it were in free space, but with a modified mass. However, as the energy is reduced to the top of a forbidden band $d^2W/dk^2 \to 0$, which suggests that $m^* \to \infty$ and corresponds to $u \to 0$, since it is not possible to move an infinitely large mass. The concept of effective mass in the allowed band *below* a forbidden band is difficult to grasp, since d^2W/dk^2 is negative at the top of the allowed band, zero in the middle and positive again at the bottom, and in these regions the interaction between the periodic lattice and the electron becomes very complicated.

The foregoing theory suggests that in an insulator the wavelength of electrons is related to the atomic spacing by eqn. (A.2.29) at energies in the forbidden band below the conduction band. Thus standing waves are set up and no electron flow can occur. However, in a metal the wavelength of electrons at the top of the valence band is not related to the atomic spacing by this equation, so that the electron waves can travel freely through the crystal.

APPENDIX 3

Density of Energy Levels in a Semiconductor

In this appendix a derivation is given for a relationship which was expressed in eqn. (2.28) for the density of the energy levels in the conduction and valence bands of a conductor:

$$(N_cN_v)^{1/2} = GT^{3/2}$$

where G is a constant. We need first an expression for the *number* of energy levels in a crystal, which is a function of energy and is directly proportional to the volume of the crystal. Thus the number of energy levels per unit volume, $\delta(W)$, is independent of volume and is known as the *density of states* function. If there are δS energy levels in a small energy range δW, then

$$S(W) = \frac{\delta S}{\delta W} \rightarrow \frac{dS}{dW} \tag{A.3.1}$$

as $\delta W \rightarrow 0$.

It is convenient to express the kinetic energy of an electron in terms of its momentum, p, which has direction as well as magnitude, and in a 3-dimensional crystal it may be expressed in terms of components p_x, p_y and p_z. These are measured along the x-, y- and z-axes respectively to form a 3-dimensional *momentum space* (Fig. A.3.1). Each point in this space represents the momentum of an electron, which can be equally well represented by a vector of length p drawn from the origin to the point. Then all the electrons with energies between 0 and W are represented by points lying within a sphere of radius p, given by

$$p = \sqrt{(2m_eW)} \tag{A.3.2}$$

which is obtained from eqn. (A.2.6), m_e being the effective mass of the electron. We are assuming here that each electron sees a uniform potential field set up by the atomic cores and the other electrons. This corresponds to uniform potential energy, so that $V = 0$ within the crystal and the energy of an electron is only kinetic.

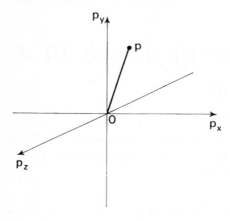

FIG. A.3.1 Momentum space

The surface area of the sphere is $4\pi p^2$, so that for a small increment of momentum δp the number of energy levels within a momentum shell of volume $4\pi p^2 \, \delta p$ may be considered in order to obtain the density of states. From quantum theory the minimum momentum-distance product is h (page 5), so that the minimum momentum-volume product is h^3. Then the number of separate energy levels in the momentum shell is, in the limit,

$$dS = 2 \times \frac{4\pi p^2}{h^3} \, dp \tag{A.3.3}$$

since each value of momentum may be associated with *two* electrons having opposite spins. From eqn. (A.3.2),

$$dp = \tfrac{1}{2}\sqrt{\frac{2m_e}{W}} \, dW \tag{A.3.4}$$

so that, on substitution in eqn. (A.3.3),

$$\frac{dS}{dW} = \frac{8\pi\sqrt{2}m_e^{3/2}W^{1/2}}{h^3} \tag{A.3.5}$$

This equation may be applied to electrons in the conduction band by measuring their energy with respect to W_c at the bottom of the band (Fig. A.3.2), so that, in the conduction band,

$$\frac{dS}{dW} = S(W) = \frac{8\pi\sqrt{2}m_e^{3/2}(W - W_c)^{1/2}}{h^3} \tag{A.3.6}$$

Similarly the energy of the holes in the valence band can be measured *down* from the energy W_v at the top of the band, so that, in the valence band,

$$S(W) = \frac{8\pi\sqrt{2}m_h^{3/2}(W_v - W)^{1/2}}{h^3} \tag{A.3.7}$$

where m_h is the effective mass of a hole. These expressions are illustrated in Fig. A.3.2.

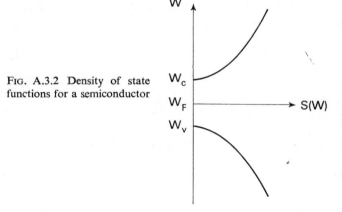

FIG. A.3.2 Density of state functions for a semiconductor

The number density of electrons dn within an energy range dW will depend on both the density of available states dS and the probability that each state is occupied, $p_F(W)$. Thus

$$dn = p_F(W)\,dS \tag{A.3.8}$$

and for the whole range of energy being considered the number density of electrons is

$$n = \int p_F(W)\frac{dS}{dW}\,dW \tag{A.3.9}$$

For a crystal $p_F(W)$ is given by the Fermi function (eqn. (2.17)), which for electrons in a semiconductor reduces to

$$p_F(W) = \exp\left(-\frac{W - W_F}{kT}\right) \tag{A.3.10}$$

Then, substituting in eqn. (A.3.9) and integrating from the bottom of the conduction band,

$$n = \frac{8\pi\sqrt{2}m_e^{3/2}}{h^3} \int_{W_c}^{\infty} (W - W_c)^{1/2} \exp\left(-\frac{W - W_F}{kT}\right) dW$$

$$= \frac{8\pi\sqrt{2}m_e^{3/2}}{h^3} \exp\left(-\frac{W_c - W_F}{kT}\right)$$

$$\int_{W_c}^{\infty} (W - W_c)^{1/2} \exp\left(-\frac{W - W_c}{kT}\right) dW \tag{A.3.11}$$

The integral in this equation can be expressed in a standard form if we make $y^2 = (W - W_c)/kT$. Then the integration becomes of the type

$$\int_{0}^{\infty} y^2 \exp(-y^2)\, dy = \frac{\sqrt{\pi}}{2} \tag{A.3.12}$$

so that

$$n = \frac{2}{h^3} (2\pi m_e kT)^{3/2} \exp\left(-\frac{W_c - W_F}{kT}\right) \tag{A.3.13}$$

which, by comparison with eqn. (2.21), can be written

$$n = N_c \exp\left(-\frac{W_c - W_F}{kT}\right) \tag{A.3.14}$$

where

$$N_c = \frac{2}{h^3} (2\pi m_e k)^{3/2} T^{3/2} \tag{A.3.15}$$

Thus N_c can be conveniently regarded as the density of states lying at the energy W_c which gives the same answer as integration over the whole conduction band of energies. By a similar analysis the density of states lying at the energy W_v in the valence band is

$$N_v = \frac{2}{h^3} (2\pi m_h kT)^{3/2} \tag{A.3.16}$$

using equation 2.34,

$$p = N_v \exp\left(-\frac{W_F - W_v}{kT}\right) \qquad (A.3.17)$$

The product of N_c and N_v is then obtained from eqns. (A.3.15) and (A.3.16), so that

$$(N_c N_v)^{1/2} = \frac{2}{h^3} (2\pi k)^{3/2} (m_e m_h)^{3/4} T^{3/2} = GT^{3/2} \qquad (A.3.18)$$

Thus the value of G depends on the value of the effective masses of electrons and holes. If it is assumed that $m_e = m_h = m$, the mass of an electron in free space, then $G = 4.83 \times 10^{21}/\text{m}^3\text{-K}$ electrons or holes. However, in practice the numerical value of G is 1.76×10^{22} for germanium and 3.87×10^{22} for silicon and it is assumed that the discrepancy is due to the effective masses being unequal to m.

In an intrinsic semiconductor $n = p$, so that the position of the Fermi level may be found by equating eqn. (A.3.13) to eqn. (A.3.16) combined with eqn. (A.3.17), which gives

$$m_e^{3/2} \exp\left(-\frac{W_c - W_F}{kT}\right) = m_h^{3/2} \exp\left(-\frac{W_F - W_v}{kT}\right) \qquad (A.3.19)$$

and

$$W_F = \frac{W_c + W_v}{2} + \tfrac{3}{4}kT \log_e \frac{m_h}{m_e} \qquad (A.3.20)$$

This shows that if $m_e = m_h$ the Fermi level lies in the middle of the energy gap for all temperatures, but that if $m_e \neq m_h$ it is only exactly in the middle of the energy gap for $T = 0\,\text{K}$.

For intrinsic semiconductors it has already been shown (page 43) that the position of the Fermi level is a function of both the doping density and the temperature, the effective masses being involved through N_c and N_v. The carrier densities for intrinsic and n-type semiconductors as a function of energy are illustrated in Fig. A.3.3. The values of $p(W)$ and $1 - p(W)$ are both one-half at the Fermi level, but the exponential variation of these probabilities is much more rapid than shown in these diagrams. However, they illustrate that the variation of the electron density close to W_c and the variation of the hole density close to W_v are

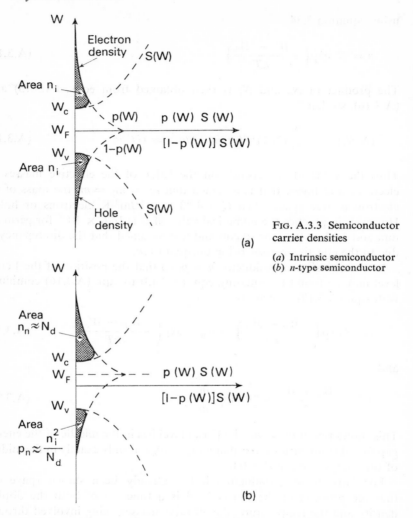

FIG. A.3.3 Semiconductor
carrier densities

(*a*) Intrinsic semiconductor
(*b*) *n*-type semiconductor

each controlled mainly by the corresponding $S(W)$ function. The variations of the carrier density at energies above W_c and below W_v are controlled by the respective probability functions. Thus there is an energy in the conduction band at which the electron density is a maximum and an energy in the valence band at which the hole density is a maximum. The area under the electron density curve must correspond to the total number of electrons, and the area under the hole density curve must correspond to the total number of holes. In an intrinsic semiconductor these areas are equal since the Fermi level is in the centre of the conduction band. In an *n*-type semiconductor the Fermi level is nearer the bottom of the conduction band, so that the total number of electrons is much greater than the total number of holes. In a *p*-type semiconductor the Fermi level is nearer the top of the valence band so that the situation is reversed and the number of holes greatly exceeds the number of electrons.

APPENDIX 4

Relativistic Mass Increase

When charged particles are accelerated by electric fields they rapidly acquire very high velocities. This may be seen by applying eqn. (6.12) to electrons, with their very small mass, which shows that even a 1 kV electron moves at 1.2×10^7 m/s. Thus their velocities approach that of light, and in this region the theory of relativity applies, which is the science of high speeds.

A consequence of the theory is that mass and energy are related by eqn. (A.2.1): $W = mc^2$. This equation states that mass is a form of energy, so that when mass is destroyed energy is released. This is the energy binding together the nucleus of an atom and released when atomic energy is produced. For example, if 1 mg of matter is destroyed the energy released is 9×10^{10} J (25 000 kWh).

In relativity dynamics, force is defined as

$$F = \frac{d}{dt}(mu) \tag{A.4.1}$$

which still leaves $dW = F\,dx$. But $dW = c^2\,dm$, when mass is allowed to vary, so that

$$c^2\,dm = \frac{d}{dt}(mu)\,dx \tag{A.4.2}$$

$$= \left(m\frac{du}{dt} + u\frac{dm}{dt} \right) dx \tag{A.4.3}$$

$$= \left(m\frac{du}{dx}\frac{dx}{dt} + u\frac{dm}{dx}\frac{dx}{dt} \right) dx \tag{A.4.4}$$

Thus

$$c^2 \, dm = mu \, du + u^2 \, dm \tag{A.4.5}$$

and

$$\frac{dm}{m} = \frac{u}{c^2 - u^2} \, du \tag{A.4.6}$$

Integrating,

$$\log_e m = \log_e (c^2 - u^2)^{-1/2} + \log_e C \tag{A.4.7}$$

and

$$m = \frac{C}{(c^2 - u^2)^{1/2}} \tag{A.4.8}$$

When $u = 0$, $m = m_0$, the *rest mass*, so that $C = m_0 c$. Hence

$$m = \frac{m_0}{\left(1 - \dfrac{u^2}{c^2}\right)^{1/2}} \tag{A.4.10}$$

and the corresponding energy is

$$W = \frac{m_0 c^2}{\left(1 - \dfrac{u^2}{c^2}\right)^{1/2}} = mc^2 \tag{A.4.11}$$

Expanding this equation by means of the binomial theorem for the case when $u/c \ll 1$,

$$W = m_0 c^2 + \tfrac{1}{2} m_0 u^2 \tag{A.4.12}$$

$m_0 c^2$ is the kinetic energy for zero velocity relative to the observer; this is known as the *rest energy*, W_0, and is about $5 \cdot 1 \times 10^5$ eV for an electron. Thus the *increase* in energy due to an applied voltage V is given by

$$W - W_0 = eV \tag{A.4.13}$$

and at velocities much less than the velocity of light,

$$W - W_0 = \tfrac{1}{2} m_0 u^2 \tag{A.4.14}$$

This shows that the relativity theory gives the same result as the theory developed on page 236 for low velocities.

At high velocities, from eqn. (A.4.11),

$$\frac{u}{c} = \left[1 - \left(\frac{W_0}{W} \right)^2 \right]^{1/2}$$

(A.4.15)

or, substituting from eqn. (A.4.13),

$$\frac{u}{c} = \left[1 - \left(\frac{W_0}{W_0 + eV} \right)^2 \right]^{1/2}$$

(A.4.16)

This equation is plotted in Fig. 6.22 and yields the same value of velocity as eqn. (6.12) for voltages up to about 5 kV. Above 10 MeV the velocity is practically equal to that of light, which it can never exceed, and the energy is increased above this level almost entirely by the relativistic increase in mass.

APPENDIX 5

Electric Field between a Cylinder and a Coaxial Wire

Consider a cylinder of radius r_2 and a coaxial wire of radius r_1, each 1 m long, with a charge $-Q$ on the wire due to a voltage V on the cylinder

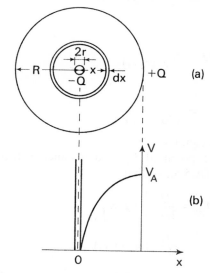

Fig. A.5.1 Coaxial wire and cylinder system

(Fig. A.5.1(a)). At distance x from the centre a surface of area $2\pi x \times 1$ square metres encloses the charge $-Q$ so that the electric flux density is

$$D = \frac{-Q}{2\pi x} = E\epsilon_0 \tag{A.5.1}$$

where E is the electric field, so that

$$E = \frac{-Q}{2\pi x \epsilon_0} \tag{A.5.2}$$

The potential difference across an element of thickness dx is

$$dV = -E\,dx = \frac{Q\,dx}{2\pi x \epsilon_0} \tag{A.5.3}$$

Hence the total voltage between wire and cylinder is

$$V = \int_{r_1}^{r_2} \frac{Q\,dx}{2\pi x \epsilon_0} \tag{A.5.4}$$

$$= \frac{Q}{2\pi \epsilon_0} \left[\log_e x \right]_{r_1}^{r_2} \tag{A.5.5}$$

$$= \frac{Q}{2\pi \epsilon_0} \log_e (r_2/r_1) \tag{A.5.6}$$

Thus

$$Q = \frac{2\pi \epsilon_0 V}{\log_e (r_2/r_1)} \tag{A.5.7}$$

and

$$E = \frac{-V}{x \log_e (r_2/r_1)} \tag{A.5.8}$$

from eqn. (A.5.2).

The potential V_x at a distance x from the centre is obtained from eqn. (A.5.5), so that

$$V_x = \frac{Q}{2\pi \epsilon_0} \left[\log_e x \right]_{r_1}^{x} \tag{A.5.9}$$

$$= \frac{Q}{2\pi \epsilon_0} \log_e (x/r_1) \tag{A.5.10}$$

Then, using eqn. (A.5.7),

$$V_x = \frac{V}{\log_e (r_2/r_1)} \log_e (x/r_1) \tag{A.5.11}$$

This equation is illustrated in Fig. A.5.1(*b*).

In the case of a corona discharge (page 320), if the critical field for breakdown at the surface of the wire is E_B the critical voltage between cylinder and wire is V_B, where

$$V_B = E_B r \log_e (r_2/r_1) \tag{A.5.12}$$

from eqn. (A.5.8), neglecting the minus sign.

Symbols

a	grade constant	N	number density of atoms
A	amplification	p	momentum
B	bandwidth		number density of holes
	magnetic flux density		probability
C	capacitance		quantum number $l = 1$
	capacitance per unit area	P	power
d	distance		pressure
	quantum number $l = 2$	q, Q	charge
D	charge per unit area	r	distance
	diffusion coefficient		radius
	penetration factor	R	resistance
E	electric field	s	distance
f	frequency		quantum number $l = 0$
	quantum number $l = 3$		spin quantum number
g	conductance	S	surface area
	generation rate	t	time
h	hybrid parameter	T	temperature
j	$\sqrt{-1}$	u	speed
J	current density	$\bar{u}$	mean speed
k	momentum vector	w, W	width
	triode constant	W	energy
K	kinetic energy	Z	atomic number
	stability factor		impedance
	thermal conductance		thermoelectric figure of merit
K_H	Hall coefficient	α	common-base current gain
l	angular momentum quantum number		ionizing power
		α_P	Peltier coefficient
	free path	α_S	Seebeck coefficient
	length	β	common-emitter current gain
$\bar{l}$	mean free path	γ	emitter efficiency
L	diffusion length		secondary emission coefficient (in a gas)
	length		space-charge reduction factor
m	magnetic quantum number	δ	base transport factor
m_o	rest mass of electron		secondary emission coefficient (in a vacuum)
m^*	effective mass of electron		small change
M	current multiplication factor	Δ	large change
n	number density of free particles		transistor determinant
	principal quantum number		
N	density of energy levels		

ϵ	permittivity		ρ	resistivity
ζ	quantum yield		σ	conductivity
κ	thermal conductivity		τ	time-constant
λ	wavelength		ϕ	work function
μ	amplification factor		Φ	magnetic flux
	feedback factor		χ	depth of conduction band
	mobility		ψ	diffusion potential,
	permeability			probability function
μ_B	Bohr magneton			
ρ	charge density		ω	angular frequency

SUBSCRIPTS

a, A	anode		m	mutual
a	acceptor		M	Maxwell
b, B	base			maintaining
B	breakdown		n	electron
BO	breakover			noise
c	conduction band			n-type
c, C	collector		oc	open-circuit
d	delay		p	hole
	donor			p-type
	drift			pump
d, D	drain		r, R	retarding
D	domain			reverse
e	effective		r	rise
	electron		s, S	saturated
e, E	emitter			signal
f	fall			source
f, F	forward			storage
F	Fermi		sc	semiconductor
g	gap			short-circuit
g, G	gate		T	threshold
h	high		V	valley
	hole		v	valence band
	horizontal			vertical
i	idler		0	initial
	input			leakage (current)
	ion			mid-band
j	junction			

VOLTAGE AND CURRENT SYMBOLS

I_A	direct anode current		V_A	direct anode voltage
I_B	direct base current		V_C	direct collector voltage
I_C	direct collector current			(microwave valves)
I_E	direct emitter current		V_H	direct helix voltage
I_G	direct grid current		V_K	direct cathode voltage

V_M	maintaining voltage		V_{AA}	anode supply voltage
V_P	direct probe voltage		V_{BB}	base supply voltage
V_R	direct reflector voltage		V_{CC}	collector supply voltage
	retarding voltage			

The symbols for alternating quantities are illustrated below in terms of collector current. Similar types of symbol and subscript are used for voltages and for other electrodes.

i_c	incremental (instantaneous) value		I_{cm}	peak value of varying component
	of varying component		I_c	r.m.s. value of varying component
i_C	total (direct plus incremental) value			

ABBREVIATIONS FOR NAMES OF UNITS (UNIT SYMBOLS)

Derivation

A	ampere	
C	coulomb	As
cd	candela	
F	farad	C/V
H	henry	Vs/A
Hz	hertz	cycle per second
J	joule	$kg\,m^2/s^2$
K	kelvin	
kg	kilogramme	
lm	lumen	cd-sr
lx	lux	lm/m^2
m	metre	
N	newton	J/m
rad	radian (plane angle)	
s	second	
sr	steradian (solid angle)	
S	siemens	(mho, or reciprocal ohm)
T	tesla	Wb/m^2
V	volt	W/A
W	watt	J/s
Wb	weber	Vs
Ω	ohm	V/A

		Value
eV	electronvolt	$1.602 \times 10^{-19}\,J$
°C	degree Celsius	K + 273

Prefixes

T	tera-	10^{12}		m	milli-	10^{-3}
G	giga-	10^9		μ	micro-	10^{-6}
M	mega-	10^6		n	nano-	10^{-9}
k	kilo-	10^3		p	pico-	10^{-12}
d	deci-	10^{-1}		f	femto-	10^{-15}
c	centi-	10^{-2}		a	atto-	10^{-18}

PHYSICAL CONSTANTS

Charge of electron, $e = 1{\cdot}602 \times 10^{-19}\,$C
Mass of electron, $m = 9{\cdot}109 \times 10^{-31}\,$kg
Specific charge of electron, $e/m = 1{\cdot}759 \times 10^{11}\,$C/kg
Speed of light, $c = 2{\cdot}998 \times 10^{8}\,$m/s
Permittivity of vacuum, $\epsilon_0 = 8{\cdot}854 \times 10^{-12}\,$F/m
Permeability of vacuum, $\mu_0 = 4\pi \times 10^{-7}\,$H/m
Planck constant, $h = 6{\cdot}626 \times 10^{-34}\,$Js
Boltzmann constant, $k = 1{\cdot}380 \times 10^{-23}\,$J/K
Avogadro constant, $N_A = 6{\cdot}023 \times 10^{23}$/mol
Relative permittivity, ϵ_r:
  germanium, 16; silicon, 12

Index

Abnormal glow discharge, 319
Abrupt p–n junction, 90
  capacitance, 110
  charge density, 106
  field, 110
  voltage, 108
  width, 109, 194
Accelerating region, 340
Acceptor atom, 31
Allowed band, 25, 410
Amplification factor, μ, 286
Amplifier—
  klystron, 334
  parametric, 360, 366, 387
  transistor, 155, 200
  travelling wave, 345
  triode, 288
  tunnel diode, 350
Angular momentum, orbital, 6, 12
  quantum number, l, 12
Angular momentum, spin, 15
  quantum number, s, 15
Anode, 215, 237, 339
  virtual, 304
Anode fall, 304
Anode slope resistance, r_a, 288
Applegate diagram, 332, 337
Arc discharge, 319
Atomic energy, 420
Atomic number, Z, 11
Avalanche breakdown, 103, 312
  temperature coefficient of, 105
Avogadro number, N_A, 25, 429

Backward wave, 344
Balmer, 8
Band, energy, 24
Barium, 244

Barrier, energy, 75
Base, 121
  charge, q_B, 127, 146, 186
  graded, 121, 183, 209
  spreading resistance, $r_{bb'}$, 162
  transport factor, δ, 126
  uniform, 121
Beam tetrode, 291
Bias—
  forward, 80, 94, 139
  reverse, 81, 99, 101
  zero, 78, 91
Binding energy, 22
Bipolar transistor (*see* Transistor, junction)
Blocking condition, 215
Bohr, 3, 406
Bohr magneton, 15
Boltzmann constant, 34
Boltzmann factor, 37
Bonds, covalent and metallic, 22
Boron, 31
Breakdown diode, 105
Breakdown, reverse, 103
Breakover, 215, 219
Brewster angle, 393
Brillouin zone, 352
Bunch, electron, 341, 343
Buncher, 332

Candela, 224
Capacitance—
  depletion layer, 110, 163, 221, 372
  diffusion, 162
  interelectrode, 291
Capacitor—
  integrated circuit, 212
  thin film, 214

Carbon, 19
Catcher, 334
Cathode, 215, 237, 339
  oxide, 245
  virtual, 282
Cathode fall, 304
Cavity magnetron, 339
Cavity, resonant, 333, 350, 371, 394
Charge—
  balance equation, 45
  control equation, 142
  control parameters, 142
Child's law, 254
Coherent emission, 387
Collector, 121, 334
  efficiency, M, 128
  leakage current, I_{CBO}, 122
  leakage current, I_{CEO}, 137
  maximum temperature, 182
  time constant, τ_C, 127, 152, 157, 162
Collision cross-section, σ, 299
Colour, 228
Concentration gradient, 37
Conduction—
  electrical, 27
  in insulators, 27
  in metals, 27, 46
  in semiconductors, 27
  intrinsic, 28
  thermal, 89
Conduction band, 25, 352
Conductivity, electrical, 54, 90
Conductivity modulation, 63
Confocal mirrors, 394
Contact—
  metal-to-semiconductor, 77
  ohmic, 77
  rectifying, 81
Contact potential, 77, 282
Copper, 19
Core levels, 25
Corona, 320
Covalent bond, 22
Current carriers, 27
  excess, 62
Current density, J, 53
Cylindrical magnetron, 269

Davisson, 9
De Broglie, 9, 278
Decibel, 170
Defect, crystal, 62
Deflection of electron beam—
  electromagnetic, 266
  electrostatic, 262
Decelerating region, 340
Degenerate parametric amplifier, 362
Degenerate semiconductor, 46, 230, 345
Delay time, t_d, 152
Density of states, 41, 413
Depletion layer, 78, 91, 106
  capacitance, 110, 114, 163, 221, 372
Diffraction, electron, 409
Diffusion, 37
  capacitance, 162
  coefficient, 37, 67
  length, 68
  potential, 78, 91, 93
Digital computer, 208
Diode—
  electroluminescent, 227
  equivalent, 285
  characteristic, 84, 98
  gas-filled, 304
  Gunn, 352
  hot carrier, 81
  IMPATT (Read), 369
  isolation, 209
  photo-, 223
  p–i–n, 372
  p–n–i–n, 370
  Schottky, 81
  step recovery, 369
  thermionic, 249
  tunnel, 345
  varactor, 367
Dipole layer, 246
Direct gap, 229
Discharge—
  abnormal glow, 319
  cold cathode, 317
  corona, 320
  hot cathode, 300
  normal glow, 317
  self-sustained, 315

Domain, 354
Donor atom, 30
Doping, 30
Downconverter, 366
Drain, 194
Drift velocity, 50, 354
Durchgriff, D, 286
Dust precipitator, 320

Effective mass, 31, 353, 411
Einstein, 241, 380, 402
Elastic collisions, 301
Elastic waves, 49
Electrical conduction, 27
Electrical noise, 372
Electromagnetic radiation, 3
Electron, 2, 267
  affinity, 319
  avalanche, 103, 312
  beam, 254
  bunch, 341, 343
  cloud, 23
  diffraction, 409
  diffusion coefficient, 67
  emission, 234
  gas, 34
  gun, 254, 277
  microscope, 277
  mass, 2, 414
  mobility, 51
  motion in electric field, 236
  motion in magnetic field, 261, 265
  optics, 273
  sheath, 305
  temperature, T_e, 306
  trap, 63
  tunnelling, 349, 407
  valence, 17, 22
  wave properties, 9, 278, 402
Electronvolt, 3
Elliptical orbit, 8
Emission—
  coherent, 387
  electron, 2, 34
  field, 105, 236, 256
  incoherent, 387
  photo-, 2, 236, 240, 243

Emission—(*contd*).
  secondary, 236, 257, 260, 315, 341
  spontaneous, 381
  stimulated, 380
  thermionic, 2, 236, 243, 244
  velocity, 241, 280
Emitter, 121
  efficiency, γ, 124, 190
  resistance, r_ε, 158
Energy—
  bands, 24
  barrier, 75
  binding, 22
  density, 380
  exchange, 328
  gap, 27, 221
  kinetic, 5, 403
  ionization, 8, 301
  level, 5, 24
  optical, 4, 228
  potential, 5, 25, 403
  thermal, 28
Epitaxial layer, 186
Epitaxial planar transistor, 185
Equivalent diode, 285
Excess base charge—
  graded-base transistor, 186
  uniform-base transistor, 127, 146
Excess carriers, 62
Excess current, 62
Excitation, 7, 301
Excited state, 6
Exclusion principle, 15
Extrinsic semiconductor, 30

Fabry–Perot interferometer, 394
Fall time, t_f, 150
Fermi–Dirac function, 38,
Fermi level, 38, 75, 417
Field effect, 197
Field-effect transistor, FET, 194
  (*see also* Transistors)
Field emission, 105, 236, 256
  microscope, 256
Flicker noise, 374
Forbidden band, 25, 223, 229, 411

Forward bias, 80, 94, 139
  breakover voltage, 215, 219
  characteristics, 118
Frequency dependence of h_{fe}, 170
Frequency multiplier, 369
Frequency response, 172, 189

Gain-bandwidth product, f_T, 169, 202
Gallium, 31
  arsenide, 228
  phosphide, 228
Gamma rays, 4
Gap, energy, 27, 221
  direct, 229
  indirect, 229
Gas diode—
  cold-cathode, 317
  hot-cathode, 304
  (*see also* Discharge)
Gas, electron, 34
Gas inert, 16
Gas laser, 392
Gate, 194, 204, 215, 219
Germanium, 19, 228
Germer, 9
Giant pulse, 391
Gold, 2
Grade constant, 112
Graded base, 121
Graded junction, 111
  capacitance, 114
  charge density, 112
  field, 114
  voltage, 114
  width, 114
Grid, 283
Grid current, 283
Ground state, 6, 301
Guiding wave, 8
Gun, electron, 277, 254
Gunn diode, 352

Hall effect, 58
Heat sink, 182
Heisenberg, 10
Helium, 16, 299, 392
Helix, 342, 344

High-frequency equivalent circuit, 161, 164
High-level injection, 116
Holding current, I_H, 215
Hole, 28, 31
  current, 30
  diffusion coefficient, 67
  mass, 30, 415
  mobility, 51
  trap, 63
Hot carrier diode, 81
Hydrogen atom, 2

Idler circuit, 363
Illumination, 224
Impact avalanche transit time (IMPATT) diode, 369
Impurity atom, 30
Imref, 100
Incoherent emission, 387
Indirect gap, 229
Indium, 31
Induced currents, 239, 328
Inelastic collision, 301
Inelastic reflection, 259
Inert gases, 16
Infrared radiation, 4
Insulated-gate field-effect transistor (IGFET) (*see* Transistor, metal-oxide-semiconductor)
Injection—
  high level, 116
  low level, 95
Injection laser, 396
Insulators, 27
Integrated circuit, 208, 214
Interelectrode capacitances, 291
Intrinsic conduction, 28
Inversion layer, 205
Ion, 302, 304
Ionization, 8
  energy, 8, 301
  potential, 301
Ionizing power, α, 313
Isolation diode, 209

Junction diode (see *p–n* junction diode)

Index

Junction-gate field-effect transistor (*see* Transistor, junction-gate field-effect)

Junction, metallurgical, 90

Junction transistor (*see* Transistor, junction)

Kelvin's law, 86

Kerr cell, 392

Kinetic energy, 5, 403

Klystron—
  amplifier, 334
  oscillator, 336

Langmuir probe, 306

Laser—
  applications, 396
  gas, 392
  injection, 396
  modes, 395
  ruby, 389
  semiconductor, 396

Lattice scattering, 48

Leakage current, I_0, 81, 99, 217

Lifetime, 62, 66, 144, 147

Limited space-charge accumulation (LSA) mode, 358

Linear integrated circuit, 214

Lithium, 16

Load line, 141

Loschmidt number, 34

Low-level injection, 95

Lumen, 224

Luminous flux, 224

Lumped components, 325

Lux, 224

Lyman, 8

Magnetic deflection, 266

Magnetic focusing, 271, 326, 345

Magnetic lens, 274

Magnetic moment, 13

Magnetic quantum number, m, 13

Magneton, 15

Magnetron—
  cavity, 339
  cylindrical, 269

Majority carriers, 46

Maser, 375, 380

Mass—
  effective, 31, 411
  electron, 2, 414
  hole, 30, 415
  increase, 261, 420

Mass spectrometer, 268

Maxwell–Boltzmann—
  distribution, 37
  statistics, 41

Mean free path, l, 35, 297, 299

Measurement—
  lifetime, 65
  semiconductor properties, 58

Mendeleeff, 17

Mercury, 299, 304

Metal, 27, 46

Metallic bond, 22

Metallurgical junction, 90

Metal-oxide-semiconductor transistor (*see* Transistor, metal-oxide-semiconductor)

Metal-to-semiconductor contacts, 77

Metastable state, 302

Microcircuit, 208

Microwave frequencies, 325

Microwave valve, 326

Miller effect, 291

Millikan, 2

Mobility, 50, 321, 353

Mode—
  depletion, 197, 205
  enhancement, 197, 205
  Gunn diode, 358
  laser, 395
  LSA, 358
  reflex klystron, 337
  π, 339

Momentum—
  angular, or moment of, 6
  of quantum, 9, 230, 402
  space, 413
  vector, k, 411

Monolithic integrated circuit, 214

Multichip integrated circuit, 214

Multiplication, avalanche, 103

Multiplier—
frequency, 369
signal, 61
Mutual conductance, g_m, 167, 200, 288

n-channel, 194
n-type semiconductor, 30
Negative resistance—
arc discharge, 319
Gunn diode, 354
IMPATT diode, 371
parametric amplifier, 362
tetrode, 292
tunnel diode, 345, 349
Neon, 299, 392
Neutron, 11
Nitrogen, 299
Noise, electrical, 372
Noise factor, 374
Noise input temperature, T_i, 375
Nondegenerate parametric amplifier, 365
Nucleating centre, 355
Nucleus, 2, 22

Ohm's law, 63
Ohmic contact, 77
Open-channel conductance, 201
Orbit, 2
elliptical, 8
preferred, 5
$s\,p\,d\,f$ classification, 16
Orbital, 12
Oscillation, 351, 362
Oscillator—
Gunn, 352
klystron, 336
tunnel diode, 351
Overdrive (of junction transistor), 146, 150
Oxide cathode, 245
Oxide layer, 257

Parametric amplifier, 360, 387
Partition noise, 374
Paschen, 8, 317
Pauli, 15
p-channel, 194
Peak voltage, 347

Peltier coefficient, α_P, 85
Peltier effect, 85
Penetration factor, D, 286
Pentode, 291
Periodic table, 17
Perveance, 254
Phonon, 49, 230
Photocell—
gas-filled, 312
vacuum, 243
Photoconduction, 264
Photocurrent I_P, 222
Photodiode, 223
Photoemission, 2, 236, 240, 243
Photoetching, 184
Photomultiplier, 260
Photon, 3, 63, 221, 230, 240
Photoresist, 184
Phototransistor, 225
p–i–n diode, 372
Pinch-off voltage, V_P, 196, 205
Planar technology, 131
Planar transistor, 183, 209
Plasma, 303
Plasma potential, 303, 307
p–n–i–n diode, 370
p–n junction
in FET, 194, 205
in junction transistor, 121
in photodiode, 222
in thyristor, 215
in tunnel diode, 345
in varactor diode, 360
(*see also* Abrupt; Graded)
p–n junction diode—
forward bias, 94, 116
forward voltage, 118
practical characteristic, 117
rectifying properties, 84
reverse bias, 101
reverse current, 99
temperature effects, 102
Population inversion, 385
Positive ion, 304
Positive ion sheath, 305
Potential energy, 5, 25, 403
Potential well, 6, 404

Index

Precession, 14
Principal quantum number, n, 12
Probability, 38, 403
  distribution, 11

Q-factor, 333
Q-switching, 391
Quantum, 3, 402
Quantum numbers, 12
Quantum yield, 224
Quasi-Fermi levels, 100
Quiescent point, 142, 289

Radiation, electromagnetic, 2, 3
  coefficient of absorption, B_{12}, 382
  coefficient of spontaneous emission, A_{21}, 382
  coefficient of stimulated emission, B_{21}, 383
Radio waves, 4
Ramsauer, 298
Read (IMPATT) diode, 369
Recombination, 62, 68, 97, 305
  centre, 62
  noise, 374
Rectifying contact, 81
Reflector, 336
Refractive index, 276
Relativistic change of mass, 261, 420
Resistors, 210
Resolving power, 277
Resonance condition, 333, 349, 368
Resonant transfer, 393
Rest mass, m_0, 261, 421
Reverse bias, 81, 99, 101
Reverse breakdown, 103
Reverse current, I_0, 81, 99, 217
Richardson, 245
Rise time, t_r, 146
Ruby laser, 389
Ruby maser, 375, 380

Satellite communication, 366
Saturated current, 245
Saturation charge, q_{BS}, 146
Schottky diode, 81
Schottky effect, 247

Schrödinger, 10, 402
Secondary emission, 236, 257, 260, 315, 341
  coefficient δ, 257
  coefficient γ, 315
Seebeck coefficient α_S, 86
Seebeck effect, 85
Self-sustained gas discharge, 315
Semiconductor—
  degenerate, 46, 345
  intrinsic, 29, 56, 417
  lamp, 227, 231
  laser, 396
  n-type, 30, 419
  p-type, 31, 419
Shot noise, 373
Silicon, 19, 184, 228
Solar cell, 226
Sommerfeld, 8
Source, 173, 194
Sparking potential, 316
$s\,p\,d\,f$ classification, 16
Spectrometer, mass, 268
Spectroscope, 8
Spontaneous emission of radiation, 381
Spreading velocity, 220
Stability factor, S, 179, 180
Stability, thermal, 178
Standing wave, 406
Stefan–Boltzmann constant, σ, 224
Step recovery diode, 369
Stimulated emission of radiation, 380
Straight amplifier, 366
Striking voltage, 309
Sulphur hexafluoride, 319
Surface effects, 129
Surface leakage, 117
Switch, 142, 203, 208, 351

Temperature, effect on—
  breakdown voltage, V_B, 105
  electron-hole pairs, 43
  forward voltage, V_F, 102
  metals, 47
  mobility, 52
  n- and p-type semiconductors, 47
  resistivity, 54
  reverse leakage current I_0, 102

Temperature-limited current, 245
Tetrode, 291
Thermal conductance, 89
Thermal energy, 28
Thermal noise, 373
Thermionic diode, 249
Thermionic emission, 2, 236, 243
Thermistor, 56
Thermoelectric cooling, 88
    figure of merit, Z, 90
Thermoelectricity, 85
Thin-film circuit, 214
Thomson, 2, 267
Thomson effect, 87
Thoriated tungsten, 244
Thorium, 246
Threshold wavelength, 221, 241
Thyratron, 209
Thyristor, 215, 217
    blocking condition, 215
    forward breakover voltage, V_{BO}, 215,
        219
    holding current, I_H, 215
    reverse breakdown voltage, V_{RA}, 215
    spreading velocity, 220
    static characteristics, 217
Townsend, 311
Transconductance (*see* Mutual conduct-
    ance)
Transistor, field effect (FET), 194
Transistor, insulated-gate field-effect
    (IGFET) (*see* Transistor, metal-oxide-
    semiconductor)
Transistor, junction, 121
    active region, 138
    amplifier, 155, 158, 200
    breakdown, 135, 138, 190
    capacitances, 162, 163
    charge control, 142
    current gain, 177
        h_{FE}, 137
        h_{fe}, 169, 175
        α_0, 124
        β_0, 136
    determinant, 177
    at high currents, 129
    at high frequencies, 155

Transistor—(*contd.*)
    f_T, 169
    f_β, 170
    hybrid parameters, 174, 175
    hybrid-π equivalent circuit, 168
    input resistance, 177
    leakage current, I_{CBO}, 122
    leakage current, I_{CEO}, 137
    operation, 140
    output resistance, 177
    reverse current, I_R, 139
    saturated region, 139
    static characteristics—
        common-base, 132, 135
        common-emitter, 139, 137
    switch, 142
    thermal stabilization, 178
    transient response, 142
    voltage gain, 177
Transistor, junction-gate field-effect
    (JUGFET), 194
    amplifier, 200
    breakdown, 200
    capacitances, 202
    current, I_{DSS}, 200
    depletion mode, 197
    enhancement mode, 197
    equivalent circuit, 202
    at high frequencies, 202
    input impedance, 194
    open-channel conductance, 201
    pinch-off voltage, V_P, 196
    static characteristics, 198, 201
        ohmic region, 196
        pinch-off region, 199
        triode region, 199
    switch, 203
    temperature effects, 200
Transistor, metal-oxide-semiconductor
    (MOST), 203, 209
    breakdown, 208
    capacitance, 208
    depletion type, 205
    enhancement type, 205
    equivalent circuit, 208
    at high frequencies, 208
    input impedance, 205, 208

Index

Transistor—(*contd.*)
   inversion layer, 205
   pinch-off voltage, V_P, 205
   static characteristics, 206
   switch, 208
Transit-time—
   cavity, 335
   mode, 358
   Read (IMPATT) diode, 371
   transistor base, 127, 155
   triode, 326
Transmission line, 162, 325
Trap, 62
Travelling wave, 9, 340, 342, 399
Travelling-wave amplifier, 345
Travelling-wave tube, 341
Triode, 283
   capacitances, 291
   equivalent circuit, 290
   equivalent diode, 285
   operation, 288
   parameters, 288
   static characteristics, 287
Tuned circuit, 325
Tungsten, 244
Tuning slug, 334
Tunnel current, 349
Tunnel diode, 105, 345
   amplifier, 350
   dynamic range, 349
   equivalent circuit, 349
   forward voltage, V_F, 347
   peak current, I_P, 347
   peak voltage, V_P, 345
   static characteristics, 346
   switch, 351
   temperature effect, 351
   valley current, I_V, 345
   valley voltage, V_V, 345

Uncertainty principle, 10
Uniform base, 121
Unipolar transistor, 197
Upconverter, 366

Valence band, 25, 30, 31
Valence electron, 17, 22
Varactor diode, 367
   equivalent circuit, 368
Velocity modulation, 331
Virtual anode, 304
Virtual cathode, 282
Voltage feedback factor, μ, 160
Voltage gain, 177

Wafer, semiconductor, 184
Wave—
   backward, 344
   equation, 399
   functions, 407
   guide, 325
   guiding, 8
   properties of an electron, 9, 402
   standing, 406
   travelling, 340, 342, 399
Wavelength, 9, 325
   threshold, 221, 241
   visible light, 228

Xenon, 299, 389
Xenon flash tube, 389
X-rays, 4

Zener breakdown, 105
   temperature coefficient of, 105
Zero bias, 78, 91, 194, 205, 285